MODERN ASPECTS OF ELECTROCHEMISTRY

No. 57

Series Editors:
Ralph E. White
Department of Chemical Engineering
University of South Carolina
Columbia, SC 29208

Constantinos G. Vayenas
Department of Chemical Engineering
University of Patras
Patras 265 00
Greece

For further volumes:
http://www.springer.com/series/6251

Previously from Modern Aspects of Electrochemistry

Modern Aspects of Electrochemistry No. 55

Biomedical Applications

Edited by Stojan S. Djokić, Professor of Chemical & Materials Engineering at the University of Alberta

Topics in Number 55 include:

- CoCrMo alloy for biomedical applications
- Electroless synthesis of metallic nanostructures for biomedical technologies
- Biodegradable Mg alloys: Corrosion, surface modification and biocompatibility
- Microcantilever sensors: Electrochemical aspects and biomedical applications
- Surface treatments with silver and its compounds for biomedical applications

Modern Aspects of Electrochemistry No. 56

Applications of Electrochemistry in Medicine

Edited by Mordechay Schlesinger, Professor Emeritus, Department of Physics, University of Windsor, Canada.

Topics in Number 56 include:

- Electrochemistry in the design and development of medical technologies and devices
- Medical devices at the interface of biology and electrochemistry
- Sensing by screen printed electrodes for medical diagnosis
- Electrochemical glucose sensors
- Electrochemistry of adhesion and spreading of lipid vesicles on electrodes
- Bio-Electrochemistry and chalcogens
- Nanoplasmonics in medicine
- Extravascular hemoglobin: aging contusions
- Modeling tumor growth and response to radiation

Stojan S. Djokić
Editor

Electrodeposition and Surface Finishing

Fundamentals and Applications

Springer

Editor
Stojan S. Djokić
Elchem Consulting Ltd.
Edmonton, AB, Canada

ISSN 0076-9924 ISSN 2197-7941 (electronic)
ISBN 978-1-4939-4830-7 ISBN 978-1-4939-0289-7 (eBook)
DOI 10.1007/978-1-4939-0289-7
Springer New York Heidelberg Dordrecht London

Printed on acid-free paper

Springer is part of Springer Science+Business Media (www.springer.com)

Dedicated to John O'M. Bockris
and Brian E. Conway

Preface

The series *Modern Aspects of Electrochemistry* has presented new developments in Electrochemistry since its inception in the early 1950s. This prestigious series contains works of many distinguished electrochemists worldwide including those of the series founders, Professors John O'M. Bockris and Brian E. Conway. The vision of Professors Bockris and Conway continues to achieve greatness with its 57th volume thanks to Springer and Dr. Kenneth Howell. I would especially like to point out that I was blessed with opportunities to study, work with, and collaborate with Professors Conway and Bockris. I owe a great deal to these two fine individuals for all I have learnt in the field of electrochemistry. Professors Bockris and Conway collaborated with many electrochemistry groups around the globe. In this way, Professors Bockris and Conway significantly influenced many electrochemists and developments in the electrochemical science and technology. I would especially like to mention the contributions of my former teacher, Professor Konstantin I. Popov, and his coworkers, who in spite of very limited equipment and finances are still achieving first class results.

This volume of *Modern Aspects of Electrochemistry* is devoted to Professors John O'M. Bockris and Brian E. Conway. Although Professor Brian E. Conway passed away in 2005 and Professor John O'M. Bockris, just recently, in July of 2013, their vision continues and will continue in the future. Significant contributions to this volume have come from individuals who were in one or another way influenced by these fine minds.

Chapter 1 by Jović et al. describes the electrodeposition of alloys and composite materials. After a historical overview of the early work, the chapter discusses the conditions for the electrodeposition of alloys from a thermodynamic point of view. The narrative then further explores the characterization of the electrodeposited alloys by electrochemical techniques. Anodic linear sweep voltammetry for the characterization of the electrodeposited alloys, e.g., eutectic, solid solution, and alloys with intermediate phases or intermetallic compounds is described and referenced in detail. In addition, the results of the analysis of electrodeposited alloys by the anodic sweep voltammetry are compared to those obtained by the X-ray diffraction. Finally, the chapter describes electrodeposition of composite materials and their mechanical and electrical properties. Important discussion is devoted to the electrodeposition of Ni–MoO_2 composite coatings as cathodes for hydrogen evolution in industrial electrolysis.

In Chap. 2, Nikolić and Popov discuss the mechanistic aspects of lead electrodeposition. Electrodeposition of lead is characterized by a relatively high exchange current densities. For a while, it was generally accepted that the electrodeposition of lead proceeds in the whole range of overpotentials and that it is diffusion controlled. The experimental results and discussion in this chapter show that electrodeposition of lead may proceed under the conditions of pure ohmic or mixed ohmic-diffusion control. The conditions for lead electrodeposition are influenced by the concentration of Pb(II) in the solution. At higher concentrations of Pb(II) in the solution, electrodeposition of lead is completely under ohmic control. The surface morphology of the electrodeposited lead is determined by the conditions of electrodeposition. While under ohmic control well-defined single crystals of lead are produced, elecetrodeposition under mixed ohmic-diffusion control leads to the formation of dendritic deposits. The formation of different shapes of dendrites of lead during the electrodeposition is further attributed to the composition of the electrolytes used in the experiments. The primary type of dendrites is produced from the simple electrolytes, while the secondary type is formed from *complexed* solutions.

In Chap. 3, Mišković-Stanković discuses the electrophoretic deposition of ceramic materials onto metal surfaces. Ceramic coatings may be useful as anticorrosion surfaces or in biomedical applications as implants. Materials presented in this chapter include alumina, boehmite, monetite, brushite, hydroxyapatite, and their combination with silver and/or lignin. The effects of the parameters of electrochemical deposition on the thickness, morphology, and structure of the deposited ceramic coatings are discussed in this chapter. Various instrumental methods are used to describe the properties of electrodeposited ceramic coatings.

Chapter 4 by Tsui and Zangari reviews the fundamentals of the electrodeposition of metal oxides for the energy conversion and storage technologies. Electrochemical growth of oxide materials and methods to control their composition, surface morphology, and crystal structure in relation to the applications for the energy conversion are discussed. The materials examined in this chapter include ZnO for solar cells, Cu_2O for photovoltaic and photoelectrochemical systems, α-Fe_2O_3 for photoelectrochemical water splitting, and MnO_2 for the supercacitor energy storage.

Chapter 5 by Stojadinović et al. is devoted to the anodization of aluminum and more particularly to an interesting phenomenon—namely, luminescence—occurring during this process. While the anodization of aluminum has been investigated for quite some time, little or no attention has been paid to the occurrence of so-called galvanoluminescence, an emission of weak optical radiation, mostly in the visible spectrum. The nature of the galvanoluminescence depends on the electrolytes used in the process, surface pretreatment, and anodizing conditions. As such, the galvanoluminescent methods are recommended for use in the determination of the oxide film thickness, growth rate, refraction index, optical constants of alumina, etc. Anodization of aluminum above the breakdown voltage leads to the formation of plasma as indicated by the presence of sparks on the surface, accompanied by the simultaneous gas evolution. As a consequence, the total luminescence increases. The spectroscopy analysis of plasma allows the determination of electron temperature and electron number density.

In Chap. 6, Cadian et al. analyze the electrochemical aspects of chemical and mechanical polishing. This process is widely used in the semiconductor manufacturing in order to generate complex three-dimensional geometries. A successful polishing of metals is based on the formation of a passive surface film. Chemical mechanical polishing performed on both tungsten and copper and the electrochemical interactions with the polishing slurry are discussed in this chapter. Electrochemical processes that lead to undesirable phenomena such as pitting, etching of the metal surface, galvanic corrosion, and re-deposition of material onto the polished surface are described in detail.

Chapter 7 by Djokić and Magagnin reviews the surface treatments prior to metallization of semiconductors, ceramics, and polymers. Classical and recently developed methods such as treatment of the ceramic and polymer surfaces with Pd (II) and Ag(I), as well as galvanic displacement reactions for the case of semiconductors, are discussed. Metallization of nonconductive surfaces is very important for many industrial applications.

I sincerely hope that this volume of the *Modern Aspects of Electrochemistry* will bring to scientists, researchers, engineers, and students review chapters related to the latest findings in the field of electrodeposition and surface finishing, The ideas discussed will have significant import for electronics, aerospace, automotive, energy devices, and biomedical applications.

Edmonton, AB, Canada

Stojan S. Djokić

Contents

Contributors

Ivan Belča Faculty of Physics, University of Belgrade, Belgrade, Serbia

K. Cadien Department of Chemical and Materials Engineering, University of Alberta, Edmonton, AB, Canada

K. Dawkins Department of Chemical and Materials Engineering, University of Alberta, Edmonton, AB, Canada

Stojan S. Djokić Elchem Consulting Ltd, Edmonton, AB, Canada

B.M. Jović Department of Materials Science, Institute for Multidisciplinary Research, Belgrade, Serbia

V.D. Jović Department of Materials Science, Institute for Multidisciplinary Research, Belgrade, Serbia

Bećko Kasalica Faculty of Physics, University of Belgrade, Belgrade, Serbia

U.Č. Lačnjevac Department of Materials Science, Institute for Multidisciplinary Research, Belgrade, Serbia

Luca Magagnin Dipartimento di Chimica, Materiali e Ingegneria Chimica 'Giulio Natta', Politecnico di Milano, Milano, Italy

Vesna B. Mišković-Stanković Faculty of Technology and Metallurgy, University of Belgrade, Belgrade, Serbia

Nebojša D. Nikolić ICTM-Institute of Electrochemistry, University of Belgrade, Belgrade, Serbia

L. Nolan Portland Technology Development, Intel Corporation, Hillsboro, OR, USA

H. Pirayesh Department of Chemical and Materials Engineering, University of Alberta, Edmonton, AB, Canada

Konstantin I. Popov ICTM-Institute of Electrochemistry, University of Belgrade, Belgrade, Serbia

Faculty of Technology and Metallurgy, University of Belgrade, Belgrade, Serbia

Stevan Stojadinović Faculty of Physics, University of Belgrade, Belgrade, Serbia

Lok-kun Tsui Department of Materials Science and Engineering, University of Virginia, Charlottesville, VA, USA

Rastko Vasilić Faculty of Physics, University of Belgrade, Belgrade, Serbia

Z. Xu Department of Chemical and Materials Engineering, University of Alberta, Edmonton, AB, Canada

Giovanni Zangari Department of Materials Science and Engineering, University of Virginia, Charlottesville, VA, USA

Ljubiša Zeković Faculty of Physics, University of Belgrade, Belgrade, Serbia

Chapter 1
Electrodeposition and Characterization of Alloys and Composite Materials

V.D. Jović, U.Č. Lačnjevac, and B.M. Jović

1.1 Introduction

It is general experience in materials science that alloy can exhibit qualities that are unobtainable with parent metals. This is particularly true for electrodeposited alloys. Some important properties of materials, such as hardness, ductility, tensile strength, Young's modulus, corrosion resistance, solderability, wear resistance, and antifriction service, may be enhanced. At the same time some properties that are not characteristic for parent metals, such as high magnetic permeability, other magnetic and electrical properties, amorphous structure, etc., can also be obtained. In some cases alloy coatings may be more suitable for subsequent electroplate overlayers and conversion chemical treatments [1].

Some alloys may be more easily obtained by electrodeposition than by metallurgical processes. This is particularly true for alloys composed of metals having large differences in melting temperatures or cannot be mixed in a liquid state. Such metals can very often be codeposited from the solutions (e.g., alloys Ag–Ni, Ag–Co, and Cd–Co). Taking into account that some metals cannot be deposited from the aqueous solutions (Ti, V, W, Nb, Zr, etc.), they could be deposited from the melts of their salts. In recent times the processes of metals and alloys deposition from the room temperature molten salts were also investigated and developed (deposition of Al–Cu, Al–Co, Al–Ni alloys from $AlCl_3$–MeEtImCl melt).

The fast-growing requirements of modern industry for materials with special qualities in the last century have given rise to increasing interest in electrodeposition of alloys, particularly in corrosion protection and in the modern electronic industry [1].

V.D. Jović (✉) • U.Č. Lačnjevac • B.M. Jović
Department of Materials Science, Institute for Multidisciplinary Research,
Kneza Višeslava 1, 11030 Belgrade, Serbia
e-mail: vladajovic@imsi.rs

S.S. Djokić (ed.), *Electrodeposition and Surface Finishing: Fundamentals and Applications*, Modern Aspects of Electrochemistry 57,
DOI 10.1007/978-1-4939-0289-7_1, © Springer Science+Business Media New York 2014

Historically, the electroplating of alloys is practically as old as electroplating of pure metals, since the electroplating of brass and bronze was performed by De Ruolz [2] in 1842, shortly after the discovery of the first cyanide baths. It is interesting to note that these baths were essentially similar to ones used nowadays, being based on the use of Cu, Sn, and Zn complexes with cyanide.

From 1842 until the end of the nineteenth century, over 180 alloys involving 40 elements have been deposited [3]. An excellent review of the achievements up to 1962 is given in the book of Brenner [4], while from practical point of view it is recommended to consider the book of Bondar, Grimina, and Pavlov [5], which contains recipes and references for more than 1,100 baths for alloy deposition.

Although the first alloys [2] were deposited in 1842, practically the first attempt at scientific approach to electrodeposition of alloys, discussing the role of cathodic potential in the deposition of brass, came rather late with the work of Spitzer [6] in 1905. In 1914 a more comprehensive attempt came from Schlötter [7], but better under-standing of the alloy deposition process by understanding the electrochemical thermodynamics and kinetics, as well as complexometry and some other fields in order to obtain clear scientific bases, had to await second part of the nineteenth century. Some attempts were made by Gorbunova and Polukarov [8], Fedoteev et al. [9], and Faust [10], but they remained at a rather elementary level, obviously oriented to help practical electroplaters.

The results obtained until 1995 are summarized in the chapter by Despić and Jović [1]. Recently, published results on electrodeposition of alloys, with a particular attention on Ni–MoO_2 composite coatings used as catalysts for the hydrogen evolution reaction (HER), are discussed in this chapter.

1.2 Electrodeposition of Alloys from Aqueous Solutions

1.2.1 Conditions for Electrodeposition of Alloys

The metals immersed in the solution of their simple salts establish the reversible potential. The values of the reversible potentials for different metals could differ for about 3 V. Electrodeposition of metals could take place only at potentials more negative than the reversible ones. Accordingly, in the solution of ions of two metals (cf. Cu^{2+} and Zn^{2+}) with one being on the positive side of the potential scale (vs. SHE) (Cu) and another one being on the negative side of the potential scale (Zn), intensive deposition of Cu could take place at potentials at which Zn would not deposit at all. Taking into account that the reversible potentials of metals could change with the presence of different anions in the solution (complexation of metal ions), and that the rates of electrodeposition of different metals are usually different, it is possible to achieve conditions for simultaneous deposition of these two metals [1].

For simultaneous deposition of two metals, A and B, their deposition potentials (E) must be identical, E(A) = E(B), i.e.,

$$E_r(\mathrm{A}) + \eta(\mathrm{A}) = E_r(\mathrm{B}) + \eta(\mathrm{B}), \tag{1.1}$$

where E_r(A) and E_r(B) are reversible potentials of metals A and B, while η(A) and η(B) correspond to the overpotentials needed for the deposition of these two metals. The reversible potential could be changed by the change of metal ions concentration in the solution and by the temperature of the solution and is defined by the Nernst's equation:

$$E_r(\mathrm{A}) = E^{\Theta}(\mathrm{A}) + \frac{RT}{pF}\ln a(\mathrm{A}^{p+}), \tag{1.2}$$

$$E_r(\mathrm{B}) = E^{\Theta}(\mathrm{B}) + \frac{RT}{qF}\ln a(\mathrm{A}^{q+}), \tag{1.3}$$

where E^{Θ}(A) and E^{Θ}(B) are standard potentials of metals A and B, a activities of corresponding metal ions in the solution, and p and q numbers of electrons to be exchanged during the process of metal deposition.

The condition defined by Eq. (1.1) could be accepted only as a first approximation, since the potential of the metal deposition is undefined quantity if the value of corresponding current density is not known. It appears that a better definition of the conditions for simultaneous deposition of two metals would be current density at which both metals deposit with approximately the same current density. More precisely, for two-components alloy to be deposited with the molar ratio of the more noble metal x and the less noble metal $(1 - x)$, assuming that the Faraday's low is obeyed, following relations should be fulfilled:

$$x = \frac{n_\mathrm{A}}{n_\mathrm{A} + n_\mathrm{B}} = \frac{\frac{j_\mathrm{A}}{p}}{\frac{j_\mathrm{A}}{p} + \frac{j_\mathrm{B}}{q}} \tag{1.4}$$

and

$$(1 - x) = \frac{n_\mathrm{B}}{n_\mathrm{A} + n_\mathrm{B}} = \frac{\frac{j_\mathrm{B}}{q}}{\frac{j_\mathrm{A}}{p} + \frac{j_\mathrm{B}}{q}}, \tag{1.5}$$

where n_A and n_B are numbers of moles of components A and B. Hence, the current density ratio for the deposition of these two metals should be defined as

$$\frac{j_\mathrm{A}}{j_\mathrm{B}} = \frac{p}{q}\frac{x}{(1-x)}. \tag{1.6}$$

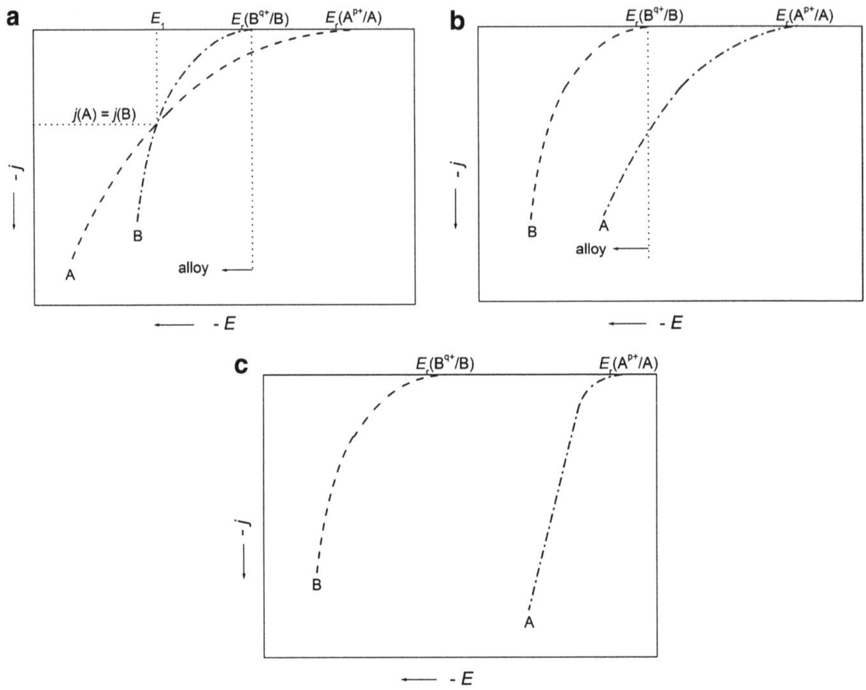

Fig. 1.1 Schematic presentation of the characteristic cases for alloy deposition

The condition defined by Eq. (1.6) could be achieved by proper adjustment of three essential variables: the concentration of the depositing ions at the electrode/solution interface (where the discharge occurs), the electrode potential, and the temperature.

In order to obtain better inside into the conditions defined by Eqs. (1.1, 1.2, 1.3, 1.4, 1.5, and 1.6), it is important to present polarization curves (current density versus potential relationships) for deposition of each metal. The characteristic cases are presented in Figs. 1.1 and 1.2. The first case is presented in Fig. 1.1a: the overpotential for deposition of the more noble metal A is higher than that for the less noble metal B. From the potentials $E_r(A^{p+}/A)$ to $E_r(B^{q+}/B)$ only more noble metal deposition occurs, while the deposition of alloy commences at the potential E_1. In the potential range from $E_r(B^{q+}/B)$ to E_1 metal A deposits with higher current density than metal B (the alloy contains more metal A than B). At the potential E_1 both metals deposit with the same current density, and the alloy contains the same amount of both metals. At the potentials more negative than E_1, the metal B deposits with higher current density and, accordingly, the alloy contains more metal B than metal A. The second case is presented in Fig. 1.1b: the overpotential for deposition of metal A is slightly lower than that for metal B, i.e., the polarization curves are almost parallel. Hence, the deposition of alloy commences at the potential $E_r(B^{q+}/B)$, while the alloy contains more metal A than B. If the difference between $E_r(A^{p+}/A)$ and $E_r(B^{q+}/B)$ is high and the overpotential for deposition of the

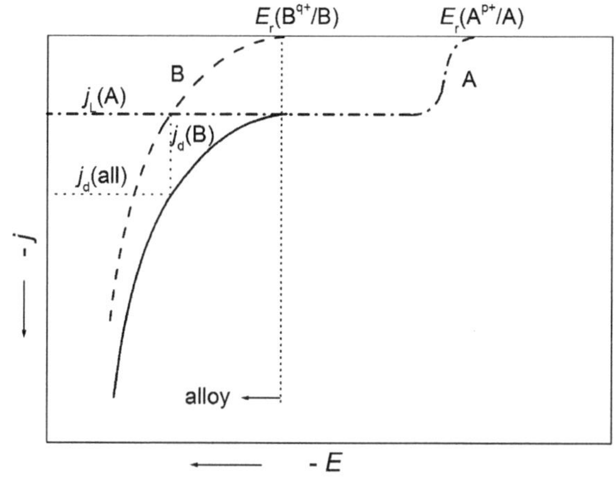

Fig. 1.2 Polarization curves for the deposition of more noble metal (A) and less noble metal (B): $j_L(A)$—diffusion limiting current density for the deposition of metal (A); $j_d(B)$—current density for the deposition of metal (B); $j_d(all)$—current density for the deposition of alloy

more noble metal A is lower than that for the less noble metal B, the third case, presented in Fig. 1.1c applies: in such a case alloy deposition is impossible. The difference between the reversible potentials of two metals could be changed (lowered) by the change of metal ions concentration (activity), and in most cases this is achieved by the complexation.

Simultaneous deposition of two metals is possible even if the difference in their reversible potentials is high if the applied current density for alloy deposition is higher than the diffusion limiting current density for the deposition of the more noble metal. Such a case is schematically presented in Fig. 1.2.

If $p = q = 2$ the molar ratios of metals (A) and (B) in the alloy are defined by the following relation:

$$\begin{aligned} x(\mathrm{A}) &= \frac{j_{\mathrm{L}}(\mathrm{A})}{j_{\mathrm{L}}(\mathrm{A}) + j_{\mathrm{d}}(\mathrm{B})} = \frac{j_{\mathrm{L}}(\mathrm{A})}{j_{\mathrm{d}}(\mathrm{all})}, \\ x(\mathrm{B}) &= \frac{j_{\mathrm{d}}(\mathrm{B})}{j_{\mathrm{L}}(\mathrm{A}) + j_{\mathrm{d}}(\mathrm{B})} = \frac{j_{\mathrm{d}}(\mathrm{B})}{j_{\mathrm{d}}(\mathrm{all})}. \end{aligned} \tag{1.7}$$

1.2.2 *Reversible Potential of Alloys in the Solution of Corresponding Ions*

According to the electroplating literature [4], when an alloy composed of metals A and B is immersed in the solution containing corresponding metal ions (A^{p+} and B^{q+}), its potential is termed as "static potential." In such a case it is desired to establish the conditions under which no net process would take place, so that the potential could be considered as the reversible potential of the alloy. Taking into account that an alloy may (and very often does) consist of several phases (intermetallic compounds), with each phase having different thermodynamic properties, it should be expected to have different reversible potential for each alloy composition. Such a case is essentially a nonequilibrium situation. Nevertheless, the thermodynamic properties of each phase, and accordingly the problem of its reversible potential, can be treated assuming that this is the only phase present in a given situation (it will be discussed later).

The problem of the reversible potential of alloys has been treated in the literature [1, 11, 12] in terms of the Nernst equation applied to the less noble metal only, while the more noble component of the alloy being considered as an inert metal matrix. Such approach is not appropriate since it yields unrealistic results. Hence, in a proper approach the thermodynamic property of the alloy cannot be assigned to an individual component of the alloy. Instead, a phase should be treated as composed of a chemical entity of stoichiometric composition corresponding to the alloy composition. Accordingly, a phase can be described as $A_xB_{(1-x)}$ and the formation of one mole of the substance characterizing this phase should be presented by the following chemical reaction:

$$xA + (1-x)B \Leftrightarrow A_xB_{(1-x)}. \tag{1.8}$$

1.2.2.1 The Influence of the Gibbs Energy of Phase Formation

The change of the Gibbs energy in the formation of the phase $A_xB_{(1-x)}$ can be described in terms of their standard partial molar Gibbs energies (standard chemical potentials) as

$$\Delta_f G^\ominus\left(A_xB_{(1-x)}\right) = \mu^\ominus\left(A_xB_{(1-x)}\right) - x\mu^\ominus(A) - (1-x)\mu^\ominus(B). \tag{1.9}$$

The phase $A_xB_{(1-x)}$ can also be formed in the electrochemical cell from the ions of both metals (A^{p+} and B^{q+}) present in the solution. In such a case the electrochemical cell is composed of an electrode made of the alloy phase as cathode and a standard hydrogen electrode as anode. The cell reaction is then

$$xA^{p+} + (1-x)B^{q+} + \left[\frac{xp + (1-x)q}{2}\right]H_2 = A_xB_{(1-x)} + [xp + (1-x)q]H^+. \tag{1.10}$$

The standard Gibbs energy change for this reaction is defined by the equation:

$$\begin{aligned}\Delta G^\ominus_{cell} &= \mu^\ominus\left(A_xB_{(1-x)}\right) + [xp + (1-x)q]\mu^\ominus(H^+) - x\mu^\ominus(A^{p+}) \\ &\quad - (1-x)\mu^\ominus(B^{q+}) - \left[\frac{xp + (1-x)q}{2}\right]\mu^\ominus(H_2).\end{aligned} \tag{1.11}$$

Taking into account that

$$[xp + (1-x)q]\mu^\ominus(H^+) - \left[\frac{xp + (1-x)q}{2}\right]\mu^\ominus(H_2) = 0, \tag{1.12}$$

it follows that the standard Gibbs energy change in the electrochemical cell is defined by the equation:

$$\Delta G^{\ominus}_{\text{cell}} = \mu^{\ominus}\left(\mathrm{A}_x\mathrm{B}_{(1-x)}\right) - x\mu^{\ominus}(\mathrm{A}^{\mathrm{p}+}) - (1-x)\mu^{\ominus}(\mathrm{B}^{\mathrm{q}+}). \tag{1.13}$$

The electromotive force of this cell, which is identical with the electrode potential of the alloy phase on the standard hydrogen scale, is given by

$$E\left(\mathrm{A}_x\mathrm{B}_{(1-x)}\right) = E^{\ominus}\left(\mathrm{A}_x\mathrm{B}_{(1-x)}\right) + \frac{RT}{[xp + (1-x)q]F} \ln a(\mathrm{A}^{\mathrm{p}+})^x a(\mathrm{B}^{\mathrm{q}+})^{(1-x)}. \tag{1.14}$$

The standard electrode potential of the alloy phase, $E^{\ominus}(\mathrm{A}_x\mathrm{B}_{(1-x)})$, is related to the standard Gibbs energy change in the cell, defined by the following equation:

$$\begin{aligned} E^{\ominus}\left(\mathrm{A}_x\mathrm{B}_{(1-x)}\right) &= \frac{-\Delta G^{\ominus}_{\text{cell}}}{[xp + (1-x)q]F} \\ &= \frac{-\mu^{\ominus}\left(\mathrm{A}_x\mathrm{B}_{(1-x)}\right) + x\mu^{\ominus}(\mathrm{A}^{\mathrm{p}+}) + (1-x)\mu^{\ominus}(\mathrm{B}^{\mathrm{q}+})}{[xp + (1-x)q]F}, \end{aligned} \tag{1.15}$$

where $[xp + (1 - x)q]F$ represents the total number of electrons exchanged in one act of the cell reaction [Eq. (1.10)].

The standard Gibbs energies of formation of the ions relative to that of the hydrogen ion (taken as zero), which are equal to the standard chemical potentials of the ions, are related to the standard potentials of the corresponding metals on the standard hydrogen scale as

$$\Delta_{\mathrm{f}}G^{\ominus}(\mathrm{A}^{\mathrm{p}+}) = \mu^{\ominus}(\mathrm{A}^{\mathrm{p}+}) = pFE^{\ominus}(\mathrm{A}^{\mathrm{p}+}/\mathrm{A}) \tag{1.16}$$

and

$$\Delta_{\mathrm{f}}G^{\ominus}(\mathrm{B}^{\mathrm{q}+}) = \mu^{\ominus}(\mathrm{B}^{\mathrm{q}+}) = pFE^{\ominus}(\mathrm{B}^{\mathrm{q}+}/\mathrm{B}). \tag{1.17}$$

Substituting Eqs. (1.16) and (1.17) into Eq. (1.15) and then substituting the resulting $E^{\ominus}(\mathrm{A}_x\mathrm{B}_{(1-x)})$ into Eq. (1.14), the reversible potential of the alloy phase is obtained as

$$\begin{aligned} E\left(\mathrm{A}_x\mathrm{B}_{(1-x)}\right) &= \frac{xp}{[xp + (1-x)q]}\left[E^{\ominus}(\mathrm{A}^{\mathrm{p}+}/\mathrm{A}) + \frac{RT}{pF} \ln a(\mathrm{A}^{\mathrm{p}+})\right] \\ &+ \frac{(1-x)q}{[xp + (1-x)q]}\left[E^{\ominus}(\mathrm{B}^{\mathrm{q}+}/\mathrm{B}) + \frac{RT}{qF} \ln a(\mathrm{B}^{\mathrm{q}+})\right] \\ &- \frac{\mu^{\ominus}\left(\mathrm{A}_x\mathrm{B}_{(1-x)}\right)}{[xp + (1-x)q]F}, \end{aligned} \tag{1.18}$$

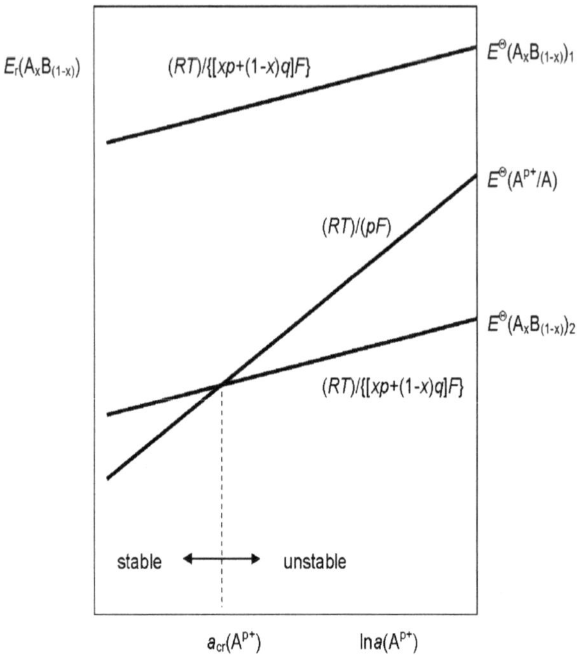

Fig. 1.3 Schematic presentation of the range of stability of phases $(A_xB_{(1-x)})_1$ and $(A_xB_{(1-x)})_2$ as a function of the concentration of more noble metal ions, ln $a(A^{p+})$ (Reprinted from [1] with the permission of Springer)

where $\mu^{\ominus}(A_xB_{(1-x)}) = \Delta_f G^{\ominus}(A_xB_{(1-x)}) - x\mu^{\ominus}(A) - (1-x)\mu^{\ominus}(B)$ is the standard chemical potential of the alloy phase relative to those of metal constituents. From Eq. (1.18) it could be concluded that: (a) the reversible potential of an alloy phase depends on the activities of ions of both metals in solution and (b) the dominant role in determining the position of the standard potential of the alloy phase relative to the pure metals is played by the standard molar Gibbs energy of the alloy phase formation [1].

1.2.2.2 Stability of Phases in the Solution of Corresponding Ions

When ions corresponding to metal constituents of an alloy phase are present in the solution, it is necessary to take into consideration their tendency to form other possible alloy phases or undergo reduction to pure metals. Accordingly, if another alloy phase or a pure metal would yield a more noble reversible potential, thermodynamic conditions for anodic dissolution of the existing alloy phase are created ("replacement reaction"). Hence, the considered alloy phase becomes unstable tending to undergo corrosive degradation. It can be shown that the instability depends on the activity of the metal ions in solution. Such a case [1] is schematically presented in Fig. 1.3.

Equation (1.14) defines the reversible potential of any alloy phase, as well as the reversible potential of the pure more noble metal ($x = 1$). Assuming a constant activity of the ions of the less noble constituent of the alloy ($a(B^{q+}) = 1$), it can be seen in Fig. 1.3 that the slope of a plot of E_r versus ln $a(A^{p+})$ is always larger for the pure more noble metal than for the alloy phase, since $\{x/[xp + (1 - x)q]\} < 1/p$ for any $x < 1$. If $E^\Theta(A_xB_{(1-x)})_1$ of the alloy phase 1 turns out to be more positive than other alloy phase 2, $E^\Theta(A_xB_{(1-x)})_2$, or of the pure more noble metal, $E^\Theta(A^{p+}/A)$, then the alloy phase will remain stable over the entire range of activities of ions of the more noble metal. If $E^\Theta(A_xB_{(1-x)})_2 < E^\Theta(A^{p+}/A)$, as shown in Fig. 1.3, there is a crossing point between the two functions at the activity $a_{cr}(A^{p+})$. At the activities larger than $a_{cr}(A^{p+})$ the metal would tend to precipitate on the account of dissolution of the alloy phase 2, i.e., the alloy phase 2 will be unstable. Conversely, at activities smaller than $a_{cr}(A^{p+})$ the alloy phase 2 will remain stable in the solution [1].

Taking into account discussion about the alloy phase stability, it is recommended not to introduce into the solution ions of the more noble metal, but only those of the less noble one. At the beginning of the nineteenth century (1907), Pushin [12] was aware of this fact and in his work the potentials of alloys were measured in the solution containing only less noble metal ions, and the same metal was used as the reference electrode.

1.2.3 *Types of Electrodeposition of Alloys*

According to Brenner [4], electrochemical codeposition of two metals to form an alloy could be: equilibrium, irregular, regular, anomalous, and induced. This classification is based on the relation between the composition of the deposited alloy (percentages of metals in the alloy) and the “metal ratio” which represents percentages of corresponding metal ions in the solution independently of their ionic form (“stoichiometric concentrations”). For the regular, irregular, and equilibrium codeposition, it is characteristic that the relative content of metals in the deposited alloy corresponds to that expected from the relation between their reversible potentials, whereas anomalous codeposition corresponds to the reverse situation. Induced codeposition is characteristic for the metals which cannot be deposited from the aqueous solutions, Mo, Ti, W, Ge, but can be codeposited with the iron-group metals (Fe, Co, Ni).

1.2.3.1 Equilibrium Codeposition

Equilibrium codeposition implies a common reversible potential for both metal constituents so that the reduction of both metal ions would take place at potentials more negative than the reversible ones. To close the gap between the reversible potentials of depositing metals, it is necessary to make the concentration of simple salts (undergoing complete dissociation) of the more noble metal impractically low and of the less noble metal impractically high. As stated in Sect. 1.2.2.1 the best way to overcome this problem could result from complexation of metal ions with different ligands. Complexation usually changes the activity of the resulting species in solution by many orders of magnitude, while keeping the total amount of one or other metal in solution sufficiently high for a good supply of plating material to the cathode. It is very often case that the ions of both metals form complexes with one and the same ligand with similar values of the stability constants, so that the change of the potential of the deposition of each metal is the same (or similar) value. Usually, in such a case, the complexation with two different ligands could result in a more pronounced change of the deposition potentials of two metals. In a further text an example for the deposition of the Ni–Sn alloy from the solution containing pyrophosphate and glycine ligands is presented [13].

According to the literature [14] standard potential of the Ni deposition is −0.23 V vs. SHE, while that for Sn is −0.1364 V vs. SHE and accordingly Ni is less noble metal. Taking into account that the overvoltage for Ni deposition [15] is much higher than that for Sn deposition, the difference between the potentials of deposition of these two metals should be larger than that of their standard potentials. In the data presented in Dean's Handbook of Chemistry [16] Ni forms two pyrophosphate complexes, $[Ni(P_2O_7)]^{2-}$ and $[Ni(P_2O_7)_2]^{6-}$, and three glycine complexes, $[Ni(NH_2CH_2COO)]^+$, $[Ni(NH_2CH_2COO)_2]$, and $[Ni(NH_2CH_2COO)_3]^-$. There are only three papers in the literature with the data for different complexes of Ni and Sn in the pyrophosphate and glycine solutions, the data proposed by Duffield et al. [17], Turyan et al. [18], and Orekhova et al. [19]. Corresponding reactions for the formation of different complexes and their formation (stability) constants are given in the work of Duffield et al. [17]. All species and their stability constants used for the calculation of the distribution of different complexes in the solution containing Sn, Ni, pyrophosphate, and glycine ions are listed in Table 1.1.

The calculation of the distribution of complexes in the solution containing pyrophosphate and glycine showed that $[Sn(P_2O_7)_2]^{6-}$ is dominant complex with Sn at pH 8.0, while two complexes of Ni dominate: complex $[Ni(P_2O_7)_2]^{6-}$ and complex $[Ni(NH_2CH_2COO)_3]^-$. This is shown in Fig. 1.4.

The values of the equilibrium potentials of prevailing complexes (E_{eq}), calculated using explanations based on the Gibbs energy change for reaction of certain complex formation [20] (assuming that the ions activities are equal to their concentrations), are also presented in Table 1.1. As can be seen, the equilibrium potential for deposition of Sn by the reduction of $[Sn(P_2O_7)_2]^{6-}$ complex is −0.847 V vs. SCE, while the equilibrium potentials for the reduction of $[Ni(P_2O_7)_2]^{6-}$ and $[Ni(NH_2CH_2COO)_3]^-$ complexes are more positive, being about −0.716 V vs. SCE, and situation becomes opposite to that for deposition from the solution of simple ions. After the complexation Ni becomes more noble metal, while Sn becomes less noble one. Hence, it could be concluded that at pH 8.0 Sn would deposit from the complex $[Sn(P_2O_7)_2]^{6-}$, while Ni would deposit simultaneously from two complexes, $[Ni(P_2O_7)_2]^{6-}$ and $[Ni(NH_2CH_2COO)_3]^-$, in the presence of both complexing anions. The equilibrium potentials for deposition

Table 1.1 All complexes present in the solution containing 0.1 M $SnCl_2$ + 0.1 M $NiCl_2$ + 0.6 M $K_4P_2O_7$ + 0.3 M NH_2CH_2COOH, their concentrations, stability constants, and equilibrium potentials of prevailing complexes

Complexes	log β	Conc. (M)	E_{eq} (V vs. SCE)
$[H(P_2O_7)]^{3-}$	8.14		
$[H_2(P_2O_7)]^{2-}$	14.01		
$[H_3(P_2O_7)]^{-}$	15.78		
$[H_4(P_2O_7)]$	16.63		
$[H(NH_2CH_2COO)]$	9.64		
$[H_2(NH_2CH_2COO)]^{+}$	12.05		
$[Sn(NH_2CH_2COO)H]^{2+}$	12.78		
$[Sn(NH_2CH_2COO)]^{+}$	10.02		
$\mathbf{[Sn(P_2O_7)]^{2-}}$	**13.05**	**0.007**	**−0.847**
$[Sn(P_2O_7)H]^{-}$	15.92		
$[Sn(P_2O_7)H_2]$	17.47		
$\mathbf{[Sn(P_2O_7)_2]^{6-}}$	**16.27**	**0.093**	**−0.847**
$[Sn(P_2O_7)_2H]^{5-}$	22.31		
$[Sn(P_2O_7)_2H_2]^{4-}$	26.79		
$[Sn(P_2O_7)_2H_3]^{3-}$	30.07		
$[Sn(P_2O_7)_2H_4]^{2-}$	31.58		
$[Sn(P_2O_7)OH]^{3-}$	5.32		
$[Sn(P_2O_7)(OH)_2]^{2-}$	−4.77		
$[Sn(P_2O_7)_2OH]^{5-}$	7.04		
$[Ni(NH_2CH_2COO)]^{+}$	5.60		
$\mathbf{[Ni(NH_2CH_2COO)_2]}$	**10.40**	**0.009**	**−0.716**
$\mathbf{[Ni(NH_2CH_2COO)_3]^{-}}$	**13.80**	**0.057**	**−0.716**
$\mathbf{[Ni(P_2O_7)]^{2-}}$	**5.80**	**0.005**	**−0.716**
$\mathbf{[Ni(P_2O_7)_2]^{6-}}$	**7.40**	**0.029**	**−0.716**

Reprinted from [13] with the permission of Electrochemical Society

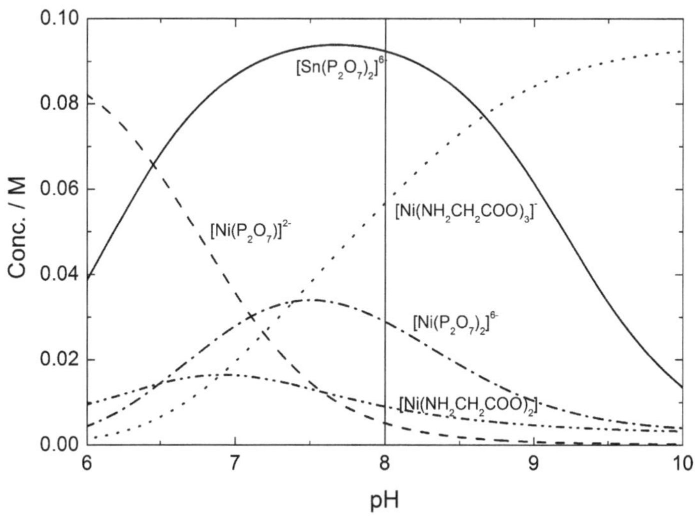

Fig. 1.4 Distribution of different complexes in the solution containing 0.1 M $SnCl_2$ + 0.1 M $NiCl_2$ + 0.6 M $K_4P_2O_7$ + 0.3 M NH_2CH_2COOH as a function of the solution pH

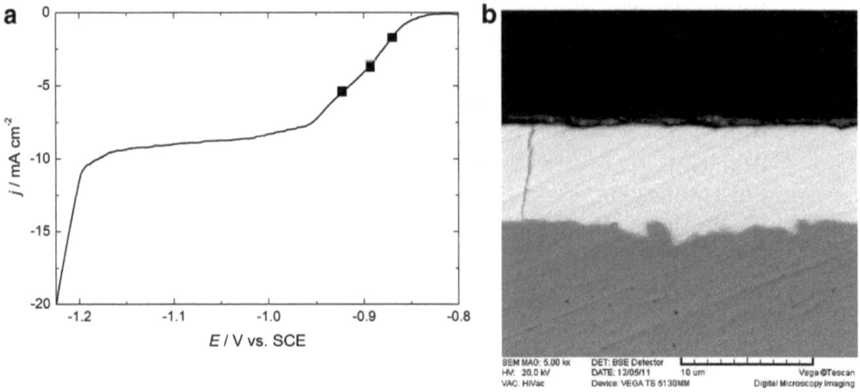

Fig. 1.5 (**a**) Polarization curve for deposition of the Ni–Sn alloy onto Ni electrode. (**b**) Typical cross section of coatings obtained at different current densities marked with solid squares in Fig. 1.5a (Reprinted from [21] with the permission of Int. J. Hydrogen Energy)

of Sn and Ni still differ for 0.131 V. As already stated, because of high overvoltage for Ni deposition [13], it could be expected that two metals possess identical, or similar, potential of deposition. This is exactly the case for these two metals in the pyrophosphate–glycine solution. The polarization curve for Ni–Sn alloy deposition onto Ni electrode is shown in Fig. 1.5a [21]. The deposition process commences at about −0.83 V vs. SCE being activation controlled down to about −0.95 V vs. SCE, while in the potential range from about −0.95 V to about −1.20 V well-defined diffusion limiting current density (−10 mA cm^{-2}) is established. In the region of the activation control (squares marked on Fig. 1.5a), Ni–Sn alloy coatings were deposited at the current densities of −2, −4, and −6 mA cm^{-2}. Flat and compact deposits were obtained in all cases, as shown in Fig. 1.5b. The composition of the coatings changed with the increase of cathodic current density from about 37 at.% Ni (for sample obtained at −2 mA cm^{-2}) to about 45 at.% Ni (for sample obtained at −6 mA cm^{-2}) [21], but in all cases both metals were present in the coating, indicating a good example for equilibrium codeposition.

1.2.3.2 Irregular Codeposition

The irregular type of codeposition is very often characterized by simultaneous influence of cathodic potential and diffusion phenomena, i.e., it mainly occurs under the activation and/or mixed control of the deposition processes. The rate of deposition in such a case is expressed by Butler–Volmer equation which is usually used for the kinetics of electrochemical processes [1]:

$$\frac{j}{j^o} = \frac{j_o}{j^o}\left[\exp\left(\frac{\alpha_a F}{RT}\eta\right) - \left(\frac{c}{c_o}\right)\exp\left(\frac{-\alpha_c F}{RT}\eta\right)\right], \tag{1.19}$$

where $\eta = E - E_r$ corresponds to the overpotential needed for the deposition of metals [see Eq. (1.1)], c and c_o represent concentrations of the discharging species at the surface of the electrode and in the bulk of the solution, respectively, j_o is "exchange current density" related to the rate constant of the deposition process, and α_a and α_c are anodic and cathodic transfer coefficients related to the mechanism of discharge (note that for a cathodic process both η and j acquire negative signs. Also note that the current densities are divided by arbitrarily chosen unit current j^o in order to obtain dimensionless values for further use).

If the discharge of depositing species is sufficiently slow so that their supply to the electrode surface occurs without difficulty, the concentration c virtually does not deviate from c_o, and such a case is termed "activation controlled" deposition with the rate-determining step being the activation energy of the discharge process. At any cathodic overpotential larger than −40 mV, the first term in Eq. (1.19) becomes negligible, so that this equation can be transformed into a simpler one, known as the Tafel equation:

$$\eta = a - b\log\left(-\frac{j}{j^o}\right), \tag{1.20}$$

where the Tafel constant a is

$$a = \frac{2.3RT}{\alpha_c F}\log\left(\frac{j_o}{j^o}\right), \tag{1.21}$$

while the slope of the linear dependence obtained from a plot η versus $\log(-j)$ (Tafel slope) is

$$b = \frac{2.3RT}{\alpha_c F}. \tag{1.22}$$

The above reasoning applies equally and independently to both metals (A) and (B), j_A and j_B, and the total current density being $j_{alloy} = j_A + j_B$.

It should be stated here that the concept of overpotential is related to the reversible potential of a pure metal in a given solution. In the case of codeposition of two metals and the formation of a phase $A_xB_{(1-x)}$, this potential has no physical meaning since it represents an arbitrary point to which j_o is related.

Typical cases of activation-controlled codeposition of the metals A (j_A) and B (j_B), presented as polarization curves for pure metals and an alloy phase (j), are shown in Fig. 1.6.

The Tafel functions, presented in Fig. 1.7, indicate linear relationships between the logarithm of the partial current densities and the electrode potential. When this

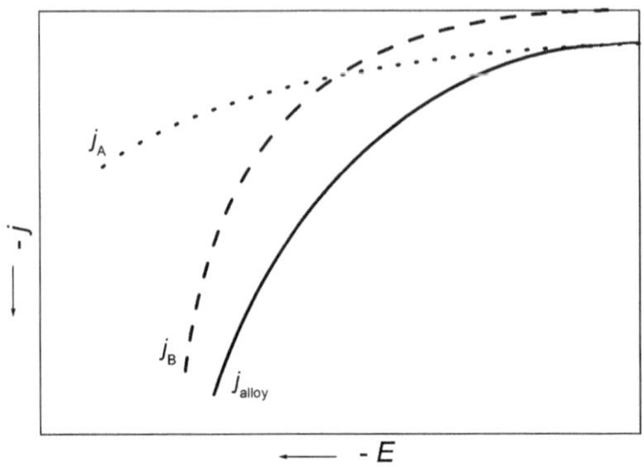

Fig. 1.6 Typical cases of activation-controlled codeposition of the metals A (j_A) and B (j_B), presented as polarization curves for pure metals and an alloy phase ($j_{alloy} = j_A + j_B$) (Reprinted from [1] with the permission of Springer)

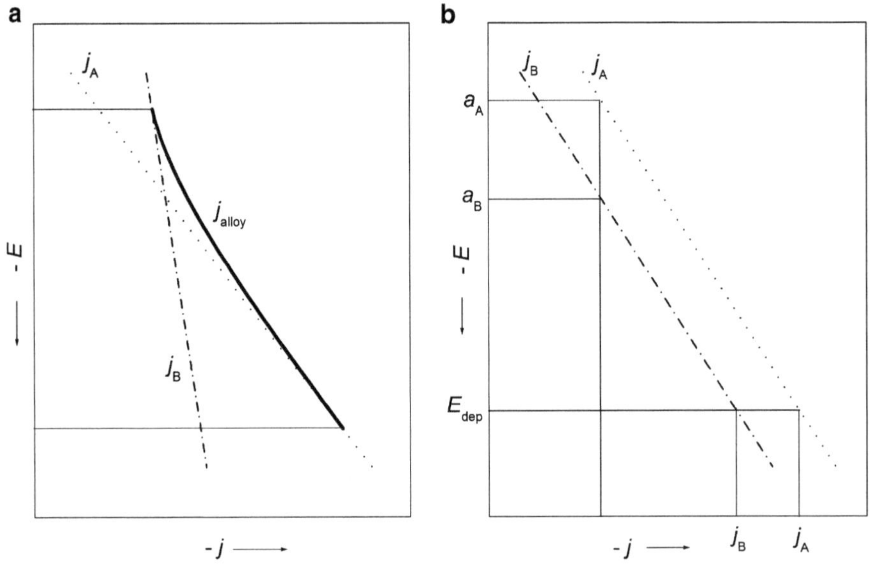

Fig. 1.7 Tafel functions for activation-controlled codeposition of the metals A (j_A) and B (j_B) and an alloy (j_{alloy}). (**a**) Different slopes of Tafel functions for pure metals deposition. (**b**) The same slopes of Tafel functions for pure metals deposition (Reprinted from [1] with the permission of Springer)

is the case, the total current density j_{alloy} cannot be a linear function of potential in the region in which the two partial current densities are comparable, since log ($j_A + j_B$) $\neq$ log j_A + log j_B (Fig. 1.7a). When the Tafel function of the total current density merges with one or the other partial current density line, then one or the other metal is obtained virtually pure. In the extreme case in which the Tafel slopes for both depositing metals are equal, as shown in Fig. 1.7b, the difference between log j_A and log j_B remains constant, i.e., the composition of the alloy is constant at all potentials. In such a case the actual composition of the alloy depends on the difference between the values of the Tafel constants (a) for the two metals.

Considering the values of the Tafel constants and Tafel slopes, it is possible to analyze the factors determining the deviation of the metal ratio in the alloy from the metal ratio in solution. At any constant potential following relation is valid:

$$a_{\mathrm{A}} - b_{\mathrm{A}}\log\left(-\frac{j_{\mathrm{A}}}{j_{\mathrm{o}}}\right) = a_{\mathrm{B}} - b_{\mathrm{B}}\log\left(-\frac{j_{\mathrm{B}}}{j_{\mathrm{o}}}\right). \tag{1.23}$$

According to Eq. (1.6) one can derive:

$$\log\frac{x_{\mathrm{A}}}{x_{\mathrm{B}}} = \log\left(\frac{pj_{\mathrm{A}}}{qj_{\mathrm{B}}}\right) = \left[\left(\frac{a_{\mathrm{A}}}{b_{\mathrm{A}}} - \frac{a_{\mathrm{B}}}{b_{\mathrm{B}}}\right) - \left(\frac{b_{\mathrm{B}} - b_{\mathrm{A}}}{b_{\mathrm{A}}b_{\mathrm{B}}}\right)E\right]\frac{p}{q}. \tag{1.24}$$

Returning to the linear coordinates one obtains:

$$\frac{x}{1-x} = \frac{x_{\mathrm{A}}}{x_{\mathrm{B}}} = \frac{(j_{\mathrm{o}})_{\mathrm{A}}pa(\mathrm{A}^{\mathrm{p}+})^{(\alpha_{\mathrm{c}})_{\mathrm{A}}/p}}{(j_{\mathrm{o}})_{\mathrm{B}}qa(\mathrm{B}^{\mathrm{q}+})^{(\alpha_{\mathrm{c}})_{\mathrm{B}}/q}}\exp\left(\frac{RT}{F}\right)$$
$$\left\{\left[(\alpha_{\mathrm{c}})_{\mathrm{A}}E^{\ominus}(\mathrm{A}^{\mathrm{p}+}/\mathrm{A}) - (\alpha_{\mathrm{c}})_{\mathrm{B}}E^{\ominus}(\mathrm{B}^{\mathrm{q}+}/\mathrm{B})\right] - \frac{(\alpha_{\mathrm{c}})_{\mathrm{A}} - (\alpha_{\mathrm{c}})_{\mathrm{B}}}{(\alpha_{\mathrm{c}})_{\mathrm{A}}^{2}(\alpha_{\mathrm{c}})_{\mathrm{B}}^{2}}\right\}. \tag{1.25}$$

Hence, it appears that the composition of the alloy follows a complex dependence on the metal ratio, involving all the thermodynamic and kinetic parameters determining activation-controlled codeposition [1].

As shown in Fig. 1.7 the Tafel lines are meant to pertain to the deposition of pure metals under the assumption that alloying does not change the Tafel constants and that the current density for alloy deposition should be a sum of partial current densities for pure metals. However, in practice, deviation from such behavior has been recorded, and an attempt to explain this phenomenon has been reported by Gorbunova and Polukarov [8] on an extreme case in which surface diffusion of A across the grains of B and nucleation of new grains of A are strongly inhibited.

An example for this type of alloy deposition is presented in Fig. 1.33 for the system Ag–Cd (cf. Sect. 1.3.3.3).

1.2.3.3 Regular Codeposition

Regular codeposition assumes transport-controlled codeposition in which diffusion of metal ions of both metals is a rate-determining step in the overall codeposition reaction.

Under steady state conditions of deposition the diffusion is governed by Fick's first law [22]:

$$\frac{c}{c_o} = \frac{j_L + j}{j_L} \tag{1.26}$$

with

$$j_L = \frac{zFD}{\delta} c_o, \tag{1.27}$$

where D is the diffusion coefficient of the depositing species and δ is the Nernst diffusion layer thickness.

Introducing Eq. (1.26) into Eq. (1.19) and rearranging following relation is obtained:

$$j = \frac{j_o\left[\exp\left(\frac{\alpha_a F}{RT}\eta\right) - \exp\left(-\frac{\alpha_c F}{RT}\eta\right)\right]}{1 + \frac{j_o}{j_L}\exp\left(-\frac{\alpha_c F}{RT}\eta\right)}. \tag{1.28}$$

At increasing values of overpotential, the second term in the denominator becomes overwhelming and the current density tends to a potential-independent diffusion limiting one, j_L.

Under the conditions of diffusion control, a plot of E versus log $[(-j_L/j) - 1]$ should be made instead of the regular Tafel plot, and the relationship should be linear with the slope of $(2.3RT/\alpha_c F)$.

One of the cases of regular codeposition is the deposition of Ag–Pd alloy from high concentration chloride (12 M LiCl) containing bath [23]. Polarization curves for the deposition of Ag–Pd alloy from the solution containing 0.005 M $PdCl_2$ + 0.05 M AgCl + 12 M LiCl + 0.1 M HCl (t = 80 °C), recorded onto rotating glassy carbon electrode under the conditions of stationary (RPM = 0) and convective (RPM = 200) diffusion are presented in Fig. 1.8a. The Pd deposition commences at about 0.07 V vs. SCE, while the codeposition of Ag begins at about −0.07 V vs. SCE. A well-defined diffusion limiting current densities for Pd (j_L(Pd)) and alloy (j_L(alloy)) deposition, representing actually the sum of j_L(Pd) and j_L(Ag), could be detected on the polarization curves, indicating that the j_L(Ag) is lower than that of Pd (taking into account that the concentration of AgCl is ten times higher than that of $PdCl_2$) and that the diffusion coefficient for Ag deposition is lower than that for Pd. This is confirmed in Fig. 1.8b. The deposition of Pd occurs by the reduction of $[PdCl_4]^{2-}$, while the deposition of Ag occurs by the reduction of $[AgCl_4]^{3-}$.

From the slopes of the linear j_L versus $\omega^{1/2}$ dependences [23] (Fig. 1.8b), confirming that the diffusion of both species obeys Levich's equation:

$$j_L = 0.62zFD^{2/3}\nu^{-1/6}c_o\omega^{1/2} \tag{1.29}$$

(ν—kinematic viscosity, ω—rotation speed), corresponding values of D for both species are obtained: $D([AgCl_4]^{3-}) = 5.5 \times 10^{-6}$ cm^2 s^{-1} and $D([PdCl_4]^{2-}) = 1.8 \times 10^{-5}$ cm^2 s^{-1}.

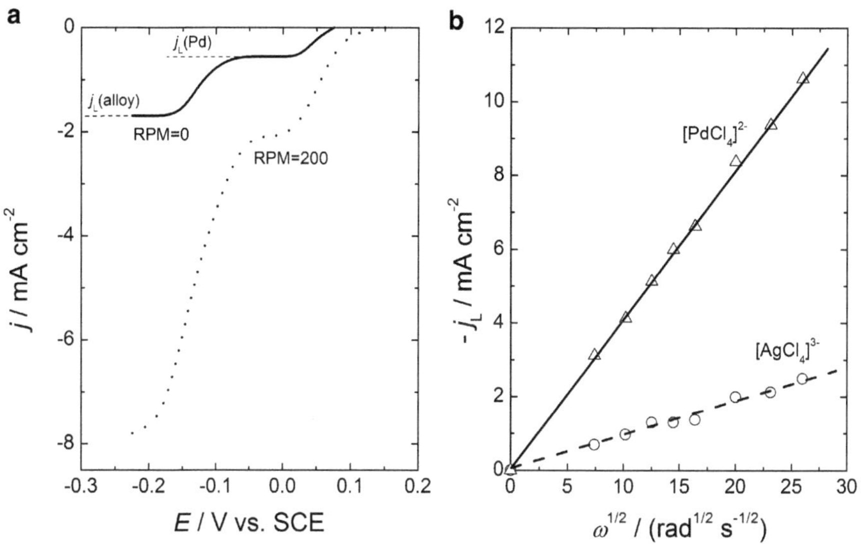

Fig. 1.8 (**a**) Polarization curves for the deposition of Ag–Pd alloy from the solution containing 0.005 M $PdCl_2$ + 0.05 M AgCl + 12 M LiCl + 0.1 M HCl (t = 80 °C), recorded onto rotating glassy carbon electrode. (**b**) The j_L versus $\omega^{1/2}$ dependences for pure Ag and pure Pd deposition onto rotating glassy carbon electrode from the solutions containing 0.005 M $PdCl_2$ + 12 M LiCl + 0.1 M HCl and 0.005 M AgCl + 12 M LiCl + 0.1 M HCl (t = 80 °C) (Reprinted from [23] with the permission of Serbian Chemical Society)

1.2.3.4 Anomalous Codeposition

According to Brenner's classification [4], anomalous codeposition is characterized by the fact that the less noble metal deposits before the more noble one as the potential is driven cathodic. As a consequence, the content of the less noble metal in the alloy is higher than its content in the solution.

An excellent example [24] is found in the case of Co–Ni alloy deposition. Pure Co and pure Ni were deposited at RPM = 1,000 onto a gold disc electrode from the solutions containing 0.2 M $CoSO_4$ + 0.2 M $Na_3C_6H_5O_7$ and 1 M $NiSO_4$ + 0.2 M $Na_3C_6H_5O_7$, respectively. For deposition of their alloys of different compositions, only the concentration of Co^{2+} has been changed, being 0.005, 0.01, 0.025, 0.05, and 0.2 M, respectively. Alloy layers of different compositions of the thickness of approximately 0.34 μm were obtained by galvanostatic deposition at two different current densities [$j = -2.5$ mA cm^{-2} (○) and $j = -28.5$ mA cm^{-2} (□)] to the same amount of charge ($Q_{dep} = -1$ C cm^{-2}). Polarization curves for pure metals deposition (Co and Ni) and Co–Ni alloy deposition (alloy) from the solution containing 0.025 M $CoSO_4$ + 1 M $NiSO_4$ + 0.2 M $Na_3C_6H_5O_7$ are presented in Fig. 1.9a (all other polarization curves for Co–Ni alloy deposition are placed between that for Ni and that for alloy—shaded area). As can be seen all polarization curves for Co–Ni alloy deposition are placed at more positive potentials than either of pure metals, which is clear indication for the anomalous codeposition of these two metals.

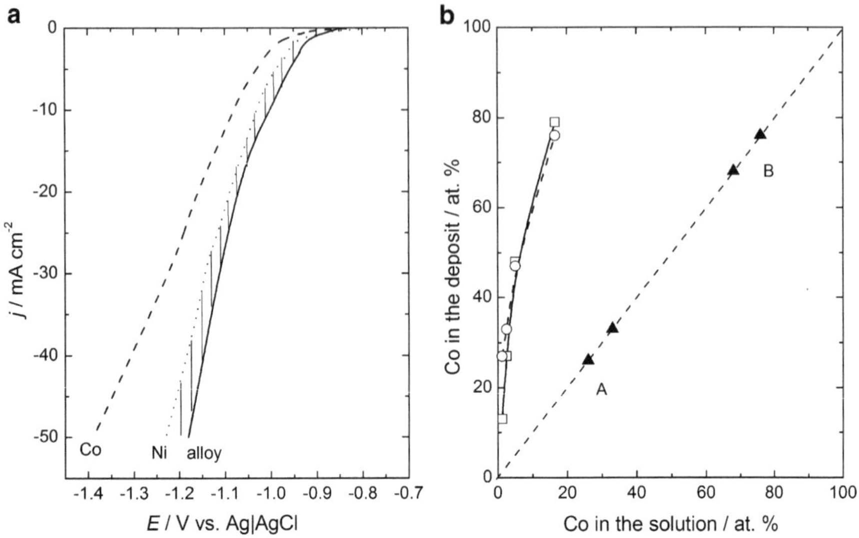

Fig. 1.9 (**a**) Polarization curves (corrected for IR drop) for the deposition of pure metals (Co and Ni) and Co–Ni alloy at the gold disc electrode (RPM = 1,000) obtained from the solutions: 0.2 M $CoSO_4$ + 0.2 M $Na_3C_6H_5O_7$ (Co); 1 M $NiSO_4$ + 0.2 M $Na_3C_6H_5O_7$ (Ni); 0.025 M $CoSO_4$ + 1 M $NiSO_4$ + 0.2 M $Na_3C_6H_5O_7$ (alloy). (**b**) Brenner's diagram for the system Co–Ni: samples obtained at $j = -2.5$ mA cm^{-2} (*open circle*) and at $j = -28.5$ mA cm^{-2} (*open square*) (Reprinted from [24] with the permission of Elsevier)

After deposition all Co–Ni alloy samples were dissolved in HCl, and the amounts of Co and Ni were determined by the AAS technique. It was found that the content of Co in the deposit increases with increasing concentration of Co^{2+} ions in the solution, changing from ~8 to ~80 at.%. The Brenner's diagram presented in Fig. 1.9b clearly shows pronounced anomalous codeposition.

Different explanations for such behavior are offered in the literature [4]. The most likely one appears to be "the hydroxide suppression mechanism" [25–28]. According to this concept, coevolution of hydrogen during the deposition causes an increase of pH at the electrode/solution interface, producing hydrolysis of less noble metal species and their precipitation as a layer of solid hydroxide. Formed hydroxide layer provides a good supply of ions of the less noble metal for their discharge and deposition, but suppresses the transport of species of the more noble metal to the cathode surface, causing anomalous codeposition.

1.2.3.5 Induced Codeposition

While it has been shown that Mo, W, Ti, and Ge cannot be separately deposited from aqueous solutions, they can be codeposited with the iron-group metals (Fe, Ni, Co) in the presence of appropriate complexing agents. This alloy-type deposition was defined by Brenner [4] as induced codeposition.

System Mo–Ni

In most of the papers concerning the process of induced codeposition of Ni and Mo, where molybdenum was present in the form of molybdate ion (MoO_4^{2-}), it was assumed that both metals were reduced to a metallic state, producing a Mo–Ni alloy coating. The most probable mechanism of alloy deposition was proposed by Podlaha and Landolt [29–31] and Marlot et al. [32]. Their investigations were performed under controlled mass transport conditions (rotating cylinder electrode). It was shown that mass transport control is an important factor for induced codeposition process. In a nickel-rich electrolyte molybdenum deposition is mass transport limited and the alloy composition is strongly influenced by the electrode rotation rate, while in a molybdenum-rich electrolyte the rate of molybdenum deposition is limited by the flux of nickel ions and alloy composition is independent of hydrodynamic effects [29]. The model assumes that nickel deposition occurs on the surface not covered by the molybdate reaction intermediate, by direct reduction of nickel species (all of them being complex of Ni^{2+} cations with the citrate anions) independently on the molybdate reaction, which can occur only in the presence of nickel species [29–32]. The model of Mo–Ni alloy deposition is described by following reduction reactions [30]:

$$[Ni(C_6H_5O_7)]^- + 2e^- \rightarrow Ni(s) + (C_6H_5O_7)^{-3} \tag{1.30}$$

$$MoO_4^{-2} + [Ni(C_6H_5O_7)]^- + 2H_2O + 2e^- \rightarrow [Ni(C_6H_5O_7)MoO_2]_{ads}^- + 4OH^- \tag{1.31}$$

$$[Ni(C_6H_5O_7)MoO_2]_{ads}^- + 2H_2O + 4e^- \rightarrow Mo(s) + [Ni(C_6H_5O_7)]^- + 4OH^- \tag{1.32}$$

$$H_2O + e^- \rightarrow ½\, H_2 + OH^- \tag{1.33}$$

assuming that the alloy deposition is always accompanied by the simultaneous hydrogen evolution [reaction (1.33)]. This model has been confirmed by in situ surface Raman spectroscopic studies, by revealing existence of adsorbed intermediate $[Ni(C_6H_5O_7)MoO_2]_{ads}^-$ [33].

It was later shown for the Mo–Ni–O powder electrodeposition [34] that this mechanism is only partially correct and that among metallic Ni and Mo, the MoO_3 phase is also deposited. Taking into account that NiO, MoO_3, and $MoNi_4$ phases were detected in as-deposited samples by transmission electron microscopy (TEM), it is obvious that the mechanism of alloy deposition similar to the one presented by Eqs. (1.30, 1.31, 1.32, and 1.33) could be applied for the formation of $MoNi_4$ phase only, since it assumes complete reduction of both metal ions. By the analogy with the mechanism proposed by Podlaha and Landolt [29–31], it was suggested that the $MoNi_4$ phase could be formed by the following mechanism [35]:

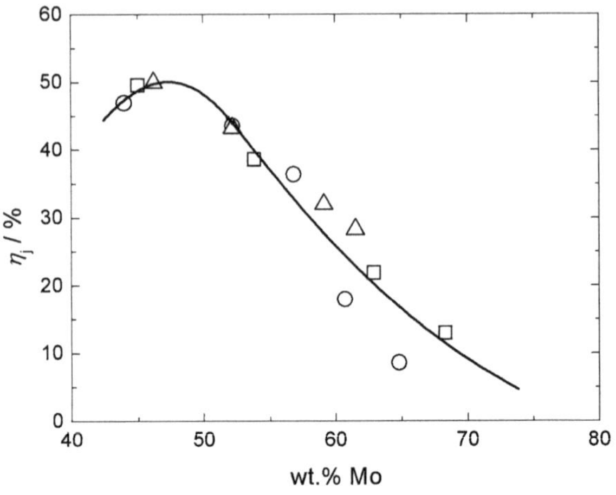

Fig. 1.10 Dependence of η_j on alloy composition (wt.% Mo); (*open square*) the influence of the $Na_4P_2O_7$ concentration; (*open circle*) the influence of the $FeCl_3$ concentration; (*open triangle*) the influence of the current density for alloy deposition (Reprinted from [36] with the permission of Serbian Chemical Society)

$$\left[Ni(NH_3)_3\right]^{2+} + 2e^- \rightarrow Ni(s) + 3NH_3 \tag{1.34}$$

$$MoO_4{}^{2-} + \left[Ni(NH_3)_3\right]^{2+} + 2H_2O + 2e^- \rightarrow \left[Ni(NH_3)_3MoO_2\right]_{ads}{}^{2+} + 4OH^- \tag{1.35}$$

$$\left[Ni(NH_3)_3MoO_2\right]_{ads}{}^{2+} + 2H_2O + 4e^- \rightarrow Mo(s) + \left[Ni(NH_3)_3\right]^{2+} + 4OH^- \tag{1.36}$$

The most probable mechanism for the MoO_3 phase formation is [34]:

$$MoO_4{}^{2-} + \left[Ni(NH_3)_3\right]^{2+} + H_2O + 2e^- \rightarrow MoO_3 + Ni(s) + 3NH_3 + 2OH^- \tag{1.37}$$

System Fe–Mo [36]

The Fe–Mo alloy coatings were deposited onto mild steel substrate from the solution containing $FeCl_3$, Na_2MoO_4, $Na_4P_2O_7$, and $NaHCO_3$, pH 9.3. Depending on the concentrations of $FeCl_3$, $Na_4P_2O_7$, and Na_2MoO_4, as well as the applied current density, the wt.% of Mo in the coatings (approximate thickness 20 μm) changed from about 45 wt.% to about 70 wt.%, while the current efficiency (η_j) varied from about 10 % to 50 %, as shown in Fig. 1.10. It should be stated that in all samples a certain (in some cases significant) amount of oxygen has been detected, but this was neglected assuming that the mechanism for induced codeposition defined by Eqs. (1.30, 1.31, 1.32, and 1.33) is operative and, accordingly, only percentages of Fe and Mo were taken into account [36].

Typical morphology of the coatings deposited at different current densities is presented in Fig. 1.11, characterized with nodular surface and large cracks. By the

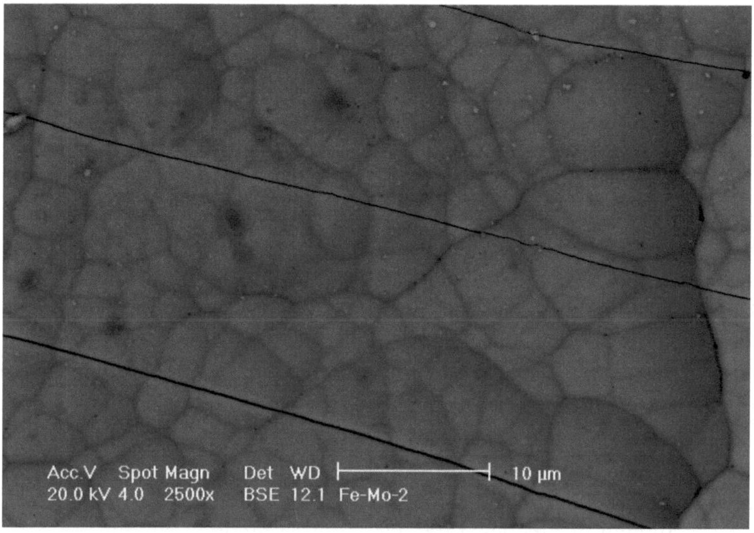

Fig. 1.11 Typical morphology of the Fe–Mo alloy coatings deposited at different current densities

Table 1.2 Composition of the Fe–Mo alloy surfaces, obtained by the EDS analysis, as a function of the deposition current density

j_{dep} (mA cm^{-2})	at.% O	at.% Mo	at.% Fe
−20	61.20	17.51	21.29
−50	34.05	27.08	38.87
−100	25.80	25.24	48.96

EDS analysis of the Fe–Mo alloy surfaces, it was found that their composition depends on the deposition current density, as shown in Table 1.2. Considering these results it appears that the amount of MoO_3 in the deposit decreases with the increase of the deposition current density (lower percentage of oxygen), indicating that the mechanism predicted by Eqs. (1.30, 1.31, and 1.32) and (1.34, 1.35, and 1.36) for Ni–Mo, prevails at higher current densities, while the mechanism for MoO_3 formation prevails at lower current densities of Fe–Mo alloy deposition.

1.3 Characterization of Electrodeposited Alloys by Electrochemical Techniques

1.3.1 Basic Concepts

The most common technique for phase identification in alloys is X-ray technique. The application of the X-ray technique on electrodeposited alloys is very often limited by the size of crystals, since in electrodeposited alloys they are often very small (lower than 10 nm) which is under the limit of detection in conventional X-ray technique. Accordingly, such crystallites appear on the diffractograms as amorphous and must be investigated by the expensive and often not available TEM technique in order to be identified. Hence, it is desirable to use some other, much simpler and less expensive technique for identification of phases in deposited alloys. It appeared that electrochemical techniques, galvanostatic dissolution and anodic linear sweep voltammetry (ALSV), could be used for successful identification of phases in electrodeposited alloys [1].

The first attempt of the use of an electrochemical technique for characterization of a deposited alloy was made in 1928 in order to determine the thickness of metallic coatings [37, 38]. Using this method it was possible to determine the thickness of the coating and that of the intermediate layer between the coating and substrate, as well as to obtain data about the phase composition of the coating. Since then, electrochemical techniques have been extensively used for investigating processes of alloy dissolution [39–48]. In all of these cases bulk alloys were investigated, and the discrepancies between the obtained results were significant.

At the beginning of nineties, three laboratories reported attempts to determine the phase composition by complete dissolution of a thin layer (up to 10 μm) of a deposited alloy [49–62, 64, 65, 72, 79, 80], using different techniques: galvanostatic [50, 52], potentiostatic [57, 58, 61], and ALSV [49, 51, 57–62, 64, 65, 72, 79, 80].

1.3.2 ALSV technique

The ALSV technique is based on the application of linear sweep voltammetry at low sweep rate of 1 mV s^{-1} during the dissolution of thin layer of deposited alloy. After deposition of a thin layer of an alloy (up to 10 μm, recommendable about 1 μm) from the solution containing ions of both metals (constituents of the alloy) onto glassy carbon, gold, silver, or platinum substrates in one cell, working electrode, together with the deposit, is transferred into another cell containing only supporting (neutral) electrolyte, and/or ions of the less noble metal of the two, in which the alloy coating is completely dissolved. If the ions of the less noble metal are present in such solution, usually the reference electrode is less noble metal in the solution of corresponding ions. In the case of neutral electrolyte any reference electrode could be used, but the starting potential for alloy dissolution should be reversible potential of the less noble metal [1].

1.3.2.1 Conditions for Phase Detection in the Thin Layer of Deposited Alloys

Following conditions must be fulfilled for successful phase detection in deposited alloys by the ALSV technique:

- It is important to dissolve the entire alloy sample, since only in such a case eventual diffusion of alloy constituents in the solid phase will not influence the process of alloy dissolution by masking of some phases by other ones. If there are some solid-state diffusion limitations, they can be overcome by using either long time polarization (galvanostatic or potentiostatic techniques) or very low sweep rate in the ALSV technique;
- The processes of passivation or replacement must be prevented during the dissolution of alloy. This can be achieved by a proper choice of the solution for alloy dissolution. The use of simple salt solutions (such as NaCl, Na_2SO_4, and $NaClO_4$) with pH values in the range $3 \leq pH \leq 6$ or simple salt solutions with addition of small concentration of the less noble metal (up to 10^{-2} M) is recommended [1];
- It is much better to use techniques in which potential is controlled since galvanostatic responses are not always well defined [50, 52]. Particularly, good results were obtained with the application of ALSV technique [51, 53, 57–62, 64, 65, 72, 79, 80].
- Concerning the process of the more noble metal deposition/dissolution, two cases can exist in the application of the ALSV technique: (1) The deposition/dissolution process of the more noble metal is reversible. Accordingly, during the dissolution of alloy the ions of the more noble metal must reprecipitate on the spot (since the potential is more negative than the reversible one of the more noble metal); (2) the deposition/dissolution of the more noble metal is irreversible and the dissolution of the more noble component starts at the potentials more positive than its reversible potential. Accordingly, no reprecipitation of the more noble metal would take place.

1.3.3 Characteristics of the ALSV Responses for Dissolution of Different Types of Alloys

The shape of the ALSV of alloy dissolution could be predicted by consideration of the phase diagram of alloys and the Gibbs energies of phases appearing in the system. Different types of alloys (eutectic, solid solution, alloys with intermediate phases, and/or intermetallic compounds) will possess different shapes of the ALSV responses.

1.3.3.1 Eutectic-Type Alloys

Since the interaction (miscibility) between the components in the solid phase of the eutectic-type alloys is absent or negligible, the Gibbs energies of each constituent should be the same as that of the corresponding pure metal. Accordingly, the reversible potential of each constituent in the alloy should be virtually the same as that of pure metal of the same grain size. In such a case, the ALSV should be

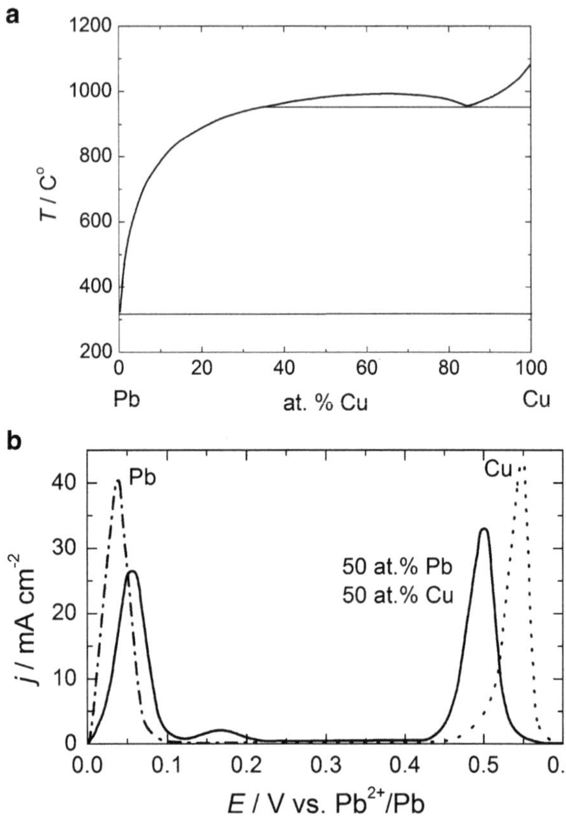

Fig. 1.12 (**a**) Phase diagram of the system Cu–Pb. (**b**) ALSV's of dissolution of pure metals (Cu—*dotted line*, Pb—*dash dot line*) and Cu–Pb alloy (*solid line*) containing approximately 50 at. % Pb and 50 at. % Cu. Conditions of dissolution: 0.01 M Pb^{2+} + 1 M HBF_4, $v = 1$ mV s^{-1}, RPM = 0 (Reprinted from [1] with the permission of Springer)

characterized by two separate dissolution peaks, each of them reflecting dissolution of the corresponding pure metal and appearing at potentials more positive than the reversible potential of that particular metal. The two peaks might overlap only if the difference between the reversible potentials of these two metals is very small and the thickness of a deposited alloy is high (over 1 μm).

System Cu–Pb [51]

The phase diagram for the system Cu–Pb [67] is shown in Fig. 1.12a, while the ALSV's of Cu, Pb, and Cu–Pb alloy dissolution, recorded in the solution containing 0.01 M Pb^{2+} + 1 M HBF_4, $v = 1$ mV s^{-1}, RPM = 0, are presented in Fig. 1.12b. Pure metals were deposited onto rotating disc electrode (RDE) made of glassy carbon (GC) at RPM = 1,000: Pb was deposited from the solution containing 0.1 M Pb^{2+} + 1 M HBF_4, $E = -0.04$ V vs. Pb^{2+}/Pb; Cu was deposited from the solution containing 0.05 M Cu^{2+} + 0.2 M H_2SO_4, $E = -0.05$ V vs. Ag/AgCl; Cu–Pb alloy was deposited from the solution containing 0.05 M Cu^{2+} + 0.01 M Pb^{2+} + 1 M HBF_4, $E = -0.52$ V vs. Ag/AgCl ($j_{dep} \approx -7.5$ mA cm^{-2}). As can be seen in

Fig. 1.12b the peaks of Cu and Pb dissolution from the alloy are well defined and separated, and their dissolution potentials are close to the dissolution potentials of pure metals [case (1) of Sect. 1.3.2.1]. Slightly, more positive peak potential for the dissolution of Pb from the alloy is due to diffusion of Pb from the bulk of alloy, while slightly more negative peak potential for the dissolution of Cu from the alloy is the consequence of the fact that, after the dissolution of all Pb from the alloy, the remaining layer of pure Cu is more porous and its dissolution takes place at more negative potentials. A small peak at about 0.15 V vs. Pb^{2+}/Pb represents the dissolution of a solid solution of Pb in Cu, which could be formed only in electrodeposited alloys [63], with up to 10 wt.% of Pb being dissolved in Cu. Such behavior indicates that deposited alloys could be characterized by the presence of some phases not indicated in the phase diagram of the system. At the same time, it is often the case that these phases cannot be detected by the X-ray technique, while they are clearly defined on the ALSV of alloy dissolution (see further text).

System Cd–Zn [64, 65]

The phase diagram for the system Cd–Zn [67] is shown in Fig. 1.13a, while the ALSV's of Cd–Zn alloys dissolution of different compositions are presented in Fig. 1.13b. The alloys were deposited at different constant current densities (to a total charge of 1 C cm^{-2}) on a RDE made of Au (RPM $=$ 1,000) from the solution containing 0.01 M $CdSO_4$ + 0.5 M $ZnSO_4$ + 0.01 M H_2SO_4. Their dissolution by the ALSV technique was performed in the solution containing 0.5 M Na_2SO_4 (pH 4), RPM $=$ 0 [64, 65]. The composition of the alloys was calculated from the ratio of current densities of both metals: more noble metal (Cd) was deposited at its diffusion limiting current density, j_L(Cd), while the difference between the total current density, j_t and j_L(Cd), was ascribed to the current density for Zn deposition, j(Zn). The current efficiency for alloy deposition was calculated from the ratio of the charge recorded during the alloy dissolution, Q_{diss}(alloy) (obtained by the integration of the surface area under the ALSV response), and that for alloy deposition, Q_t(alloy), as $\eta_j = Q_{diss}(\text{alloy})/Q_t(\text{alloy})$. Accordingly, the molar ratio of the more noble metal in the alloy is expressed as [64, 65]

$$x(\text{Cd}) = \frac{j_L(\text{Cd})}{\eta_j j_t(\text{alloy})} = \frac{Q_t(\text{Cd})}{\eta_j Q_t(\text{alloy})} = \frac{Q_t(\text{Cd})}{Q_{diss}(\text{alloy})}, \tag{1.38}$$

while that for the less noble one was

$$\begin{aligned} x(\text{Zn}) &= \frac{[\eta_j j_t(\text{alloy}) - j_L(\text{Cd})]}{\eta_j j_t(\text{alloy})} = \frac{[\eta_j Q_t(\text{alloy}) - Q_t(\text{Cd})]}{\eta_j Q_t(\text{alloy})} \\ &= \frac{[Q_{diss}(\text{alloy}) - Q_t(\text{Cd})]}{Q_{diss}(\text{alloy})}. \end{aligned} \tag{1.39}$$

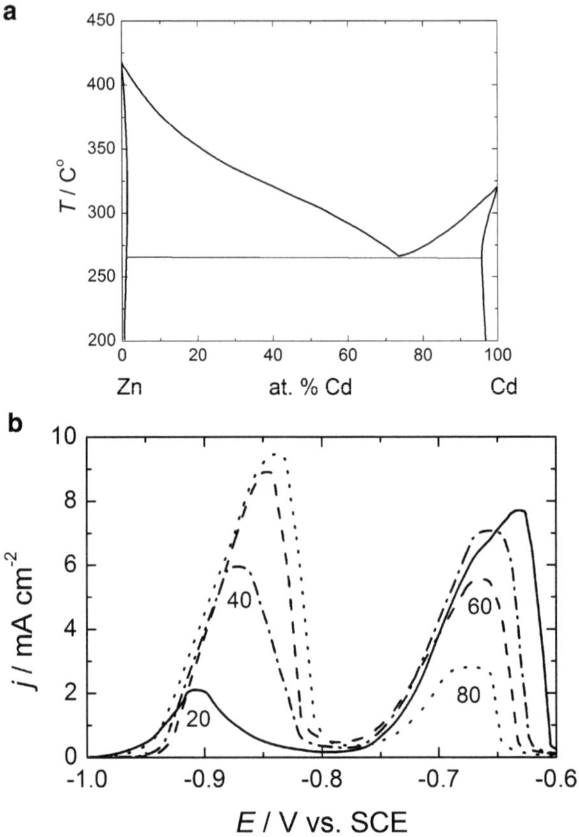

Fig. 1.13 (**a**) Phase diagram of the system Cd–Zn. (**b**) ALSV's of dissolution of Cd–Zn alloy of different compositions (at.% of Zn in the alloy marked in the figure). Conditions of dissolution: 0.5 M Na_2SO_4 (pH 4), $v = 1$ mV s^{-1}, RPM = 0 (Reprinted from [1] with the permission of Springer)

To express the molar ratio of each metal in the alloy by the ratio of their current densities for deposition, $\Delta = j_L(\mathrm{Cd})/j(\mathrm{Zn})$, it follows that

$$x(\mathrm{Cd}) = \frac{\Delta}{1+\Delta}, \tag{1.40}$$

$$x(\mathrm{Zn}) = \frac{1}{1+\Delta}. \tag{1.41}$$

The composition of alloys, i.e., the desired total current density for alloy deposition for the defined duration of the current pulse, was obtained by using Eqs. (1.38, 1.39, 1.40, and 1.41).

As can be seen in Fig. 1.13b the peaks of Zn and Cd dissolution are clearly separated [case (1) of Sect. 1.3.2.1] for all alloy compositions (at.% of Zn in the alloys are marked in the figure), as in the case of the system Cu–Pb (Fig. 1.12b). By comparing the charge under the peaks for Cd and Zn dissolution on the ALSV's and their charges for deposition (taking into account η_j), it was found that Q_{diss}(Zn) is smaller than Q_{dep}(Zn), while Q_{diss}(Cd) is higher than Q_{dep}(Cd). This finding confirms the fact that in electrodeposited Cd–Zn alloys, supersaturated solid solution of Zn in Cd can be formed [66].

System Cd–Co [72]

According to the literature data [67], several attempts were made in order to obtain phase diagram for the system Cd–Co and none of them was successful, because it was not possible to prepare alloys with less than 94 wt.% and more than 10 wt.% Co [68]. In the case of melts with up to 10 wt.%, Co cooling curves showed only eutectic arrests at 316 °C. By the structural analysis of this melt, it was stated that an intermetallic phase is present in eutectic matrix [69]. Some authors claimed that $Cd_{23}Co_3$ intermetallic compound (having structure of γ-brass) was identified in this system [70], but it could not be proven by immersion of Co in a Cd melt at 700 °C [71]. Hence, at the moment no convincing data concerning the phase diagram for the system Cd–Co are available in the literature.

The Cd–Co alloy coatings [72] were deposited from sulfate electrolytes with different compositions: (1) 0.01 M $CdSO_4$ + 0.2 M $CoSO_4$ + 0.2 M H_3BO_3, (2) 0.02 M $CdSO_4$ + 0.2 M $CoSO_4$ + 0.2 M H_3BO_3 and (3) 0.02 M $CdSO_4$ + 0.4 M $CoSO_4$ + 0.2 M H_3BO_3. Pure Co was deposited in the solution containing 0.2 M $CoSO_4$ + 0.2 M H_3BO_3 and pure Cd from the solution (2) at the potential of pure Cd deposition (−0.75 V vs. SCE). All deposits were obtained under the conditions of convective diffusion, i.e., RPM = 1,000, while all ALSV's were recorded at the sweep rate of 1 mV s^{-1} in the 1 M NaCl solution (pH 2) at RPM = 0 (stationary electrode).

In Fig. 1.14a are compared ALSVs of pure Cd, pure Co, and Cd–Co alloy of approximate composition 50 at.% Cd – 50 at.% Co, all deposited to a constant charge of 1 C cm^{-2} (case (2) of Sect. 1.3.2.1). It appears that deposited Cd–Co alloy does not belong to the eutectic-type alloys since some new, unknown phase (Un), presented by the additional ALSV peak at about −0.32 V, has been formed. At a certain solution composition (0.02 M Cd^{2+} + 0.2 M Co^{2+} + 0.2 M H_3BO_3), certain deposition potential (−1.5 V), and certain amount of deposited alloy, the most pronounced peak is the peak of the unknown phase, Fig. 1.14b. Hence, it is possible to obtain the deposited Cd–Co alloy with the highest amount of the unknown phase in the system Cd–Co [72]. It should also be emphasized that the peak of Co dissolution is placed at significantly more positive potentials than the potential of its deposition (about −1.0 V). Accordingly, Co^{2+} ions cannot reprecipitate [case (2) of Sect. 1.3.2.1], as it was the case for Cu–Pb [51] and Cd–Zn [64, 65] systems.

Corresponding charges for the deposition of Cd and Co, obtained by the calculation of the Cd deposition charge using its diffusion limiting current density ($Q_{Cd} = j_L(Cd) \cdot t$) and Co deposition charge as $Q_{Co} = Q_{tot} - Q_{Cd}$, are presented

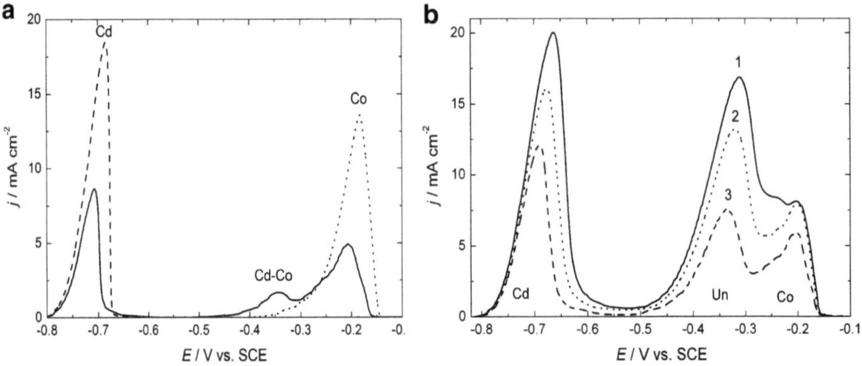

Fig. 1.14 (**a**) ALSV's recorded at $v = 1$ mV s^{-1} and RPM = 0 in the electrolyte of 1 M NaCl (pH 2) for pure Cd (*dashed line*), pure Co (*dotted line*), and Cd–Co alloy (*solid line*) deposited to a constant charge of 1 C cm^{-2} in the electrolyte containing 0.01 M $CdSO_4$ + 0.2 M $CoSO_4$ + 0.2 M H_3BO_3 at RPM = 1,000. (**b**) ALSV's recorded under the same conditions as those in (**a**) for deposits obtained at a potential of −1.5 V to the different charges of 14.8 C cm^{-2} (1), 10.7 C cm^{-2} (2), and 6.7 C cm^{-2} (3) in the electrolyte containing 0.02 M $CdSO_4$ + 0.2 M $CoSO_4$ + 0.2 M H_3BO_3 at RPM = 1,000 (Reprinted from [72] with the permission of Elsevier)

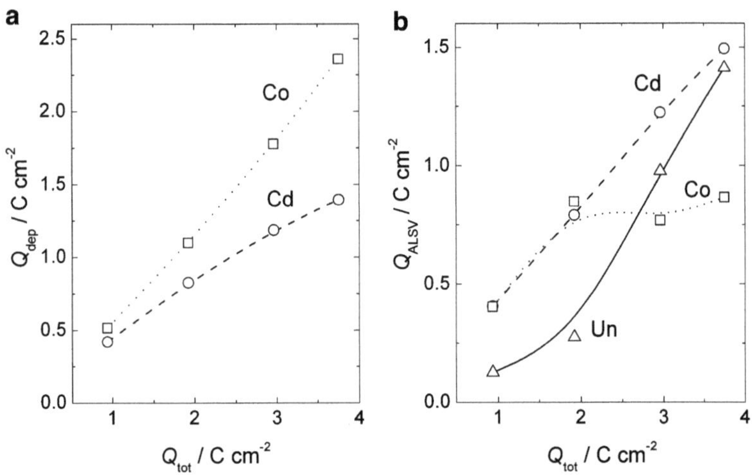

Fig. 1.15 (**a**) Deposition charges for Cd and Co as a function of the total charge for the alloys deposition. (**b**) Dissolution charges for Cd, Co, and unknown phase (Un), obtained by the integration of the charge under corresponding ALSV peaks, as a function of total charge for the alloys deposition (Reprinted from [72] with the permission of Elsevier)

in Fig. 1.15a. As can be seen with the increase of the total charge for alloy deposition, Q_{tot}, charges for both metals increase almost linearly. By the integration of the charge under corresponding peaks on the ALSV's, the charges for the dissolution of Cd, Co, and unknown phase (Un) are obtained. These charges, as a function of the Q_{tot}, are presented in Fig. 1.15b. It appears that the charge (amount) for the unknown phase sharply increases with increasing Q_{tot} for the applied potential of alloy deposition (−1.3 V), while the charge for Co increases for $Q_{tot} \leq 2$ C cm^{-2} reaching its limiting value of about 0.8 C cm^{-2} at higher values of Q_{tot}. Such behavior indicates that the amount of the unknown phase increases on the account of the amount of both metals up to $Q_{tot} \approx 2$ C cm^{-2}, while at higher values of Q_{tot} its increase is mainly on the account of deposited Co, but certain limiting amount of Cd must be deposited for the unknown phase to be formed and seen on the ALSV. Taking into account these results, it is obvious that the unknown phase detected in the deposited alloy cannot be ascribed to the $Cd_{23}Co_3$ intermetallic compound [70], since it contains higher amount of Co than Cd and its peak is close to the peak of pure Co dissolution [72].

1.3.3.2 Solid Solution Type Alloys

In the case of solid solution type alloys, the prediction of the ALSV shape could be problematic, i.e., the dissolution can proceed through two separate peaks or through one peak. According to the literature [73] metal atoms in the ideal solid solutions should be randomly arranged. In reality, such alloys can also contain some short- and long-range ordered structures. Some alloys whose metals are totally miscible in the solid state, such as Au–Cu solid solution, possess "superlattice" structures ($AuCu_3$, AuCu, and Au_3Cu), and each of these types of "superlattices" could have different dissolution characteristics. On the other hand, if the Gibbs energy of mixing is not very high and if the potential sweep is sufficiently slow [1, 51], the less noble metal should dissolve completely at more negative potentials than the potential of dissolution of the more noble metal. In such a case at least two ALSV peaks should be expected, corresponding to the dissolution of both pure metals. The preceding peaks should represent the dissolution of the less noble metal only, assuming that, even if the two metals dissolved simultaneously in the preceding peaks (as is likely to happen with a solid solution), the more noble metal should immediately reprecipitate since its ions are formed at potentials more negative than the reversible potential of the more noble metal. Such theory, predicting two ALSV peaks, has been proposed in the literature [1, 51].

System Co–Ni, Ideal Solid Solution [24]

The phase diagram for the system Co–Ni [67] is shown in Fig. 1.16a, while the ALSV's of Co–Ni alloys dissolution of different compositions are presented in Fig. 1.16b. Considering deposition/dissolution characteristics of both metals (both are irreversible), this system cannot be assigned to either of the cases (1) or (2) of Sect. 1.3.2.1.

Pure Co and pure Ni were deposited onto gold disc electrode at RPM = 1,000 from the solution containing 0.2 M $CoSO_4$ + 0.2 M $Na_3C_6H_5O_7$ and 1 M $NiSO_4$ + 0.2 M $Na_3C_6H_5O_7$, respectively. For deposition of their alloys of different

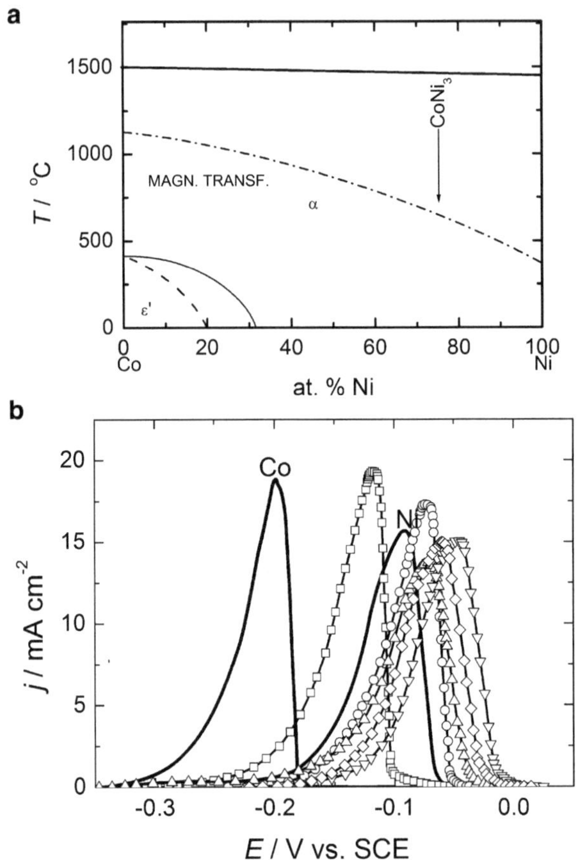

Fig. 1.16 (**a**) Phase diagram of the system Co–Ni. (**b**) ALSV's of dissolution of Co–Ni alloy of different compositions: *solid lines*—pure Co and pure Ni; *open square*—78.3 at.% Co; *open circle*—47.5 at.% Co; *open triangle*—25.2 at.% Co; *open inverted triangle*—12.9 at.% Co; *open diamond*—8.0 at.% Co. Conditions for dissolution: 1 M NaCl (pH 2), $v = 1$ mV s^{-1} (Reprinted from [24] with the permission of Elsevier)

compositions, only the concentration of Co^{2+} has been changed, being 0.005 M, 0.01 M, 0.025 M, 0.05 M, and 0.2 M, respectively (see Fig. 1.9). Alloy layers of different compositions of the thickness of approximately 0.34 μm were obtained by galvanostatic deposition at different constant current densities, but at the same amount of charge of 1 C cm^{-2}. All samples were dissolved in the solution 1 M NaCl (pH 2) by application of the ALSV technique on the stationary Au electrode (RPM = 0). The composition of alloy samples deposited at a constant current density to the amount of 1 mg and galvanostatically dissolved in 10 ml of 1 M HCl has been determined by the AAS technique [24].

Considering ALSV results presented in Fig. 1.16b, it could be concluded that the peak potentials of dissolution of pure metals are very close to each other (the difference being about 0.1 V) and that all Co–Ni alloy samples dissolve through one ALSV peak only, indicating simultaneous dissolution of both metals from their alloy. Such behavior during the alloy dissolution is the consequence of the fact that the potentials of dissolution are much more positive than the potentials of Ni deposition (reduction of Ni^{2+} ions) so that it is not possible for Ni^{2+} ions to reprecipitate. It is interesting to note that the peaks of dissolution of Co–Ni alloys with less than 50 at.% Co are placed at potentials more positive than that of the more noble metal (Ni). Most of the data concerning the phase diagram of the system Co–Ni [74] state that these two metals form solid solutions at all compositions, while in one work the possibility of the formation of ordered $CoNi_3$ structure [75] has been predicted. For a new phase in the system, one would expect additional ALSV peak (as it is in the case in the system Ag–Pd, see text below), but in the case of Co–Ni alloys dissolution this is not the case.

Let us consider the standard potential of the phase $Co_xNi_{(1-x)}$. According to the existing literature [1], standard potential of such phase should be

$$E^{\Theta}\left(Co_xNi_{(1-x)}\right) = \frac{E^{\Theta}\left(Co^{2+}/Co\right) + E^{\Theta}\left(Ni^{2+}/Ni\right)}{2} - \frac{\mu^{\Theta}\left(Co_xNi_{(1-x)}\right)}{2F}. \quad (1.42)$$

In the case of the ideal solid solutions, the Gibbs energy is entirely defined by the entropy of mixing (the enthalpy, $\Delta H^{\Theta}{}_{i} = 0$), while the entropy of mixing depends on the composition of solid solution and the temperature

$$\mu_i^{\Theta}\left(Co_xNi_{(1-x)}\right) = -RT[x \ln x + (1-x) \ln (1-x)]. \quad (1.43)$$

In real systems this behavior is rather exceptional and usually mixing is endothermic or exothermic, i.e., ΔH^{Θ} is included and explained by the quasi-chemical model assuming that the heat of mixing (ΔH^{Θ}) is only due to the bond energies between adjacent atoms [73]. In such a case the Gibbs energy change upon mixing of a so-called regular solid solution is given by

$$\mu_i^{\Theta}\left(Co_xNi_{(1-x)}\right) = \Delta H^{\Theta}\left(Co_xNi_{(1-x)}\right) - RT[x \ln x + (1-x) \ln (1-x)], \quad (1.44)$$

where the heat of mixing $\Delta H^{\Theta}(Co_xNi_{(1-x)})$ is determined by the interatomic bonds and is given by the equation:

$$\Delta H^{\Theta}\left(Co_xNi_{(1-x)}\right) = N_a z \varepsilon x(1-x) \quad (1.45)$$

with N_a being Avogadro's constant, z is the number of bonds per atom, and ε is the bond energy.

This model is also rather limited, and in the real systems the actual arrangement of atoms will be a compromise that gives the lowest internal energy consistent with sufficient entropy, or randomness, to achieve the minimum Gibbs energy. Such a tendency can lead to a formation of ordered structures known as "superlattices,"

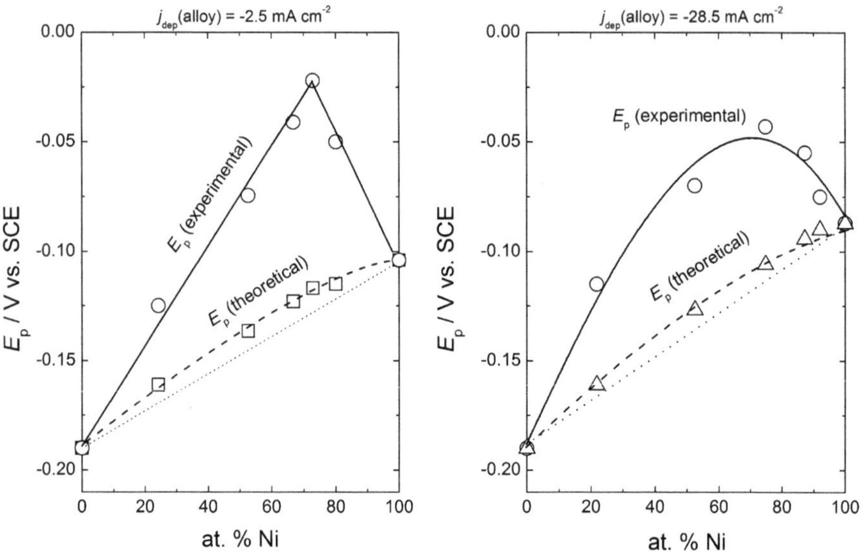

Fig. 1.17 Dependences of the ALSV peak potentials as a function of alloy composition for Co–Ni alloys deposited at different current densities (marked in the figure) (Reprinted from [24] with the permission of Elsevier)

which are clearly detected [73] in the system Au–Cu as Au_3Cu, AuCu, and $AuCu_3$. All the above considerations assume thermally prepared alloys obtained under equilibrium conditions. Since electrodeposited alloys are practically never obtained under equilibrium conditions [1], one can expect that the effects of the heat and entropy of mixing will be even more pronounced in the case of deposited alloys.

Using Eq. (1.42) the highest value of the entropy of 1.717 kJ mol^{-1} is obtained for $x = 0.5$ (for temperature of 25 °C), while the corresponding change of the standard potential of the $Co_xNi_{(1-x)}$ phase is only 9 mV. Hence, in such a case the standard potential of this phase should change linearly from that of Co, $E^{\Theta}(Co^{2+}/Co)$, to that of Ni, $E^{\Theta}(Ni^{2+}/Ni)$, while the contribution of $\mu^{\Theta}{}_{i}$ is very small and causes very small deviation in the change of standard potential from linearity (see Fig. 1.17). Taking into account that it is not possible to determine reversible potentials of Co, Ni, and Co–Ni alloys from the dissolution characteristics represented by their ALSV's in Fig. 1.16b, and bearing in mind high irreversibility of the deposition/dissolution processes for both metals, an attempt was made to use ALSV peak potentials instead of the reversible potentials for dissolution of all components of the system [24]. In Fig. 1.17 are presented dependences of the ALSV peak potentials (E_p) as a function of the alloy composition (at.% Ni) for the alloys and pure metals deposited at different current densities. As can be seen theoretical and experimental dependences are different, with the most positive experimentally determined peak potentials recorded for the composition of approximately 75 at.% Ni ($CoNi_3$). The peak potential change is much better defined for the samples deposited at lower current density, j_{dep}(alloy) $= -2.5$ mA cm^{-2}.

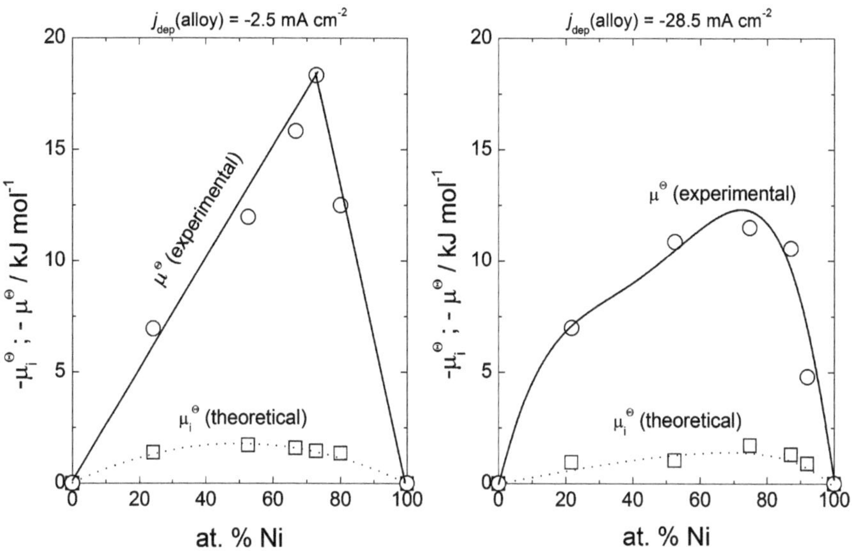

Fig. 1.18 Dependences of the μ^{Θ} and $\mu^{\Theta}{}_i$ as a function of alloy composition for Co–Ni alloys deposited at different current densities (j_{dep}(alloy)—marked in the figure) (Reprinted from [24] with the permission of Elsevier)

By expressing μ^{Θ} using E_p instead of E^{Θ} values in Eq. (1.5), it is possible to plot experimental and theoretical values for μ^{Θ} as a function of alloy composition, as shown in Fig. 1.18.

Hence, it could be concluded that even in the absence of additional ALSV peak for dissolution of "ordered structure" ($CoNi_3$), it is possible to predict the existence of such structure in the solid solution by proper analysis of the ALSV response.

System Ag–Pd [62]

The Ag–Pd alloys were deposited to a constant charge from the solution containing 0.005 M $PdCl_2$ + 0.04 M AgCl + 12 M LiCl + 0.1 M HCl (t = 50 °C) onto rotating Au disc electrode (RPM = 1,000) at constant current densities defined by the desired alloy composition [see Eqs. (1.38, 1.39, 1.40, and 1.41)]. Pure metals were deposited under the same conditions from two different solutions: Pd from the solution 0.001 M $PdCl_2$ + 12 M LiCl + 0.1 M HCl at the constant current density of −0.5 mA cm^{-2}; Ag from the solution 0.04 M AgCl + 12 M LiCl + 0.1 M HCl at the constant current density of −10.0 mA cm^{-2}. All deposited samples were dissolved by the ALSV technique in the solution 0.001 M AgCl + 12 M LiCl + 0.1 M HCl (RPM = 0).

The ALSV's of pure metals dissolution (Ag—dotted line, Pd—dashed line) and an alloy containing approximately 50 at.% Ag and 50 at.% Pd (solid line) are presented in Fig. 1.19a [case (1) of Sect. 1.3.2.1].

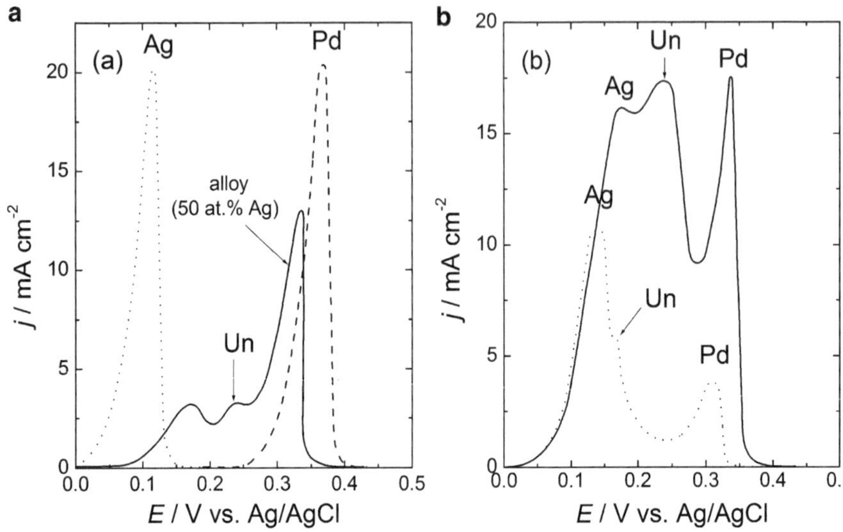

Fig. 1.19 (**a**) ALSV's recorded at $v = 1$ mV s^{-1} and RPM = 0 in the solution 0.001 M AgCl + 12 M LiCl + 0.1 M HCl for pure Ag (*dotted line*), pure Pd (*dashed line*), and Ag–Pd alloy (approximately, 50 at.% Ag and 50 at.% Pd—*solid line*) deposited to a constant charge of 1 C cm^{-2}. (**b**) ALSV's recorded under the same conditions as those in (**a**) for alloy containing 80 at.% Ag and 20 at.% Pd deposited to the different charges of 1.0 C cm^{-2} (*dotted line*) and 4.0 C cm^{-2} (*solid line*) (Reprinted from [62] with the permission of Elsevier)

As can be seen Ag and Pd dissolution peaks are well separated. The ALSV of an alloy containing approximately 50 at.% Ag and 50 at.% Pd is characterized by the presence of additional peak at about 0.24 V vs. Ag/AgCl (Un), while peak of pure Ag dissolution from the alloy is placed at more positive potentials (due to diffusion through the alloy matrix) and peak of Pd dissolution is positioned at more negative potentials (due to its porous structure—consequence of Pd reprecipitation). The peak of unknown phase (Un) is seen to depend on the amount of deposited alloy, as well as on the alloy composition, Fig. 1.19b. In the alloy with 80 at.% Ag deposited to the lower charge (1 C cm^{-2}, thinner deposit—dotted line) this peak is small and not well defined, while in the same alloy with higher thickness (4 C cm^{-2}—solid line) this peak is much bigger, indicating higher amount of this phase in the alloy. This finding [62] indicates that deposited Ag–Pd alloy is not a simple homogeneous solid solution [76, 77], as predicted by the phase diagram [67]. Confirmation of this finding is in accordance with the measurement of the microhardness of deposited Ag–Pd alloys, which was found to increase with increasing Ag content in the alloy [78]. Such behavior could only be explained by the presence of some intermetallic compound that possesses a higher value of microhardness than that of pure Pd. Since the microhardness of Pd is three times higher than that of Ag, the microhardness of their solid solutions should be somewhere in between. Hence, the unknown phase is most likely an intermetallic compound, formed only in the deposited Ag–Pd alloys.

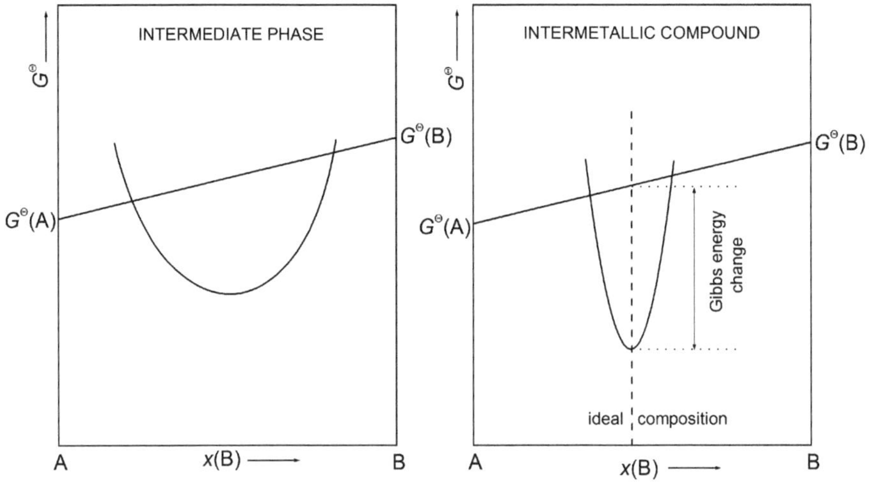

Fig. 1.20 Schematic representation of the change of Gibbs energy of intermediate phase and intermetallic compound as a function of alloy composition (Reprinted from [1] with the permission of Springer)

1.3.3.3 Alloys Characterized with Intermediate Phases and/or Intermetallic Compounds

It is known that the configuration of atoms that has minimum Gibbs energy after the mixing of two metals often does not have the same crystal structure as either of the pure metals. Such structure may be either an intermediate phase or intermetallic compound. The difference between them is not in their structures but in the change of their Gibbs energies with composition [73]. The changes of their Gibbs energies with the composition are schematically represented in Fig. 1.20. Since the change of the Gibbs energy of an intermetallic compound is very sharp, it seems reasonable to assume that its reversible potential should be a singular point. Accordingly, the dissolution peak of an intermetallic compound on the ALSV should be better defined than that of an intermediate phase [51], but this is not necessarily true since either of them should diffuse through the alloy matrix, and their diffusion could influence the shape of the ALSV peak. Hence, one can expect that the ALSV peaks of intermediate phases or intermetallic compounds which appear at the beginning of the ALSV (at the most negative potentials) could be influenced by the diffusion through the matrix of alloy (being wider), while those which are placed at the most positive potentials should be sharper.

In studying the phase composition of alloys containing intermediate phases and/or intermetallic compounds by the ALSV technique several systems were investigated. The best examples of the power of the ALSV technique are found in the work described below for the systems: Cd–Cu [51, 57, 58, 61], Cd–Ni [23], Ag–In [79], and Ag–Cd [80].

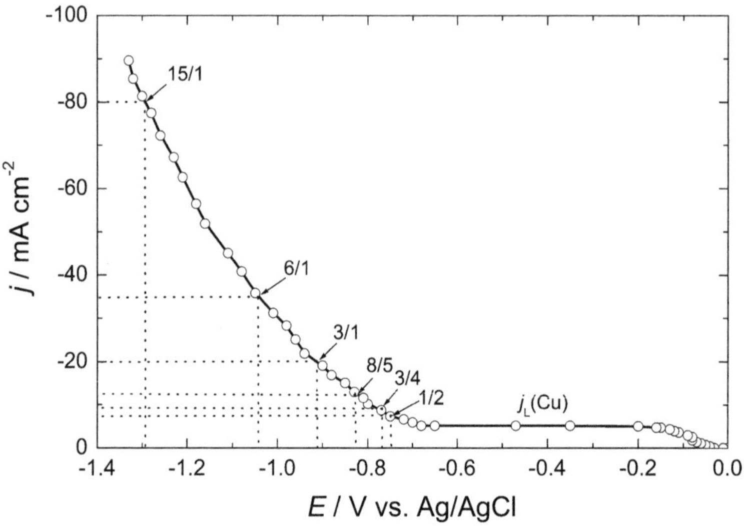

Fig. 1.21 Polarization diagrams for Cd–Cu alloy deposition at defined (marked in the figure) ratios of current densities, $j(Cd)/j_L(Cu)$. RPM = 1,000, solution: 0.01 M $CuSO_4$ + 0.5 M $CdSO_4$ + 0.2 M Na_2SO_4 + 0.01 M H_2SO_4 (Reprinted from [1] with the permission of Springer)

System Cd–Cu [58]

According to the phase diagram of the system Cd–Cu, this alloy is characterized by the presence of four intermetallic compounds [67] (Fig. 1.22a): Cd_3Cu, Cd_8Cu_5, Cd_3Cu_4, and $CdCu_2$. In order to obtain alloys of certain compositions, corresponding to four intermetallic compounds, four samples were deposited with the current density ratios 3/1, 8/5, 3/4, and 1/2. Among these, alloys with high amount of Cd were also deposited at the current density ratios 6/1 and 15/1, as shown in Fig. 1.21. Deposition of such alloys was possible due to the fact that deposition of Cu onto RDE in the investigated solution is diffusion-controlled, taking place at the diffusion limiting current density $j_L(Cu)$ (see Figs. 1.2 and 1.21).

All alloys were deposited at the gold or silver RDEs at RPM = 1,000 from the solution containing 0.01 M $CuSO_4$ + 0.5 M $CdSO_4$ + 0.2 M Na_2SO_4 + 0.01 M H_2SO_4. As can be seen in Fig. 1.21 down to about −0.7 V vs. Ag/AgCl only Cu is deposited, while at more cathodic potentials Cd deposition took place. Hence, Cu is deposited under diffusion control, while Cd is deposited under activation control and in all cases flat deposits were obtained.

Deposited alloys were dissolved by the ALSV technique (RPM = 0, v = 1 mV s^{-1}) in the solution 0.01 M $CdSO_4$ + 0.2 M Na_2SO_4 + 0.01 M H_2SO_4. Cadmium wire immersed in the solution of Cd^{2+} ions was used as the reference electrode (Cd^{2+}/Cd) and dissolution of alloy started from the reversible potential of this electrode [case (1) of Sect. 1.3.2.1].

In Fig. 1.22 are shown phase diagram of the Cd–Cu system [67] (a) and the ALSV of dissolution of alloy containing 75 at.% Cd and 25 at.% Cu (solid line),

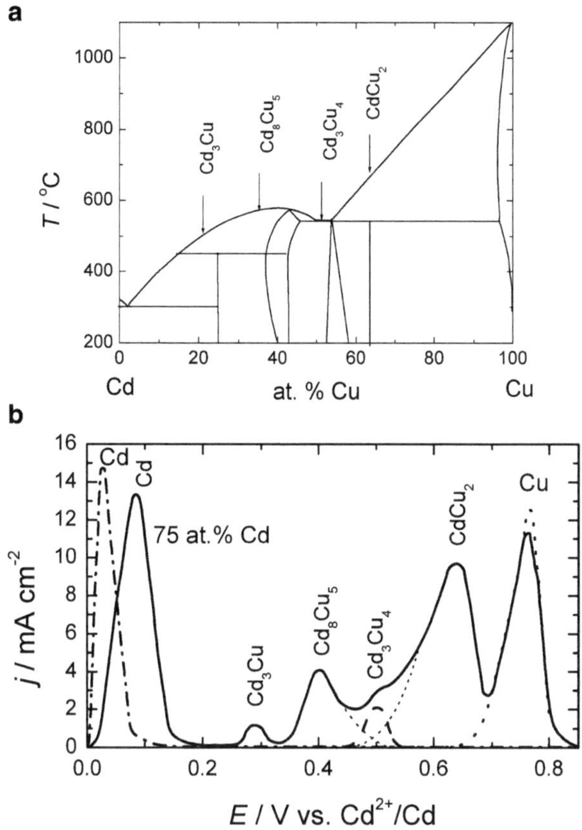

Fig. 1.22 (**a**) Phase diagram of the system Cd–Cu and (**b**) ALSV for dissolution of alloy containing 75 at.% Cd and 25 at.% Cu (*solid line*), pure Cd (*dash-dot line*), and pure Cu (*dotted line*). Solution: 0.01 M $CdSO_4$ + 0.2 M Na_2SO_4 + 0.01 M H_2SO_4; RPM = 0; $v = 1$ mV s^{-1} (Reprinted from [1] with the permission of Springer)

pure Cd (dash-dot line), and pure Cu (dotted line) (b). The peaks of dissolution of certain intermetallic compounds, as well as pure metals, are marked in the figure. As can be seen all intermetallic compounds could be detected in the deposit obtained at the current density ratio $j_{dep}(Cd)/j_{dep}(Cu) = 3/1$ (solid line). The peaks of dissolution of all intermetallic compounds are well defined, except the peak of Cd_3Cu_4 which is covered by the peak of $CdCu_2$ to a large extent. By simple deconvolution (dotted lines) of the ALSV at this position, the peak of Cd_3Cu_4 could be clearly detected (dashed line).

The shapes of ALSV's depend on the alloy composition, i.e., on the current density ratio for their deposition. In Fig. 1.23 are shown ALSV's of dissolution of Cd–Cu alloys deposited at different values of $j_{dep}(Cd)/j_{dep}(Cu)$ to a constant charge of 5 C cm^{-2}. With the increase of Cd content in the alloy the number of peaks increases. For the alloy deposited at the current density ratio 1/2 only two peaks

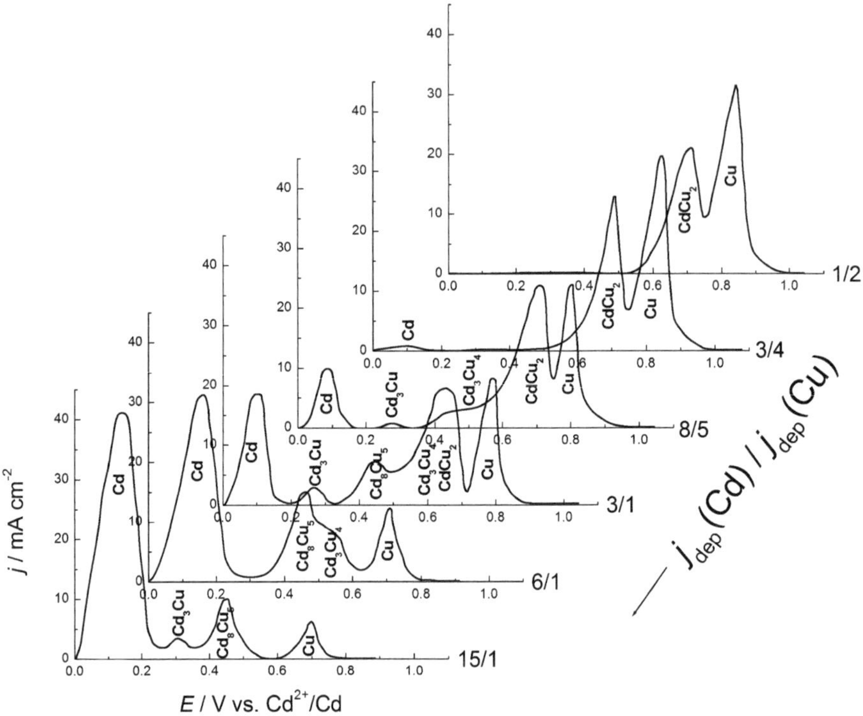

Fig. 1.23 The ALSV's for dissolution of Cd–Cu alloys deposited at different current density ratios (marked in the figure, 1/2, 3/4, 8/5, 3/1, 6/1, and 15/1). Solution: 0.01 M $CdSO_4$ + 0.2 M Na_2SO_4 + 0.01 M H_2SO_4; RPM = 0; v = 1 mV s^{-1} (Reprinted from [1] with the permission of Springer)

appear on the ALSV corresponding to $CdCu_2$ and Cu, whereas for alloys deposited at higher values of current density ratio the number of peaks increases with the peaks corresponding to other intermetallic compounds (marked in the figure).

Considering presented ALSV's it appears that Cd_3Cu_4 and $CdCu_2$ cannot be detected separately [57, 58] in any of deposited alloys and that only for the ALSV of alloy deposited at $j_{dep}(Cd)/j_{dep}(Cu) = 3/1$ the Cd_3Cu_4 could be detected after ALSV deconvolution.

It is obvious from the presented results that in spite of the adjustment of the current density ratio for the formation of particular intermetallic compound, in most cases more than one compound has been detected, indicating that single-phase equilibrium alloys could rarely be obtained, which is the fact well known in electrochemical practice [7, 8].

In order to confirm results obtained by the ALSV technique, an attempt was made to detect intermetallic compounds in alloy layers of higher thickness (>20 μm) deposited at different current density ratios, using X-ray technique. Unfortunately, the X-ray results revealed practically amorphous deposits, indicating that the grain sizes of intermetallic compounds were less than that required for

Table 1.3 Comparison of the results of the Guinier method of phase structure analysis for Cd–Cu alloys deposited at different current density ratios and submitted to subsequent thermal treatment and the results of the ALSV analysis of as-deposited samples

$j_{dep}(Cd)/j_{dep}(Cu)$	Guinier analysis results	ALSV analysis results
3/4	$CdCu_2$ Cd_3Cu_4 Cu	Cu $CdCu_2$
8/5	$CdCu_2$ Cd_3Cu_4 Cd_8Cu_5	Cu $CdCu_2$ Cd_3Cu_4 Cd_3Cu Cd
3/1	Cd_8Cu_5 Cd_3Cu Cd	Cu $CdCu_2$ Cd_3Cu_4 Cd_3Cu Cd
6/1	Cd_8Cu_5	Cu Cd_3Cu_4 Cd_8Cu_5 Cd
15/1	Cd_8Cu_5 Cd_3Cu Cd	Cu Cd_8Cu_5 Cd_3Cu Cd

Reprinted from [1] with the permission of Springer

discernible X-ray peaks. Hence, to increase the grain size of intermetallic compounds thicker samples (50 μm) were deposited onto glassy carbon disc, peeled off from the substrate surface, and submitted to thermal treatment at a temperature of 200 °C for 50 h in vacuum. The results obtained using Guinier analysis after the thermal treatment [58] are presented in Table 1.3. In spite of the fact that the X-ray analysis was performed on samples thermally treated after deposition (some solid-state reaction might take place), comparison of the ALSV results with those of the Guinier phase structure analysis shows relatively good agreement between them. The tendency for reduction in the number of intermetallic compounds for thermally treated samples could be explained in terms of the tendency towards the equilibrium [58].

From the presented results it could be inferred that deposited alloys are not in an equilibrium state. Considering ALSV's presented in Figs. 1.22b and 1.23, the question arises as to whether the ALSV peaks represent phases (intermetallic compounds) that are present in the alloy from the start or reflect the transformation of initially present phases into compounds ever richer in Cu during the process of dissolution [50]? If this would be the case (only Cd_3Cu existed at the start and is subsequently transformed successively into compounds richer in Cu), the charges under the subsequent peaks should follow a rational sequence defined by the stoichiometry. Thus, by calculation, the charges under subsequent peaks should be equal to 0.607, 0.178, and 0.357 of that under the first (most negative) peak, respectively. This is obviously not so. Indeed, the more likely event would be the

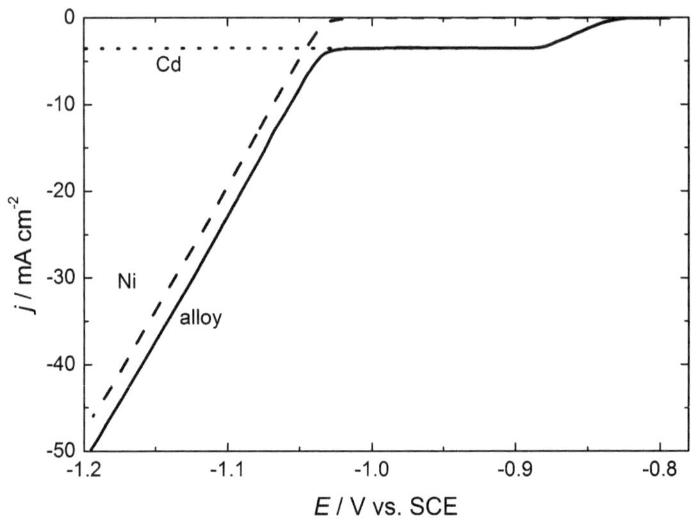

Fig. 1.24 Polarization diagrams for Cd (*dotted line*), Ni (*dashed line*), and Cd–Ni alloy (*solid line*) deposition recorded onto gold RDE at the sweep rate of 1 mV s^{-1} and RPM = 1,000 in the solution containing 0.002 M $CdSO_4$ + 0.2 M $Na_3C_6H_5O_7$ + 2.0 M $NiSO_4$ (Reprinted from [15] with the permission of Elsevier)

formation of different compounds in accordance with the standard Gibbs energy per atom sequence, but this is not the case either. Hence, it can be concluded [58] that all the compounds were formed simultaneously. This conclusion is in accordance with the results of the Guinier phase structure analysis.

System Cd–Ni [15]

System Cd–Ni is one of the most suitable binary alloy systems for the application of ALSV technique since the process of deposition/dissolution of Ni is irreversible. Although its standard potential is −0.474 V vs. SCE ($E^{\ominus}_{Cd} = -0.647$ V vs. SCE) and in this system Ni represents the more noble component, its deposition, due to high overpotential for Ni deposition [15, 24], commences at about −1.04 V vs. SCE in the solution for alloy deposition (0.002 M $CdSO_4$ + 0.2 M $Na_3C_6H_5O_7$ + 2.0 M $NiSO_4$, Fig. 1.24), while the ALSV peak for Ni dissolution in 1 M NaCl (pH 2) is positioned at about −0.10 V vs. SCE (Fig. 1.25). Accordingly, the Ni^{2+} ions formed during the dissolution of certain intermediate phase in the potential range from −0.80 V vs. SCE to −0.20 V vs. SCE (Fig. 1.25) cannot reprecipitate on the spot [case (2) of Sect. 1.3.2.1], and the ALSV peak of dissolution of pure Ni could be seen at the most positive potentials on the ALSV only if the whole amount of deposited Ni has not been dissolved during the dissolution of all phases and the deposit contains some amount of free Ni. On the other side, due to its reversible deposition/dissolution Cd starts to deposit at more positive potential than Ni of

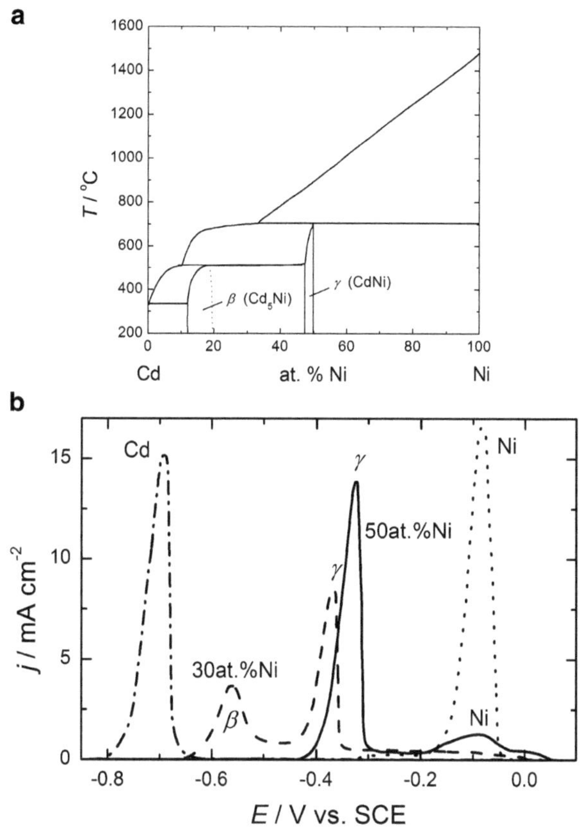

Fig. 1.25 (**a**) Phase diagram of the system Cd–Ni and (**b**) ALSV's for dissolution of pure Cd (*dash-dot line*), pure Ni (*dotted line*), alloy containing 30 at.% Ni (*dashed line*), and alloy containing 50 at.% Ni (*solid line*). Conditions of dissolution: 1 M NaCl, pH 2; RPM = 0; $v = 1$ mV s^{-1} (Reprinted from [15] with the permission of Elsevier)

about −0.82 V vs. SCE (Fig. 1.24), and its dissolution commences at about −0.80 V vs. SCE (Fig. 1.25). At the same time it is very easy to calculate alloy composition by the analysis of charge consumed during the alloy deposition, the total charge for alloy dissolution (charge under the ALSV), and the charge for deposition of Cd [see Eq. (1.38)]. The current efficiency for alloy deposition is given by the relation.

$$\eta_j = \frac{Q_{\text{diss}}(\text{alloy})}{Q_{\text{dep}}(\text{alloy})}. \tag{1.46}$$

In Fig. 1.24 are presented polarization curves for Cd, Ni, and Cd–Ni alloy deposition recorded at the sweep rate of 1 mV s^{-1} and RPM = 1,000 in the solution containing 0.002 M $CdSO_4$ + 0.2 M $Na_3C_6H_5O_7$ + 2.0 M $NiSO_4$.

The phase diagram of the system Cd–Ni [67] is presented in Fig. 1.25a. As can be seen two intermediate phases, β (Cd_5Ni) and γ(CdNi), exist in a narrow range of composition. Hence, one would expect two peaks for these phases on the ALSV of Cd–Ni alloy dissolution, placed at potentials more positive than that for pure Cd dissolution (in the case that the whole amount of Ni is not consumed in these two

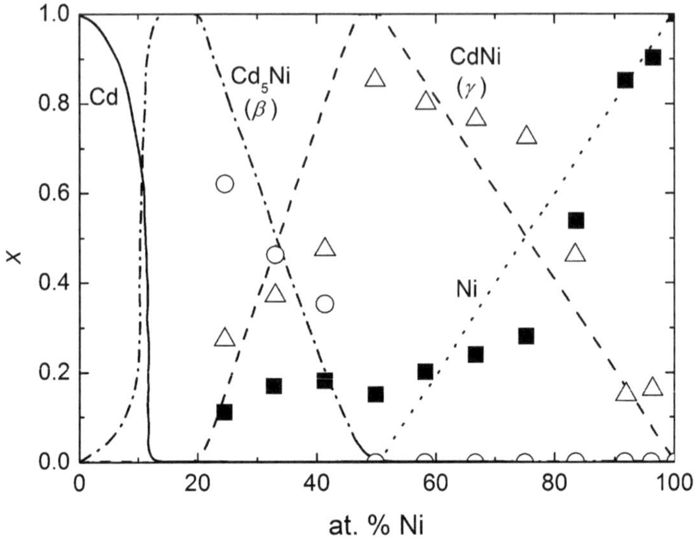

Fig. 1.26 Equilibrium molar fractions of Cd (*solid line*), β phase–Cd_5Ni (*dash-dot line*), γ phase–CdNi (*dashed line*), and Ni (*dotted line*). Molar fractions of β phase–Cd_5Ni (*open circle*), γ phase–CdNi (*open triangle*), and Ni (*filled square*) obtained by the analysis of the ALSV's (Reprinted from [15] with the permission of Elsevier)

phases, being dissolved under these peaks, additional peak for Ni dissolution might appear at the most positive potentials). Considering the ALSV's for dissolution of pure Cd (dash-dot line), pure Ni (dotted line), Cd–Ni alloy containing 30 at.% Ni (dashed line), and Cd–Ni alloy containing 50 at.% Ni (solid line), presented in Fig. 1.25b it is obvious that expected ALSV response has been obtained [15]. The Cd–Ni alloy containing 30 at.% Ni is characterized by the presence of both intermediate phases (β and γ) and no free Ni, while the alloy containing 50 at.% Ni is characterized by the presence of only one phase (γ) and small amount of free Ni.

In order to confirm agreement between the results obtained by the ALSV analysis and those obtained by the analysis of the phase diagram [67] (equilibrium ones) of the system Cd–Ni, the equilibrium molar fraction of each phase in the system (Cd, Ni, β, and γ) has been compared with that obtained by the integration of corresponding ALSV peaks. For this to be done the Cd–Ni alloys with following compositions were deposited and dissolved using ALSV technique: 24, 32, 41, 50, 58, 67, 75, 83, 92, and 97 at.% Ni. The results of this analysis are presented in Fig. 1.26. As can be seen good agreement between the equilibrium molar fractions and molar fractions obtained by the analysis of the ALSV's has been obtained for phases β and γ, while slightly higher amount of Ni (in comparison with the equilibrium one) has been obtained for alloys containing up to 50 at.% Ni. Hence, although it is well known in electrochemical practice that equilibrium alloys could rarely be obtained by electrodeposition, in some cases (as is Cd–Ni alloy) this is possible.

Taking into account that the alloys were dissolved in the electrolyte (1 M NaCl, pH 2) containing no ions of corresponding metals (Cd^{2+} and Ni^{2+}), instead of using standard potentials of metals [Eq. (1.18)] for determining the standard Gibbs energies of intermediate phases β and γ, the reversible potentials (E_r) for dissolution of Cd, Ni, β, and γ could be used. These potentials could be relatively precisely determined from the ALSV's of their dissolution as the potentials at which the dissolution current density starts to rise on the ALSV. In such a way Eq. (1.18) could be written as [15]

$$E\left(\mathrm{A}_x\mathrm{B}_{(1-x)}\right) = \frac{xp}{[xp + (1-x)q]}E_\mathrm{r}(\mathrm{A}) + \frac{(1-x)q}{[xp + (1-x)q]}E_\mathrm{r}(\mathrm{B}) - \frac{\mu^\Theta\left(\mathrm{A}_x\mathrm{B}_{(1-x)}\right)}{[xp + (1-x)q]F}. \tag{1.47}$$

Applying this equation on the intermediate phases β (Cd_5Ni) and γ (CdNi), following relations are obtained:

$$E_\mathrm{r}(\beta) = 0.167E_\mathrm{r}(\mathrm{Ni}) + 0.833E_\mathrm{r}(\mathrm{Cd}) - \frac{\mu^\Theta(\beta)}{2F}, \tag{1.48}$$

$$E_\mathrm{r}(\gamma) = 0.5E_\mathrm{r}(\mathrm{Ni}) + 0.5E_\mathrm{r}(\mathrm{Cd}) - \frac{\mu^\Theta(\gamma)}{2F}. \tag{1.49}$$

From Eqs. (1.48) and (1.49) it follows

$$\mu^\Theta(\beta) = 2F[0.167E_\mathrm{r}(\mathrm{Ni}) + 0.833E_\mathrm{r}(\mathrm{Cd}) - E_\mathrm{r}(\beta)], \tag{1.50}$$

$$\mu^\Theta(\gamma) = 2F[0.5E_\mathrm{r}(\mathrm{Ni}) + 0.5E_\mathrm{r}(\mathrm{Cd}) - E_\mathrm{r}(\gamma)]. \tag{1.51}$$

Using Eqs. (1.50) and (1.51) and values for E_r(Ni), E_r(Cd), $E_r(\beta)$, and $E_r(\gamma)$, the values for the standard Gibbs energies of intermediate phases were obtained as $\mu^\Theta(\beta) = -1.3$ kJ mol^{-1} and $\mu^\Theta(\gamma) = -28.9$ kJ mol^{-1}.

Hence, taking into account that the obtained values for $\mu^\Theta(\beta)$ and $\mu^\Theta(\gamma)$ are negative, this is in good accordance with the fact that already at 20 at.% of Ni there is practically no free Cd in the deposited alloys. Such a behavior indicates very strong interaction between Cd and Ni in order to form both phases, with the γ phase being present in the alloy at all alloy compositions, i.e., with the γ phase being the most stable one in the alloy.

System Ag–In [79]

The phase composition of the Ag–In alloy has been investigated by different authors [67, 81, 82], and the most cited paper is that of Campbell et al. [82]. Due to its complex structure the phase diagram is commonly presented in two parts: up

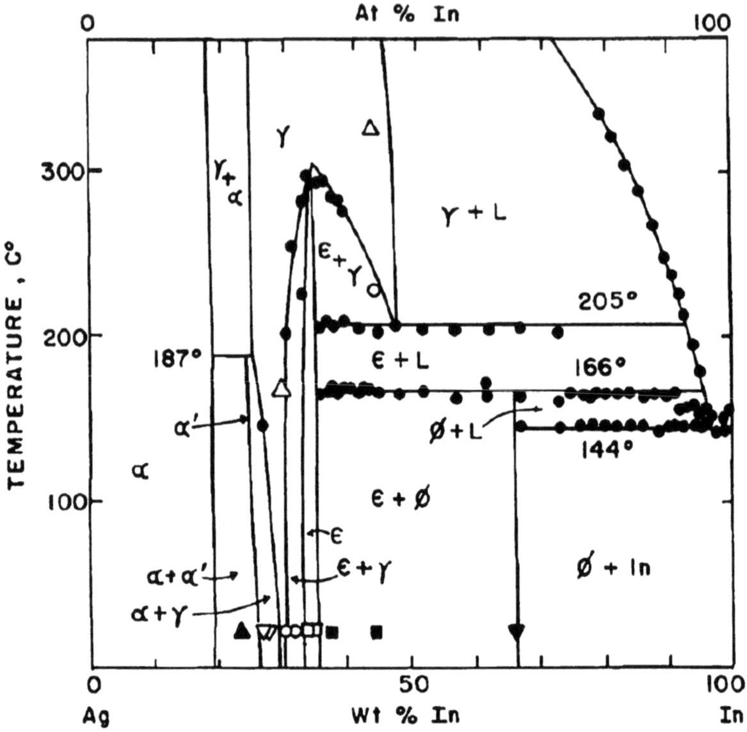

Fig. 1.27 Phase diagram of Ag–In system below 400 °C. γ (Ag_3In)—hexagonal phase; ε (Ag_9In_4)—cubic phase; ϕ ($AgIn_2$)—tetragonal phase (Reprinted from [79] with the permission of Elsevier)

to 400 °C and above. A part of the phase diagram below 400 °C is presented in Fig. 1.27. This system is characterized by the presence of three intermediate phases, ϕ ($AgIn_2$), ε (Ag_9In_4), and γ (Ag_3In), while considering deposition/dissolution characteristics of the more noble metal it belongs to the case (2) of Sect. 1.3.2.1.

The Ag–In alloy coatings were deposited onto stationary Pt electrode from cyanide electrolyte [79] with the composition: 0.2 M $InCl_3$ + 0.04 M KAg $(CN)_2$ + 0.1 M D(+) Glucose + 1 M KCN, pH 8.4. Cyclic voltammogram for alloy deposition/dissolution ($v = 1$ mV s^{-1}) is presented in Fig. 1.28, while in the inset is shown cathodic part only, with the current densities for alloy deposition marked with the arrows (four different alloy compositions). Three potential regions could be detected on the cathodic part presented in the inset: only Ag is deposited down to the inflection point A, while the alloy deposition takes place at more negative potentials (from point A to the second inflection point B) and from point B simultaneous alloy deposition and hydrogen evolution take place. Already at the current density of −2 mA cm^{-2}, Ag–In alloy coating with small amount of In is deposited, while higher amounts of In are obtained at higher cathodic current densities (−4, −6, and −8 mA cm^{-2}).

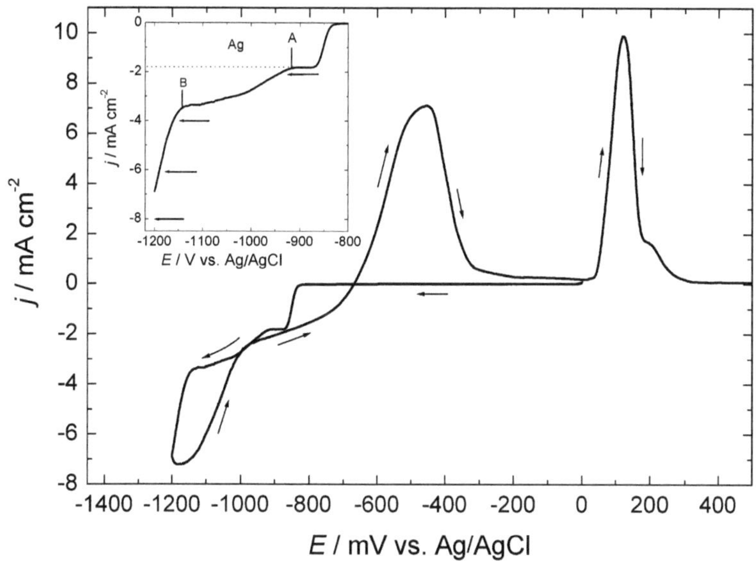

Fig. 1.28 Cyclic voltammogram for alloy deposition/dissolution (v = 1 mV s^{-1}) recorded onto Pt stationary electrode in the solution 0.2 M $InCl_3$ + 0.04 M $KAg(CN)_2$ + 0.1 M D(+) Glucose + 1 M KCN, pH 8.4. In the inset is shown cathodic part only, with the current densities for alloy deposition marked with the *arrows*

The ALSV curves of the dissolution of Ag–In coatings deposited from cyanide electrolyte at different current densities (−2, −4, −6, and −8 mA cm^{-2}) to a constant charge (Q = 1.25 C cm^{-2}) recorded in the solution of 12 M LiCl + 0.1 M HCl (v = 1 mV s^{-1}) are presented in Fig. 1.29. The ALSV curve of the alloy coating deposited at a current density of −2 mA cm^{-2} shows only the presence of one peak E, probably due to the α-phase of the alloy which is a solid solution of In in Ag that exists up to about 19 wt.% of In in the alloy (average amount of In in this coating is found to be about 4 wt.% [83, 84]) and/or pure Ag (see Fig. 1.27), since this peak is situated at the same potential as the peak of pure Ag dissolution. As shown earlier [85] the solid solution appears at the same potential as pure metal (in this case Ag) and cannot be detected as a separate peak. The ALSV curve of the alloy coating deposited at a current density of −4 mA cm^{-2} is characterized by the presence of three additional peaks, A, B, and D. These peaks become more pronounced in the coating deposited at a current density of −6 mA cm^{-2}, while new peak C appears in the coating deposited at a current density of −8 mA cm^{-2}. Hence, these peaks should be attributed to the presence of different phases in the alloy coatings: A—In, B—$AgIn_2$, C—Ag_9In_4, D—Ag_3In, and E—Ag. With the increase of the amount of Ag richer phases, the coating becomes very heterogeneous and some periodic spatiotemporal structures can be observed on its surface [79, 83, 84].

It is interesting to note that the effect of aging of the alloy coatings has clearly been demonstrated for this alloy. In Fig. 1.30 are shown ALSV's (a) and diffractograms (b) for freshly deposited alloy samples (dotted lines) and samples

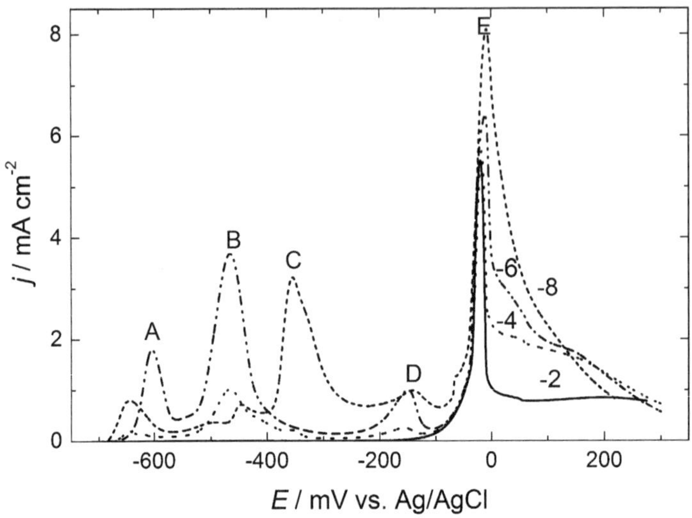

Fig. 1.29 ALSV's of Ag–In alloy coatings dissolution ($v = 1$ mV s^{-1}, solution 12 M LiCl + 0.1 M HCl) deposited at different current densities (marked in the figure in mA cm^{-2}). The ALSV peaks A, B, C, D, and E correspond to In, $AgIn_2$, Ag_9In_4, Ag_3In, and Ag, respectively (Reprinted from [79] with the permission of Elsevier)

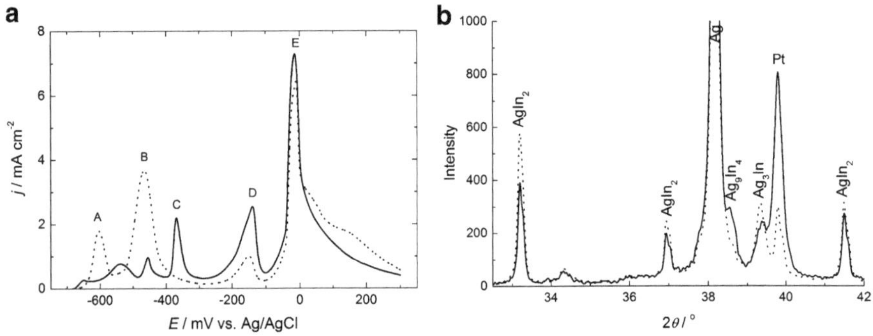

Fig. 1.30 (**a**) ALSV's of Ag–In alloy coating dissolution ($v = 1$ mV s^{-1}, solution 12 M LiCl + 0.1 M HCl) deposited at $j = -6$ mA cm^{-2} recorded immediately after deposition (*dotted line*) and after aging for 20 h at the room temperature. (**b**) Diffractograms of Ag–In alloy coating deposited at $j = -6$ mA cm^{-2} recorded immediately after deposition (*dotted line*) and after aging for 20 h at the room temperature (Reprinted from [79] with the permission of Elsevier)

kept under vacuum in a desiccator during 20 h at the room temperature (solid lines). As can be seen in these figures in a sample analyzed immediately after deposition one phase, Ag_9In_4—peak C, is missing. It is reasonable to expect that during the solid-state reaction, which takes place at the room temperature very slowly, more noble phases (Ag_9In_4 and Ag_3In) are formed on the account of the less noble ones (In and $AgIn_2$), as shown in Fig. 1.30. The same effect has also been detected for the system Cd–Cu [86].

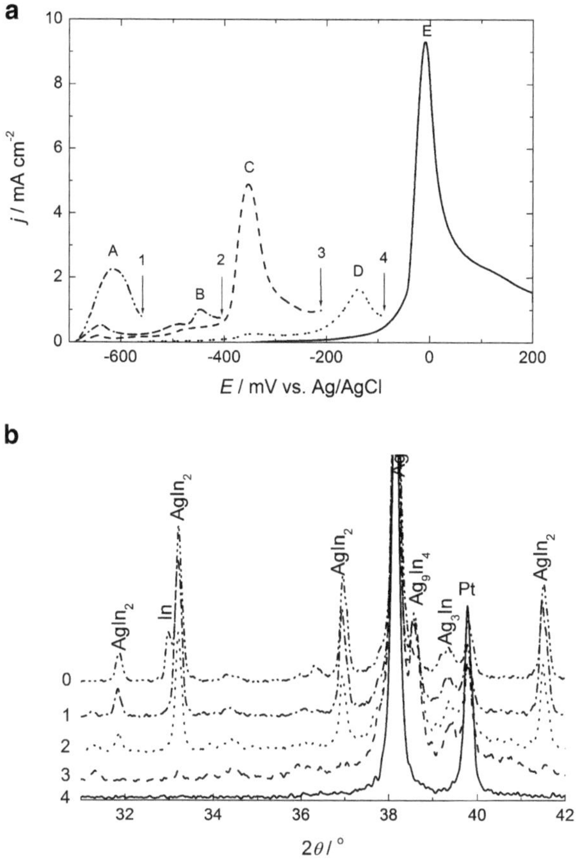

Fig. 1.31 (a) ALSV's of Ag–In alloy coating dissolution ($v = 1$ mV s^{-1}, solution 12 M LiCl + 0.1 M HCl) deposited at $j = -8$ mA cm^{-2}: *1* first dissolution sequence; *2* second dissolution sequence; *3* third dissolution sequence; *4* fourth dissolution sequence. (**b**) Diffractograms of the same Ag–In alloy coating: *0* before the first dissolution sequence; *1* after the first dissolution sequence; *2* after the second dissolution sequence; *3* after the third dissolution sequence; *4* after the fourth dissolution sequence (Reprinted from [79] with the permission of Elsevier)

In order to confirm the statement about the presence of phases mentioned above, following experiment was performed: an alloy coating containing all phases was submitted to the ALSV experiment in such a way that during the first sweep only the phase under the first peak (A) was dissolved; in the second sweep, starting from the same potential (−800 mV), the first (A) and second (B) peaks were dissolved; in the third sweep A–C peaks were dissolved; in the fourth sweep A–D peaks were dissolved; and finally during the last sweep the rest of the alloy was dissolved. The result of this experiment is shown in Fig. 1.31a. After the dissolution of the peak A (up to the potential limit 1, marked with the arrow), during the second run (up to the potential limit 2) very small peak A appears on the ALSV curve, while peak B is

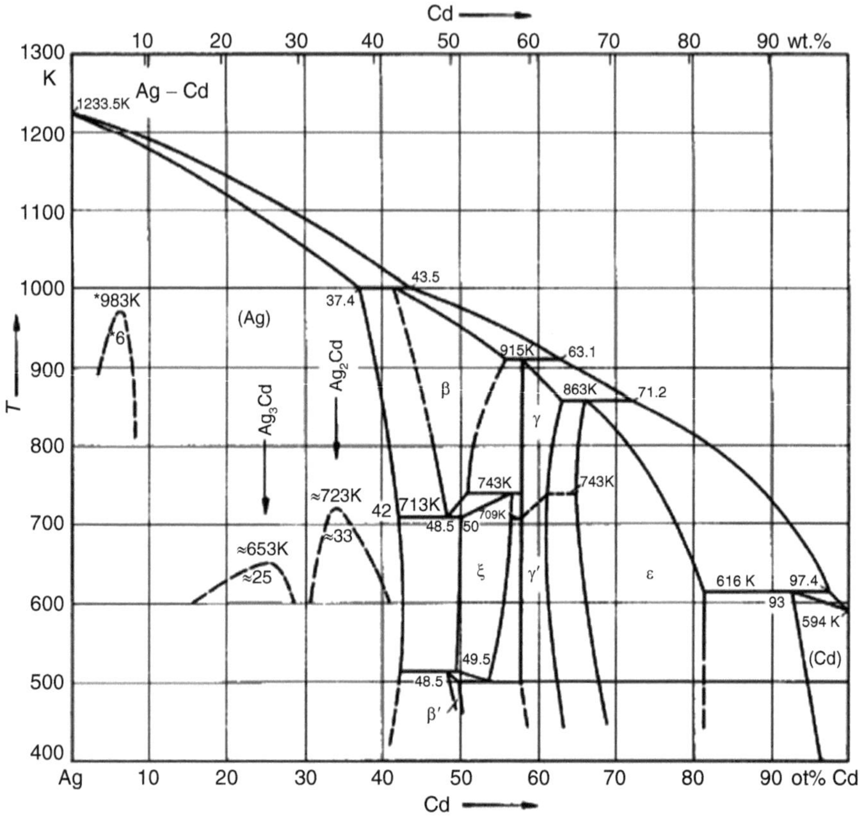

Fig. 1.32 Phase diagram of Ag–Cd system

almost the same as one recorded on the ALSV curve shown in Fig. 1.29. During the next run these two peaks are still slightly visible, while the peak C is well defined (as in Fig. 1.29). After dissolving a phase corresponding to the peak C, during the next run, peaks A and B disappear, peak C become very small, while peak D is the same as in Fig. 1.29. Finally, during the last run no peaks, except the peak E, are seen on the ALSV. After each ALSV run samples were investigated by means of X-ray diffraction, and corresponding diffractograms are shown in Fig. 1.31b. As can be seen excellent agreement between ALSV and X-ray diffraction results were obtained, confirming above statement concerning the presence of different phases.

System Ag–Cd [80]

Phase diagram of the Ag–Cd system is presented in Fig. 1.32. The phases observed in the Ag–Cd alloy system are described in the monograph of Hansen and Anderko [67] in the following way: The solubility of Cd in Ag (α-phase) reaches 42 at.%.

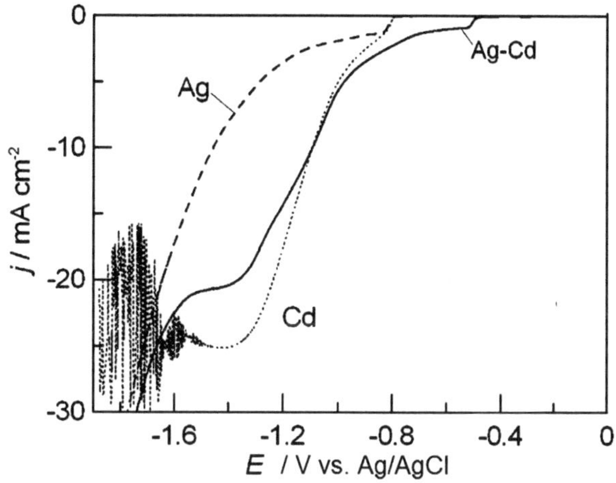

Fig. 1.33 Polarization curves recorded at a sweep rate of 1 mV s^{-1} in the electrolyte containing 0.032 M KAg $(CN)_2$ + 0.56 M KCN (Ag), 0.14 M $CdSO_4 \cdot 8/3H_2O$ + 0.56 M KCN (Cd), and 0.032 M KAg $(CN)_2$ + 0.14 M $CdSO_4 \cdot 8/3H_2O$ + 0.56 M KCN (Ag–Cd) (Reprinted from [80] with the permission of Elsevier)

In the composition range between 26 and 29 at.% Cd, the high-temperature Ag_3Cd phase could be observed. In the composition range between 42 and 60 at.% Cd, three phases (β, β' and ξ) are established in the vicinity of 50 at.% Cd at different temperatures. The β and β' phases are characterized by the *b.c.c.* lattices and the ξ phase by *h.c.p.* lattice. These phases correspond to the AgCd compound. The existence of several phases (γ, γ', ε, and η) is reported in the composition range between 60 and 100 at.% Cd. The first one (γ) is a hexagonal phase, and the second one (γ') possesses a *b.c.c.* lattice. They correspond to the Ag_5Cd_8 compound. The pure hexagonal ε phase exists in the range 68–81.4 at.% Cd and corresponds to the Cd_3Ag compound. In the composition range 81.4–96.9 at.% Cd, ε phase exists together with the η phase, which is a solid solution of Ag in Cd. Pure η phase is observed in the interval 96.9–100 at.% Cd (100 °C).

The Ag, Cd, and the Ag–Cd alloy coatings were deposited onto stationary Pt electrode from cyanide electrolytes [80]. Pure Cd from the solution containing 0.14 M $CdSO_4 \cdot 8/3H_2O$ + 0.56 M KCN, pure Ag from the solution containing 0.032 M $KAg(CN)_2$ + 0.56 M KCN, and Ag–Cd alloys from the solution containing 0.032 M $KAg(CN)_2$ + 0.14 M $CdSO_4 \cdot 8/3H_2O$ + 0.56 M KCN. Figure 1.33 shows polarization curves recorded at a sweep rate of 1 mV s^{-1} in the electrolyte containing both metals separately or together. The deposition of Ag (dashed line) is characterized by the cathodic shoulder, which is most likely indication of the diffusion limiting current density (taking into account small concentration of Ag) at a potential of about −0.82 V, with the deposition starting at about −0.79 V. The increase of the cathodic current density at potentials more negative than −1.2 V during the Ag deposition is the consequence of simultaneous hydrogen evolution. Pure Cd deposition (dotted line) starts at the same potential as that of Ag. At potentials more negative than −1.6 V, current oscillations with amplitude higher than 200 mV have been observed. In the case of Ag–Cd alloy deposition, a shoulder representing diffusion controlled deposition of Ag (solid line) appears at about −0.51 V (for about 0.3 V more positive than the shoulder of pure Ag). Hence, in the solution containing both metal ions, deposition of

Table 1.4 Concentration of different Ag complexes and their equilibrium potentials

	Concentration (%)			E_{eq}/V vs. Ag/AgCl	
Solution composition	$[Ag(CN)_2]^-$	$[Ag(CN)_3]^{2-}$	$[Ag(CN)_4]^{3-}$	$[Ag(CN)_3]^{2-}$	$[Ag(CN)_2]^-$
0.032 M $KAg(CN)_2$ + 0.56 M KCN	35.9	61.2	2.6	−0.737	
0.032 M $KAg(CN)_2$ + 0.14 M $CdSO_4$ + 0.56 M KCN	98.1	1.9	–		−0.479

Reprinted from [80] with the permission of Elsevier

Ag commences at more positive potential. In order to find out the reason for such behavior, the analysis of Ag complexes with cyanide has been performed. The results are presented in Table 1.4. As can be seen, when pure Ag is present in the KCN solution the dominant complex is $[Ag(CN)_3]^{2-}$, with the equilibrium potential of −0.737 V. In the presence of Cd ions most of the CN^- anions are consumed in Cd–CN complexes ($[Cd(CN)]^+$, $[Cd(CN)_2]$, $[Cd(CN)_3]^-$, and $[Cd(CN)_4]^{2-}$), and the dominant Ag–CN complex becomes $[Ag(CN)_2]^-$, with the equilibrium potential of −0.479 V. Hence, from the presented analysis it is obvious that the first shoulder on a solid curve in Fig. 1.33 for Ag–Cd alloy deposition corresponds to the deposition of pure Ag from $[Ag(CN)_2]^-$ complex [87].

Five alloy samples were deposited to the total charge of 1.3 C cm^{-2} at different potentials (−1.0, −1.4, −1.6, −1.8, and −2.0 V). Their ALSV's and diffractograms are presented in Fig. 1.34a, b, respectively. As in a previous case the results of the ALSV analysis and the X-ray analysis are in good agreement. Accordingly, the ALSV peaks detected in Fig. 1.34a could be ascribed to following intermediate phases: the peak A corresponds to the pure Cd phase; the peak B to the $AgCd_3$ phase; the peak C to the AgCd phase; the peak D to the Ag_3Cd phase; and the peak E to the pure Ag.

It is interesting to note that neither ALSV nor X-ray analysis is sensitive to the presence of different lattices for certain compounds at room temperature analysis: (β, β', and ξ)—AgCd compound; (γ, γ', ε, and η)—Ag_5Cd_8 compound.

There are two more systems with intermediate phases investigated by the ALSV technique application, Ni–Zn [87–90] and Co–Zn [91]. As in above presented results, ALSV and X-ray analysis showed good agreement for both systems.

1.3.4 Mechanism of Dissolution of Thin Layers of Electrodeposited Alloys

When a single-phase binary alloy is anodically dissolved in an aqueous solution, assuming that all the necessary conditions are fulfilled (cf. Sect. 1.3.2.1), the mechanism of dissolution may be [1, 40]:

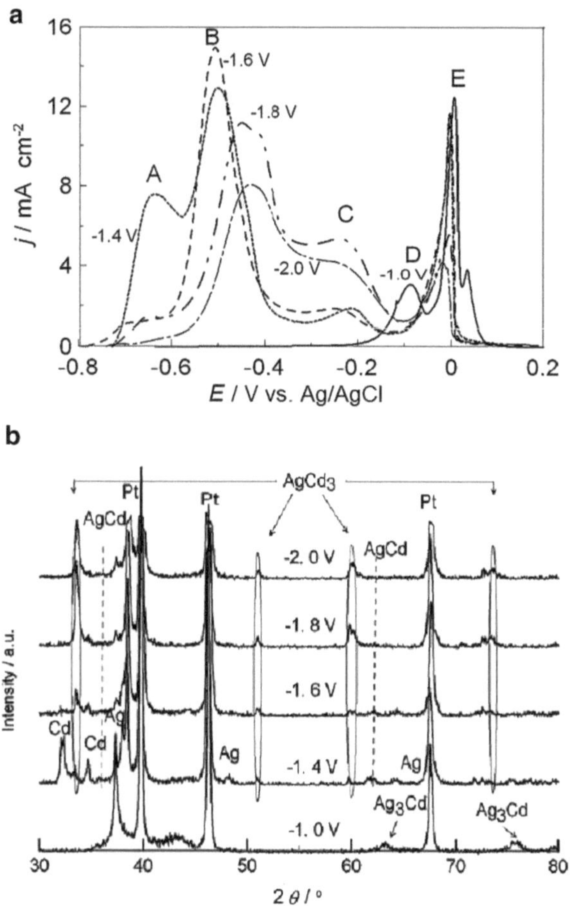

Fig. 1.34 (**a**) ALSV's of Ag–Cd alloy coatings dissolution ($v = 1$ mV s^{-1}, solution 12 M LiCl + 0.1 M HCl) deposited at different potentials (marked in the figure). (**b**) Diffractograms of corresponding alloy coatings (Reprinted from [80] with the permission of Elsevier)

1. Simultaneous dissolution of both components of the alloy, with the more noble one being reprecipitated on the spot;
2. Partial dissolution (ionization) of the less noble component, causing restructuring of the remaining matrix into a phase richer in the more noble component;
3. Complete dissolution (ionization) of the less noble component, with the atoms of the more noble one aggregating by surface diffusion (formation of patches and monolayers of the more noble component with the possibility of formation of small three-dimensional crystals, which may prevent further dissolution of the less noble component);
4. Complete dissolution (ionization) of the less noble component, leaving atoms of the more noble one in an unstable matrix. Atoms of both metals then move in the solid phase by volume diffusion, one kind dissolving, and the other kind restructuring the matrix so that it corresponds to the crystal lattice of the pure metal.

On the basis of diffusivities extrapolated from high-temperature measurements, the participation of the fourth mechanism should be negligible. However, this extrapolation assumes that the equilibrium concentration of vacancies is established and that the contribution of diffusion along small grain boundaries and dislocations is negligible. This is not the case in reality, since the diffusion coefficients at room temperature may be considerably larger than the extrapolated ones (in the case of Cu, monovacancies cannot account for this mechanism, but divacancies, with $D = 1.3 \times 10^{-12}\ \mathrm{cm^2\ s^{-1}}$, may make the mechanism operative) [40].

There is a significant inconsistency in the results reported earlier from investigations of the mechanism of alloy dissolution, particularly for results obtained with bulk alloy samples [39–48]. For example, for one and the same system (Cu–Ni) it was found both that components dissolve simultaneously [44] and that Ni dissolves preferentially [43]. In the case of Cu–Au and Cu–Zn systems, two different mechanisms were observed, preferential dissolution of the less noble component and simultaneous dissolution of both components [40–43]. Actually, it was concluded that in the Cu–Zn system, enrichment of the surface with copper is a necessary condition for simultaneous dissolution of both components to start. Hence, there exists an induction period of disproportionate dissolution of the less noble component, preceding simultaneous dissolution [43].

It is likely that the mechanism depends on the composition of the alloy. Thus, at low contents of the more noble metal, the dissolution of the less noble one leaves the atoms of the former loose and ready to oxidize. Conversely, at high contents of the more noble metal, the atoms of the less noble one are likely to tend to squeeze out of the lattice without producing much change in the latter.

It is obvious that in bulk alloy samples, diffusional limitations (diffusion of metal atoms from the bulk of the alloy to the surface) influence the process of anodic dissolution. In the case of dissolution of thin layers of alloys, where the entire amount of the alloy is dissolved, such limitations (if they exist) can be overcome (cf. Sect. 1.3.2.1). However, two phenomena should be noted:

(a) Simultaneous dissolution creates locally an increased concentration of the positive constituent, corresponding to a more positive reversible potential of the pure metal phase than that attained at a particular moment in the course of the anodic sweep. Hence, that potential turns out to be cathodic with respect to the reversible potential, so that this metal constituent must deposit on the spot either in the form of another alloy phase with a more positive reversible potential or in the form of pure metal crystallites. Hence, the net recorded current reflects the dissolution of the less noble constituent only.

(b) Dissolution of one alloy phase from a mixture with other phases, or through redeposited positive constituent, must run into transport difficulties. Hence, the dissolution current comes under diffusion control, which makes the current vs. potential (time) pattern similar to that obtained in such a situation for a cathodic metal deposition process; that is, it exhibits a maximum and subsequent decay.

Considering all presented cases it appears that the alloy phase dissolves by dissolving both metal constituents. If the deposition/dissolution process of the more noble metal is reversible [case (1) of Sect. 1.3.2.1], the ions of this metal reprecipitate on the spot and the whole amount of the more noble metal becomes dissolved through the ALSV peak positioned at the most positive potential (systems Ag–Pd, Cd–Cu, Cd–Zn, Cu–Pb, etc.). If the deposition/dissolution process of the more noble metal is irreversible [case (2) of Sect. 1.3.2.1], the more noble metal dissolves together with the less noble one through the peak of each phase. In the case that the whole amount of the more noble metal is not consumed in existing phases, the rest of the more noble metal becomes dissolved through the ALSV peak positioned at the most positive potential (systems Cd–Co, Cd–Ni, Co–Ni, etc.).

Finally, it should be noted that using the ALSV technique, it is possible to identify solid-state reactions in electrodeposited alloys at room temperature, leading to the formation of more noble phases on the account of the less noble ones (as shown for the systems Ag–In [79] and Cd–Cu [85]).

1.4 Electrodeposition of Composite Metal Coatings

1.4.1 Basic Concepts

Two types of coatings obtained by electrodeposition could be ascribed to composite metal coatings: (i) laminar metal structures and (ii) metal coatings with inclusion of nonmetallic particles [1, 92].

According to the theory of composite systems [93], the existence of intermediate layer between two metals is inevitable in multilayered laminar metal structures. The thickness of the intermediate layer is usually of the order of couple of atomic layers, and the parameters of the system (concentration of individual metals, crystal structure, Young's modulus, density, coefficient of thermal expansion, etc.) change from one layer to another. Usually such intermediate layer is composed of two layers of solid solutions and one layer of intermetallic compound, as it is schematically presented in Fig. 1.35.

By using appropriate techniques for the formation of such structures (chemical of physical evaporation, electrodeposition, etc.), it is possible to achieve high value of total surface area of the intermediate layer amounting to 3,000 $cm^2\ cm^{-3}$. Accordingly, the larger the surface of the intermediate layer, the higher the difference in the properties of the multilayered laminar metal structure (higher values of Young's modulus, hardness, tensile strength, etc.) [93]. Taking into account that the electrodeposition of such structures is the cheapest technique, its advantage over the other techniques is obvious.

It should be emphasized here that multilayered laminar metal structures could be obtained by electrodeposition of individual metal layers from separate baths, which is a time consuming process. On the other side, it is possible to obtain such deposit

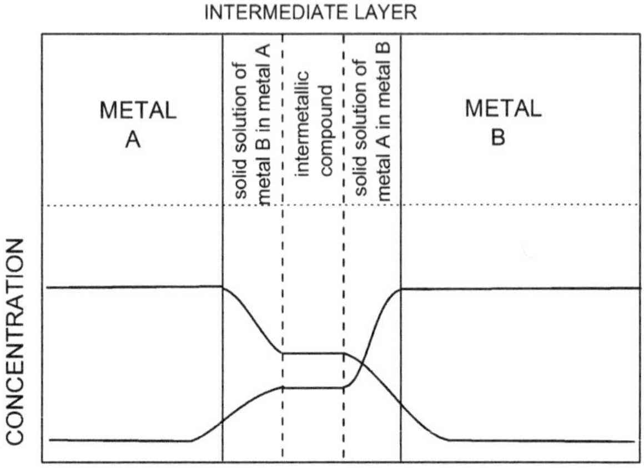

Fig. 1.35 Schematic representation of the intermediate layer in multilayered laminar metal structures

from one bath containing ions of both metals by appropriate choice of metal ions concentration and applied current density. In such a case one layer (usually first) is always pure more noble metal, while the other one (second) must be alloy of these two metals. Hence, by appropriate choice of deposition conditions, one can adjust the composition of a second layer in multilayered structures, as well as the thickness of each layer, and, accordingly, properties of such structures.

Concerning composite structures containing metals with inclusion of nonmetallic particles, despite the potential interest and many years of work in this field, only a few large-scale commercial baths have been developed, while general theoretical explanation of this phenomenon is still missing [1, 92] although some attempts were successfully applied to certain systems [1, 92].

1.4.2 *Electrodeposition of Laminar Metal Structures*

As already stated above, multilayered laminar metal structures could be obtained by electrodeposition of individual metal layers from separate baths or from one bath containing ions of both metals. Only the second case is of interest for this chapter, and such conditions will be discussed.

1.4.2.1 Spontaneous Formation of Layered Deposits

In the investigations of electrodeposition of alloys, laminar metal structures, parallel to the cathode surface, were observed in quite a few instances. Such phenomenon has been observed by Meyer and Phillips [94] during the deposition of Cu with other metals; Aotani [95] detected laminar metal structures in the case of Fe–Ni alloys deposition, while Mikhalev [96] found them in Cd–Zn deposits. This phenomenon has been accompanied with the fluctuations of potential ranging from

0.1 mV to 0.1 V depending on the system and applied current density. Any fluctuations of periodic nature (oscillations) must be the consequence of superposition of two counteracting processes. Accordingly, this phenomenon could be explained in a following way: At the beginning of the process under galvanostatic conditions, the rate of deposition of the more noble metal is sufficient for its nuclei to be formed and its grains to grow. However, after some time, due to depletion of the diffusion layer, the concentration of the ions of the more noble metal at the cathode surface becomes practically zero and the discharge of the ions of the less noble metal must compensate for the decrease of the partial current density of the more noble metal. This requires a certain increase in cathodic overpotential and nucleation and growth of the less noble metal mainly. After some time the concentration of the ions of the more noble metal at the cathode surface recovers to the point where its nucleation can start again, causing decrease of the cathodic overpotential and, accordingly, the nucleation and growth of the less noble metal subsides. Hence, the process starts all over again.

1.4.2.2 Formation of Laminar Deposits by Pulsating Current Regimes

Instead of letting laminar metal structures to be formed spontaneously during the alloy deposition, there was significant motivation to obtain well-defined multilayered structures of controlled compositions and thicknesses of individual layers. Deposition of such structures was first demonstrated by Brenner [4], while Cohen et al. [97] obtained an Ag–Pd alloy with periodically changing composition by alternating the current density, and such structure showed improved wear performance of electrical contacts. In order to obtain laminar metal structure with the thickness of individual layers (with the second layer being Cu–Ni alloy with small amount of Ni) of 300 nm in the system Cu–Ni, pulsating regimes were used [98, 99]. It was shown that such structures possessed higher tensile strength [98] and microhardness [99].

In the application of pulsating regimes [97–99] three types of pulses, schematically presented in Fig. 1.36, were used.

The first attempt in theoretical treatment of the variation of the composition of a deposit obtained by pulsating current from a bath containing two different metal cations, using complicated mathematical procedure, was presented in 1985 [100]. The quantitative theory of laminar metal deposition based on fundamental concepts of electrode kinetics, using much simpler mathematical approach, was presented in a series of papers of Despić and coworkers [101–103].

Single Current Pulse Regime

For the deposition of laminar metal coatings, two conditions must be fulfilled: (1) The reversible potentials for metals A and B must be sufficiently different so that at a given current density the less noble one (B) virtually does not deposit

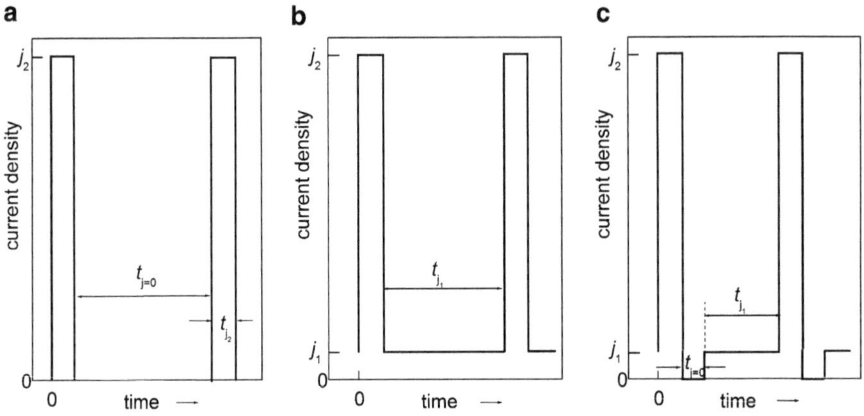

Fig. 1.36 Schematic presentation of pulsating regimes used for deposition of laminar metal structures

during the deposition of the more noble one (A) until complete concentration polarization with respect to ions of metal A takes place. (2) Within the duration of the current density pulse, Send's equation [22] for diffusional polarization is obeyed with respect to concentration change, resulting in transition from deposition of metal A to deposition of metal B after well-defined transition time.

Hence, the first layer contains pure metal A and its thickness is, according to Faraday's law, defined by

$$d(\mathrm{I}) = \left(\frac{M_\mathrm{A}}{z_\mathrm{A}\rho_\mathrm{A}F}\right) j\tau_\mathrm{A}, \tag{1.52}$$

where M_A, ρ_A, and z_A are, respectively, atomic weight, density, and charge on the ions of metal A, j is the current density in the pulse, while τ_A is the transition time with respect to the ions of metal A. Assuming that the convection and migration of ions of the electrolyte are negligible, τ_A is defined by Send's equation [22]:

$$\tau_\mathrm{A} = \frac{(z_\mathrm{A}F)^2 \pi D_\mathrm{A}}{4}\left(\frac{C_\mathrm{A}^\mathrm{o}}{j}\right)^2, \tag{1.53}$$

where $C_\mathrm{A}{}^\mathrm{o}$ and D_A are, respectively, the bulk concentration and diffusion coefficient of the ions of metal A. Introducing Eq. (1.53) into Eq. (1.52) and rearranging one obtains

$$\frac{(C_\mathrm{A}^\mathrm{o})^2}{j} = \frac{4\rho_\mathrm{A}}{z_\mathrm{A}F\pi D_\mathrm{A}M_\mathrm{A}} d(\mathrm{I}). \tag{1.54}$$

Considering Eq. (1.54) it could be concluded that desired thickness of the layer of metal A could be achieved by proper choice of metal ions concentration and the current density of the pulse.

Taking into account that Sand's equation is valid only as long as the change of concentration occurs within a stagnant layer undisturbed by convection and introducing the Nernst's diffusion layer boundary δ and hydrodynamic layer boundary Δh, the minimum current density that must be applied for the transition and deposition of the second layer to take place is given by

$$j^{\min} = \frac{z_A F \pi D_A}{2\Delta h} C_A^o \tag{1.55}$$

and, accordingly, by introducing Eq. (1.54) into Eq. (1.55), the maximum achievable thickness of the first layer (metal A) is

$$d(\mathrm{I})^{\max} = \frac{M_A C_A^o \Delta h}{2\rho_A}. \tag{1.56}$$

Deposition of the second layer starts after τ_A is reached and is accompanied by sudden change of potential to the reversible potential of metal B deposition [1].

The shape of the potential vs. time (E vs. t) response in such a case depends on the property of the alloy and its constituents in the investigated solution. Two situations could be considered: (i) replacement reaction takes place during the off-time ($j = 0$) period; (ii) replacement reaction does not take place during the off-time ($j = 0$) period.

(i) If the difference between the reversible potentials of metals A and B is sufficient, and the constituents of the alloy do not mix in the solid state (eutectic-type alloy), and/or deposition/dissolution of metal B is reversible, replacement reaction must take place during the off time ($j = 0$). Such a case is schematically presented in Fig. 1.37. The current density change is presented in (a), while corresponding potential change is presented in (b). As can be seen, after reaching τ_A, metal A continues to deposit but at a decreasing partial current density [actually by its diffusion limiting current density—$j_d(A) = j_L(A)$], while the partial current density of the metal B increases to make a constant current density of the pulse—$j_d(B)$. If the concentration of ions of metal B is much larger than that of the metal A, the growth of the second layer could be virtually unlimited and the content of metal A in the second layer will continuously decrease with the thickness of the second layer [following the decrease of $j_d(A)$]. Corresponding potential response, schematically presented in (b), is characterized by a sudden increase of cathodic potential to the value higher than $E_r(B)$ at the position of τ_A. At the position of the end of the current pulse (T) and the beginning of the current density off period ($j = 0$), replacement reaction will take place. The deposition of metal A will continue with $j_d(A) = j_L(A) = j_d(A)_r$, while, in order to keep total current density at zero, the less noble metal B will start to dissolve with the positive partial current density $j_{diss}(B)_r$ equal to that of metal A deposition. Once this process is finished,

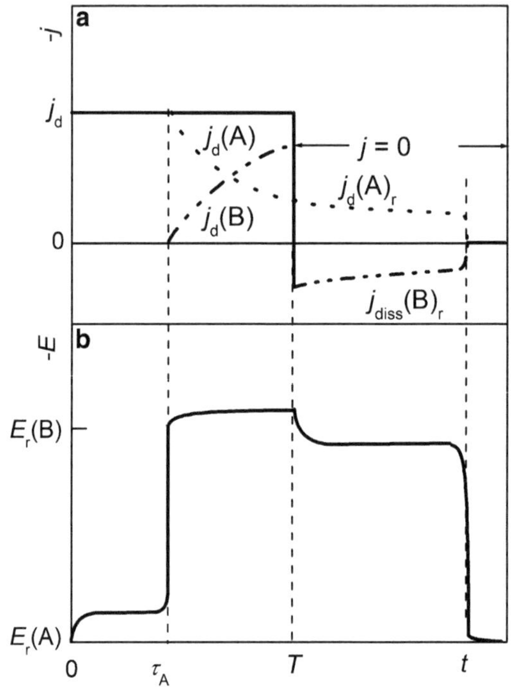

Fig. 1.37 Schematic representation of the partial current densities changes (**a**) and corresponding potential response (**b**) during the deposition of two layers of metals A and A + B by constant current density pulse (j_d) up to time T and during the replacement reaction at $j = 0$. Partial current density for deposition of metal A after reaching $\tau_A - j_d(A)$ and partial current density for deposition of metal B after reaching $\tau_A - j_d(B)$, partial current density for deposition of metal A during the replacement reaction $j_d(A)_r$, partial current density for dissolution of metal B during the replacement reaction $j_{diss}(B)_r$ (Reprinted from [1] with the permission of Springer)

either due to dissolution of the whole amount of metal B in the second layer or prevention of its dissolution for some reasons (e.g., at some places of a second layer atoms of metal B become covered with a monolayer of metal A, etc.), the current density $j_{diss}(B)_r$ will drop to zero and corresponding potential to the value of $E_r(A)$ [1].

Although the replacement reactions have extensively been studied in the literature [104], the only example for such reaction in the layered deposits was demonstrated for the system Cu–Pb [1, 101]. A two-layer structure, composed of a first layer of pure Cu and a second layer of Cu–Pb alloy, has been deposited at a constant current density $j_d = -1$ mA cm^{-2} for different times (from 60 to 300 s) on a stationary glassy carbon electrode from a solution containing 0.01 M Cu $(CH_3COO)_2$ + 0.01 M $Pb(CH_3COO)_2$ + 1 M HBF_4 [101]. After the deposition, zero current density ($j = 0$) was applied (positions marked in the figure with 1, 2, 3, 4, 5, 6, 7, 8, and 9) and corresponding potential responses are presented in Fig. 1.38. As can be seen after applying zero current density (positions 1–9), the

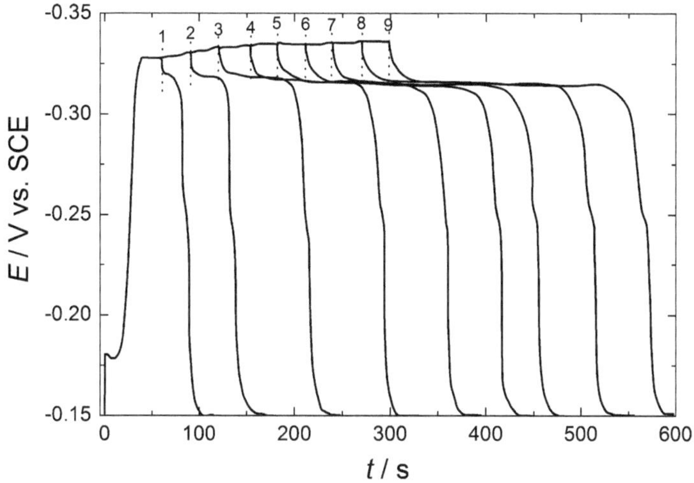

Fig. 1.38 Potential responses recorded during constant current density $j_d = -1$ mA cm^{-2} pulse trains on a stationary glassy carbon electrode from a solution containing 0.01 M $Cu(CH_3COO)_2$ + 0.01 M $Pb(CH_3COO)_2$ + 1 M HBF_4. After the deposition, zero current density ($j = 0$) was applied (positions marked in the figure with 1–9) (Reprinted from [1] with the permission of Springer)

potential remains for some time at a level slightly below the reversible potential of Pb, after which it falls to the potential of Cu. That time is virtually equal to duration of the current density pulse, indicating that the dissolution of Pb from a second layer occurs all the time. When the dissolution process is finished, potential returned to the reversible potential of Cu. By the EDS analysis it was confirmed that the whole amount of deposited Pb has been dissolved during the off-time periods. Hence, all conditions (i) for the replacement reaction to take place were fulfilled: Cu–Pb is eutectic-type alloy, and deposition/dissolution of Pb is reversible [101].

(ii) If the difference between the reversible potentials of metals A and B is sufficient, and the constituents of the alloy mix in the solid state forming solid solution and a metal B passivates in the electrolyte used, replacement reaction will not take place during the off-time ($j = 0$). Such a case is schematically presented in Fig. 1.39. The current density change is presented in (a), while corresponding potential change is presented in (b). During the current density pulse everything is the same as in a previous case. The absence of replacement reaction is characterized by sudden potential change from E_r(B) to E_r(A) at the position of the end of the current pulse (T) and the beginning of the current density off period ($j = 0$).

Typical example for such behavior is the system Cu–Ni [98, 99, 103, 105, 106]. This system belongs to the solid solution type alloys, and the dissolution of Ni is prevented by its passivation in the electrolyte for deposition. As shown earlier [15, 24] dissolution of the whole deposited amount of Ni could be achieved in chloride containing electrolyte of pH 2.

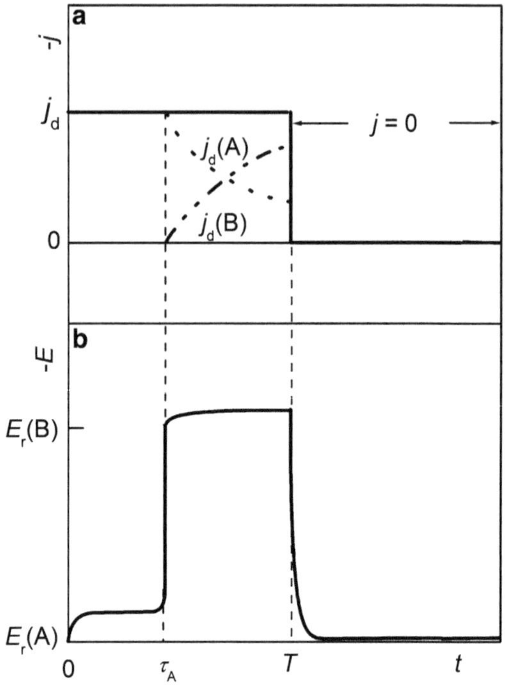

Fig. 1.39 Schematic representation of the partial current densities changes (**a**) and corresponding potential response (**b**) during the deposition of two layers of metals A and A + B by constant current density pulse (j_d) up to time T and after, in the absence of the replacement reaction at $j = 0$. Partial current density for deposition of metal A after reaching $\tau_A - j_d(A)$, partial current density for deposition of metal B after reaching $\tau_A - j_d(B)$

Dual Current Pulse Regime

Single current pulse regime imposes serious limitations on both the thickness of the layers of the more noble metal and the content of that metal in the second layer. The application of a dual current pulse scheme was found more suitable for manipulation of the thickness of each layer, as well as the content of the more noble metal in the second layer. The scheme consists of current pulses in two different intensities each. A lower current density j(I) for a certain time period T(I) is followed by a higher current density j(II) for a period T(II) before the current is interrupted (or the sequence repeated). It can be shown that with such a scheme and proper choice of parameters, any desired thickness of both layers can be achieved and the content of metal A in the second layer can be reduced to a desired level. The process can be improved, in terms of shortening the deposition time of the first layer while maintaining a desired content of metal A in the second layer and by synchronously modulating the hydrodynamic conditions. Such a process can have significant practical value for obtaining bilayers (or multilayers) [1, 102].

In the dual current pulse regime, it is desirable to have

$$C_B^o >> C_A^o \tag{1.57}$$

the best choice being that which satisfies equality

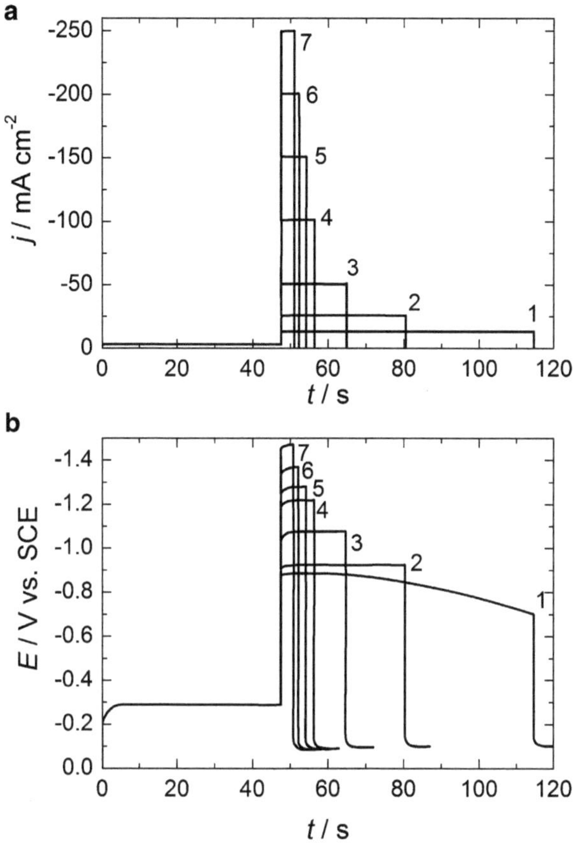

Fig. 1.40 A sequence of high current density pulses (**a**) and corresponding potential responses (**b**) (marked with numbers 1–7) during the formation of two-layer structure composed of a pure (first) Cu layer of about 2 μm and the second layer of different Cu–Ni alloy compositions of a thickness of about 12 μm. The current density ratio, c.r.—$j(\mathrm{Ni})/j(\mathrm{Cu})_\mathrm{L}$, for pulse no. 1–2.0, no. 2–5.3, no. 3–13.3, no. 4–27.6, no. 5–36.5, no. 6–56.1, no. 7–61.5 (Reprinted from [102] with the permission of Electrochemical Society)

$$\frac{C_\mathrm{B}^\mathrm{o}}{j(\mathrm{II})} = \frac{C_\mathrm{A}^\mathrm{o}}{j(\mathrm{I})}. \tag{1.58}$$

An attempt was made to deposit two-layer structure composed of pure (first) Cu layer of about 2 μm and the second layer of different Cu–Ni alloy compositions of a thickness of about 12 μm by applying dual current pulse regime presented in Fig. 1.40. The two layers were deposited onto Ag RDE (RPM = 1,000) from the solution 0.02 M $CuSO_4$ + 2 M $NiSO_4$ + 0.5 M $Na_3C_6H_5O_7$. The value of the current density in the first pulse, $j(\mathrm{I})$, was adjusted at 75 % of the value of the diffusion limiting current density for copper ions at the given concentration and rotation speed [102].

Table 1.5 The composition of the second layer of the deposits as a function of the current density ratio (c.r.)

		Composition of the second layer			
		EDS analysis		Calculated from c.r.	
Pulse no.	c.r.	at.% Cu	at.% Ni	at.% Cu	at.% Ni
1	2.0	35.4	64.6	33.3	66.7
2	5.3	20.8	79.2	16.0	84.0
3	13.3	7.9	92.1	7.0	93.0
4	27.6	4.9	95.1	3.5	96.5
5	36.5	4.7	95.3	2.7	97.3
6	56.1	3.9	96.1	1.7	98.3
7	61.5	4.2	95.8	1.6	98.4

Reprinted from [102] with the permission of Electrochemical Society

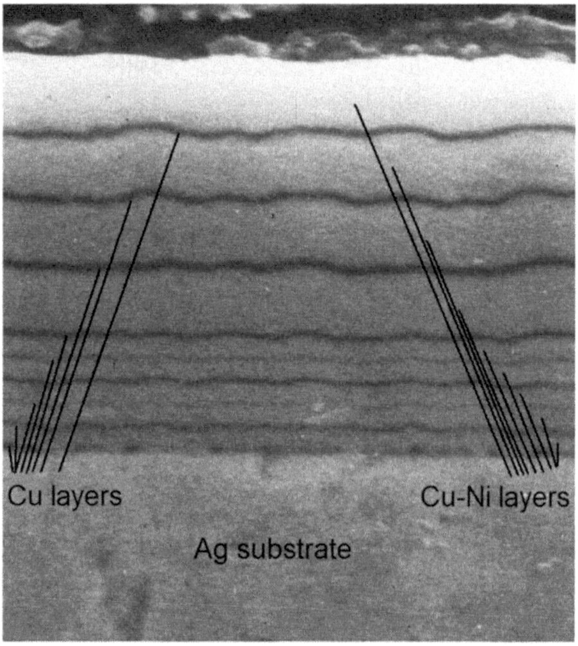

Fig. 1.41 Multilayer structure obtained by the current pulses presented in Fig. 1.42 (Reprinted from [102] with the permission of Electrochemical Society)

Relatively, smooth deposits, consisting of two well-defined layers, are obtained. Results of the EDS analysis of the composition of the second layer (Cu–Ni alloy) of the deposits obtained by the current regimes shown in Fig. 1.40 are compared in Table 1.5 with the composition of the second layer calculated from the current density ratio in the second pulse assuming 100 % current efficiency for alloy deposition, c.r.—$j(\mathrm{Ni})/j(\mathrm{Cu})_L$. Taking into account precision of the EDS analysis, it is seen that a relatively good agreement is obtained.

A multilayer structure consisting of ten relatively thick layers shown in Fig. 1.41 was obtained by the current regime presented in Fig. 1.42.

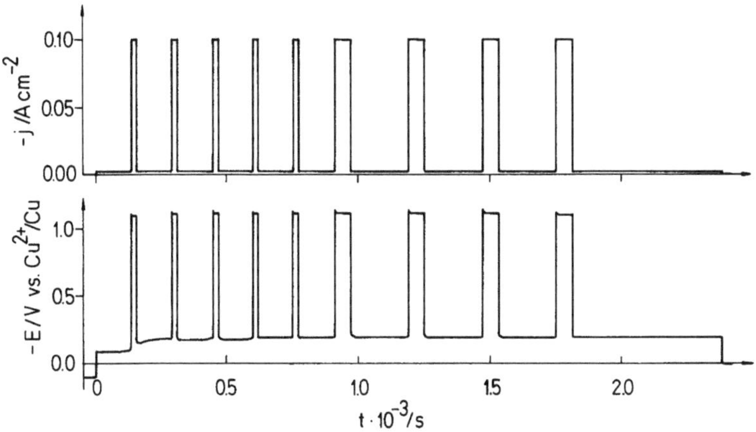

Fig. 1.42 Current density pulses and corresponding potential responses applied for the formation of multilayer structure presented in Fig. 1.41 (Reprinted from [102] with the permission of Electrochemical Society)

Presented analysis confirmed that the thickness and the composition of layered deposits could be controlled with sufficient precision [102].

1.4.2.3 Intermediate Layer Detection

Taking into account the fact that the intermediate layer in deposited laminar metal structures plays an important role in obtaining improved mechanical and magnetic properties [93], from 1980 to 2000 particular attention has been given to the pulsating regimes which would provide as thin as possible layers (of the order of nanometers) [105–109] in multilayered metal structures, and, accordingly, much higher surface area of the intermediate layer.

In theoretical consideration of multilayer deposition of Cu–Ni alloy under the conditions of convective diffusion [103], it was shown that a constant concentration of Cu in the second layer (Cu–Ni alloy) could be established at different thicknesses of the second layer, depending on the value of the current density of a second pulse and the rotation rate. As the value of the current density in the second pulse increases, the thickness at which a constant concentration of Cu could be established also increases, indicating that the thickness of the intermediate layer also increases, reaching the maximum value of about 150 nm [103]. Unfortunately, this theoretical consideration was not experimentally confirmed. Most of the experimentally confirmed theoretical treatments of the pulsed deposition of multilayer have been applied to the Cu–Ni system [110–113]. Equations for partial currents for copper and nickel deposition [19, 111] (including the evolution of hydrogen) from a single bath in the case of galvanostatic [110] and potentiostatic [111] deposition of multilayered Cu–Ni coatings under the conditions of convective diffusion were developed. The experimental results were in good agreement with the theoretical predictions. The subject of two papers [112, 113] was the process taking place during the off-time period in pulsed deposition, showing that during the off-time period copper deposits by a displacement of nickel ("replacement reaction"). These models were found to agree well with the experimental findings. Hence, the existence of an intermediate layer in these investigations has not been confirmed experimentally.

However, it was shown in the system Cd–Ni [114] that the intermediate layer could be detected by the application of the ALSV technique. Alloy samples used for the ALSV analysis were galvanostatically deposited at the amount of charge of 1 C cm^{-2} in the solution containing 2 M $NiSO_4$ + 0.2 M $Na_3C_6H_5O_7$ + 0.002 M $CdSO_4$. Pure Ni and Cd layers were also deposited galvanostatically at $j_d = -10$ mA cm^{-2} ($Q_d = -1$ C cm^{-2}) from the solutions containing 2 M Ni $(NH_2SO_3)_2$ + 0.5 M H_3BO_3 and 0.5 M $CdSO_4$ + 0.01 M H_2SO_4, respectively. The samples composed of a layer of alloy and a layer of pure Cd on top of the alloy layer, used for the analysis of the intermediate layer existence, were galvanostatically deposited from two different baths: 2 M $NiSO_4$ + 0.2 M $Na_3C_6H_5O_7$ + 0.002 M $CdSO_4$ and 2 M $Ni(NH_2SO_3)_2$ + 0.5 M H_3BO_3 + 0.002 M $CdSO_4$. All samples were dissolved in a solution of 1 M NaCl (pH 2).

Intermediate Layer Between Ni and Cd Deposited from a Separate Baths

For this purpose layers of pure Cd and pure Ni were deposited from the appropriate separate baths onto gold disc electrode. To provide normal dissolution of these metals, Ni was first deposited as a more noble metal and a layer of Cd was deposited on top of it. From such a deposit Cd dissolves at about −0.8 to −0.7 V, while Ni starts to dissolve at about −0.33 V with the peak appearing at about −0.1 V.

Hence, to detect an intermediate layer between these two metals, ALSV's of pure Ni and pure Cd dissolution were first recorded. Then, after the deposition of a layer of Cd on top of a layer of Ni, an ALSV of dissolution of a deposit consisting of two metal layers was recorded. Since these two metals make two intermediate phases [15], in the case of interaction between the two layers, one would expect one or two small ALSV peaks, pertaining to the dissolution of either of these phases, between the peaks of dissolution of pure metals. There was no indication, however, of the existence of any additional ALSV peak even after prolonged thermal treatment (deposit was kept in an atmosphere of purified nitrogen at a temperature of 50 °C for 24 h) of such deposit. Knowing that the ALSV technique is very sensitive and that it is possible to detect clearly a monolayer of metals by this technique, it seems that in the case of Ni and Cd layers, deposited from separate baths, no intermediate layer between them is formed.

Intermediate Layer Between Cd–Ni Alloy and Pure Cd Deposited from a Single Bath

In the case of Cu–Ni alloy, detection of an intermediate layer by the ALSV technique is impossible, since these two metals make a series of solid solutions

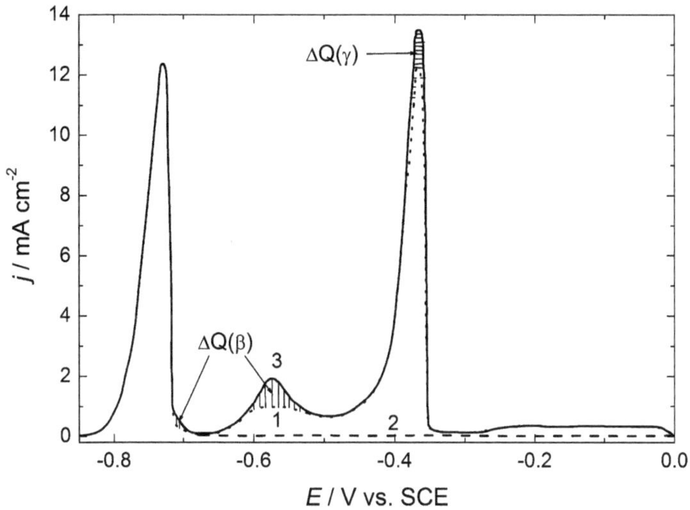

Fig. 1.43 The example of the ALSV's of Cd–Ni alloy dissolution (*dotted line*, 1), pure Cd dissolution (*dashed line*, 2) and dissolution of a two layer deposit (*solid line*, 3) composed of a layer of Cd–Ni alloy and a layer of pure Cd deposited on top of a layer of the alloy (Reprinted from [114] with the permission of Elsevier)

and the potentials of dissolution of pure metals in solution 1 M NaCl, pH 2, are very close. However, the system Cd–Ni should be very suitable for this purpose, since the ALSV's of Cd–Ni alloys dissolution are seen to be characterrized by the existence of two ALSV peaks corresponding to the dissolution of phases β and γ of the formula $NiCd_5$ and NiCd, respectively [15, 114]. Hence, if there exists the intermediate layer, it should be composed of these phases [1]. By comparing ALSV of dissolution of a single layer of the alloy with the ALSV of dissolution of the two-layers deposit, it is possible to detect an intermediate layer between these two layers. An example is shown in Fig. 1.43. Curve 1 represents the ALSV of the alloy dissolution without the Cd layer on top, curve 2 the ALSV of pure Cd dissolution, while curve 3 represents the ALSV of the two-layers deposit dissolution. As can be seen curves 1 and 3 are different, with the peaks of β- and γ-phase dissolution being somewhat higher on the curve 3, while the peak of pure Cd dissolution does not change. As the content of Ni in the alloy layer increases, the difference between these two ALSV curves becomes smaller.

Hence, the difference between curves 3 and 1 (shaded areas on Fig. 1.43) should represent the intermediate layer. By integration of the surface of shaded areas, it is possible to obtain the charges (amounts) corresponding to the increase of β and γ phases in the two-layer deposit [marked in the figure as $\Delta Q(\beta)$ and $\Delta Q(\gamma)$], which should be a consequence of the formation of an intermediate layer between a layer of alloy and a layer of pure Cd. Thus, the intermediate layer is mainly composed of β and γ phases, which is in good agreement with the theory of an intermediate layer in laminar metal structures [93].

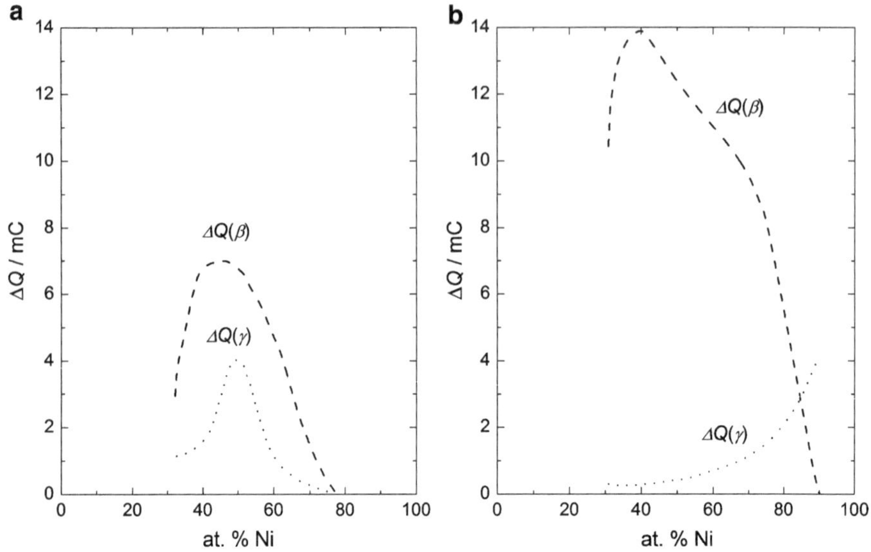

Fig. 1.44 The charges corresponding to the contribution of the β and the γ phases in the intermediate layer obtained by the analysis of the ALSV's of two-layer deposits dissolution. Alloys were deposited from the solutions containing: (**a**) 2 M $NiSO_4$ + 0.2 M $Na_3C_6H_5O_7$ + 0.002 M $CdSO_4$; (**b**) 2 M $Ni(NH_2SO_3)_2$ + 0.5 M H_3BO_3 + 0.002 M $CdSO_4$ (Reprinted from [114] with the permission of Elsevier)

These investigations were performed with the two-layers deposits obtained from two different baths (as stated above). The recorded charges $\Delta Q(\beta)$ and $\Delta Q(\gamma)$, obtained for both investigated baths for alloy deposition, are shown in Fig. 1.44 as functions of alloy layer composition. As can be seen in the figure, the intermediate layer composition is slightly different in the two cases. The amount of the β phase in the intermediate layer is seen to reach its maximum value at about the same alloy composition (30–40 at.% Ni) in both cases, while the contribution of the γ phase in the intermediate layer is different in two-layer deposits obtained from different baths. Such a behavior clearly indicates that the composition of an intermediate layer is sensitive to the composition of the electroplating bath, since the conditions of deposition were identical in both cases.

Assuming that both phases are homogeneously distributed in the intermediate layer, the intermediate layer thickness can be calculated from the charges recorded on the ALSV's. From the formula of the β phase (Cd_5Ni), one can determine the charge corresponding to Ni as 0.167 and to Cd as 0.833 of total charge $\Delta Q(\beta)$ and in a similar way for the γ phase [Ni 0.5 and Cd 0.5 of $\Delta Q(\gamma)$]. Using these charges and corresponding values of the density and molecular weights of the two metals, the thicknesses of β and γ phases in the intermediate layer can be calculated. Hence, the total thickness of the intermediate layer should represent the sum of the thickness of individual phases. Figure 1.45 shows the dependence of the thickness of β and γ phases and the total thickness of an intermediate layer (d_{il}) as a function of alloy

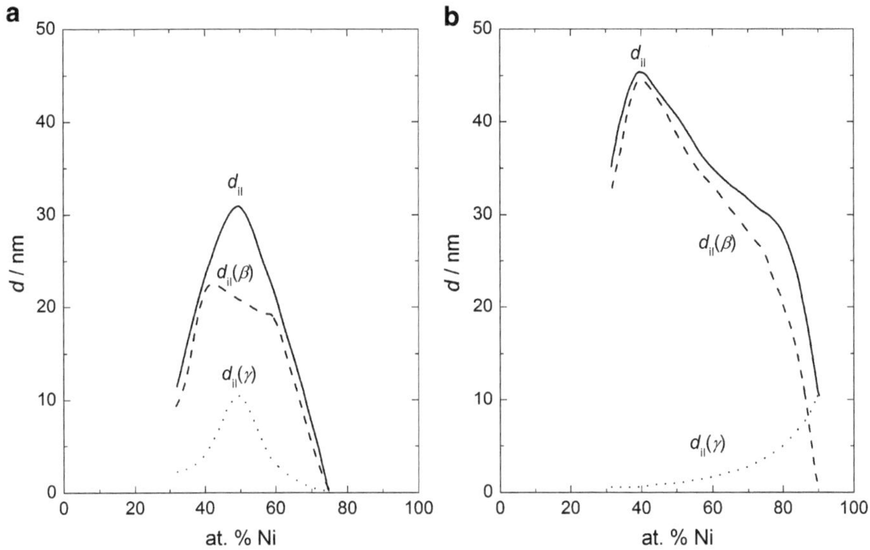

Fig. 1.45 The thickness of the individual phases, $d_{il}(\beta)$ and $d_{il}(\gamma)$, and the intermediate layer, d_{il}, obtained by the analysis of the results presented in Fig. 1.44 (Reprinted from [114] with the permission of Elsevier)

composition, obtained by the analysis of the results presented in Fig. 1.44. As can be seen the thickness of the intermediate layer depends on the solution for alloy deposition, and its maximum value is about 45 nm. Hence, in certain systems it is possible to estimate the thickness and the surface of intermediate layer by the application of ALSV technique.

1.4.2.4 Mechanical and Magnetic Properties of Multilayered Structures

Mechanical properties of multilayered structures were only investigated for deposits containing Cu/Cu–Ni layers. Following mechanical properties were mainly investigated: Young's modulus [98], hardness [115], and tensile strength [107, 115–117]. It is shown that all investigated properties depend on the thickness of the individual layers and that in all cases multilayered structures showed better properties than that of pure metals and/or their alloys.

The results of hardness investigation [115] are presented in Fig. 1.46. As can be seen maximum hardness for the deposit containing Cu/Ni–100 layers was achieved at the thickness of individual layers of 12 nm. These results are not in agreement with those detected for evaporated layers Cu/Ni–111 [116], where the maximum hardness was obtained for individual layer thickness of 1.6 nm, indicating that, most probably, the hardness depends on the thickness and the total surface area of

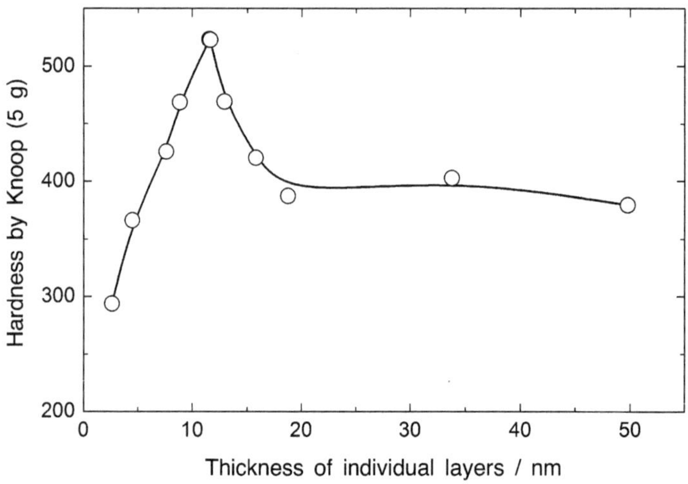

Fig. 1.46 The hardness of the deposit containing Cu/Ni – 100 layers as a function of the thickness of individual layers (Reprinted from [115] with the permission of Elsevier)

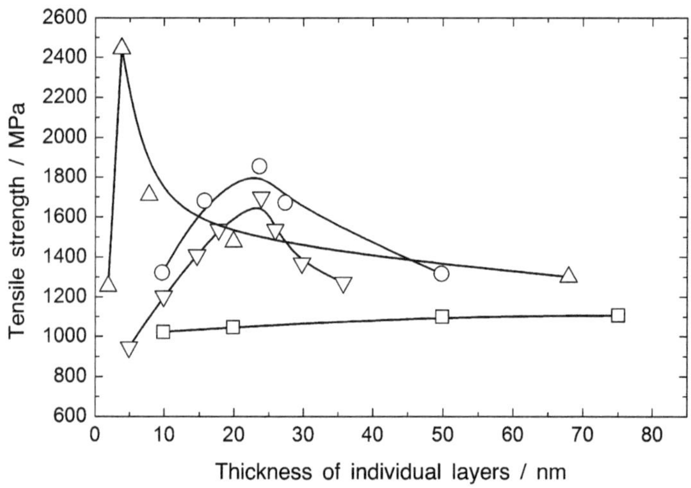

Fig. 1.47 The tensile strength of the multilayered deposits as a function of the thickness of individual layers: *open circle*—[107]; *open square*—[117]; *open triangle*—[118]; *open inverted triangle*—[108] (Reprinted from [109] with the permission of Elsevier)

the intermediate layer which is different for different techniques of multilayered structure formation.

In several investigations the influence of the individual layer thickness on the tensile strength was reported [106, 114, 116, 117]. The results of these investigations are presented in Fig. 1.47. Significant difference in the obtained results is

Table 1.6 Tensile test for electroformed 90 % Ni – 10 % Cu multilayers vs. Cu layer thickness

Cu layers (nm)	UTS (MPa)	YS (MPa)	Modulus (GPa)	Strain (%)
1.0	1,021	795	160	2.4
2.0	1,056	840	152	2.4
2.0	1,029	881	141	1.4
5.0	1,107	893	147	2.0
7.5	1,116	844	154	2.1
7.5	1,116	848	151	2.7
10.0	1,069	807	145	2.8
12.5	978	789	154	1.5
15.0	863	724	130	1.2
		Averages	148 (±18)	2.1 (±0.9)

Reprinted from [108] with the permission of Electrochemical Society

mainly the consequence of different parameters of the deposition process (bath composition, pulse regimes, temperature, etc.). A common explanation for the decrease of tensile strength with the increase of individual layer thickness (after the maximum value) is the increase of the coherence of the intermediate layer. It is concluded that with the decrease of the number of dislocations in the intermediate layer, the probability of their transfer from it to previous or subsequent layer increases, resulting in the decrease of tensile strength.

The dependence of the tensile properties of multilayered Cu–(Ni–Cu) deposits, with the nominal overall composition 90 at.% Ni–10 at.% Cu, was investigated in the work of Tench and White [108] as a function of the Cu layer thickness (varying from 1 to 15 nm). Multilayers of the nominal thickness of 50 μm were deposited from a commercial sulfamate bath with the addition of 5 mM $CuSO_4$. Pure Cu layers were plated at −0.14 V vs. SCE with cathode rotation at RPM = 750, while Ni layers (with 0.8 at.% Cu) were plated at $j = -105$ mA cm^{-2} from the stagnant bath. The results of the tensile properties (ultimate tensile strength, UTS, yield strength, YS, modulus, and strain) measurements are presented in Table 1.6. Considering X-ray data it was concluded that a decrease in the enhanced strength for such deposits, producing peak at a Cu layer thickness of about 7 nm, is associated with a decrease in the deposit (100) texture (normally established for Ni layers) and an increase in the (110) orientation preferred for Cu deposits.

Concerning the magnetic properties, giant magnetoresistance, which is a significant decrease of electric resistance in the presence of magnetic field, characteristic effect for multilayered structures, has been discovered in 1988 for the systems Fe/Cr/Fe and Fe/Cr by Grünberg and Fert (these scientists received Nobel Prize for physics in 2007). This phenomenon has also been recognized for electrodeposited multilayered structures [119–123]. The strongest effect is detected for multilayered Co–Cu/Cu [121] and Co–Ni–Cu/Cu [122, 123] structures, while less pronounced effect is detected for the multilayered Ni–Cu/Cu [121] structures. It is important to note that the presence of the intermediate layer of the thickness of 2–3 nm has been detected for the first time by high resolution transmission electron microscopy (HRTEM) for the multilayered Ni–Cu/Cu structures [119], as shown in Fig. 1.48.

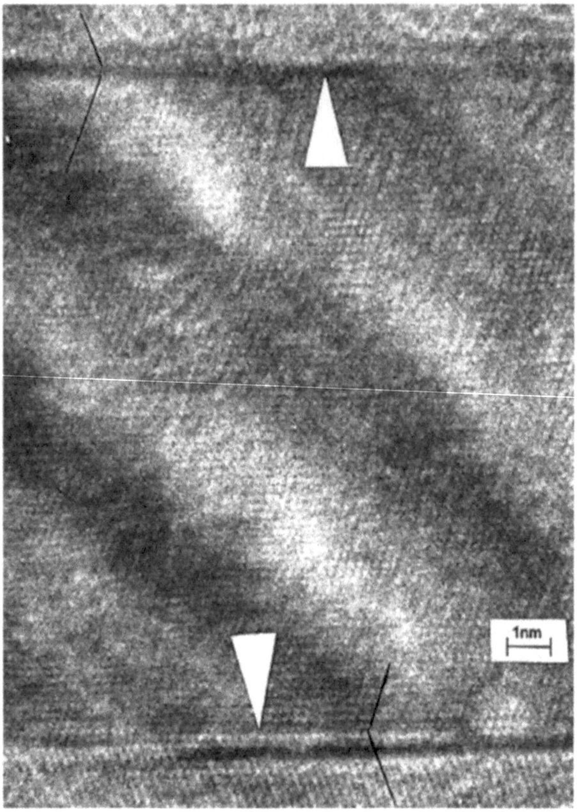

Fig. 1.48 HRTEM picture taken on the Ni–Cu (3.8 nm)/Cu(1.4 nm) sample which shows the atomic structure of a twin boundary (indicated by the *arrow*) and the magnetic/nonmagnetic layers (Reprinted from [119] with the permission of Elsevier)

1.4.3 Metal Coatings with Inclusion of Nonmetallic Particles

The basics of deposition of composite coatings containing metal and nonmetallic particles are given in the literature in chapter of Despić and Jović [1] and in a review paper of the electrochemical engineering group of Southampton University [92]. Despić and Jović [1] summarized results obtained in this field up to about 1993 in 28 available references, while review paper of Low et al. [92] presented mostly recent literature survey of the deposition of metallic coatings containing nano-sized particles citing 82 references. In this survey [92] are summarized various types of nano-structured materials obtained by electrodeposition, examples of theoretical models used to describe such a process, investigations of the inclusion of Al_2O_3, SiC, SiO_2, ZrO_2, diamond, Si_3N_4, TiO_2, etc. into metal matrix of Ni, Cu, Au, Zn, and Fe, types of current density pulses used for these investigations, effects of nano-sized particles on deposit characteristics, influence of bath agitation and cathode movement, and examples of studies using rotating cylinder electrode and other operating parameters (pH, bath temperature, type of additives and concentration of metal ions, and additives). Also, gradient distribution of particles and properties of nano-composite deposits (corrosion rates, microhardness, wear resistance, magnetic properties) as a function of the deposition parameters are discussed. In almost all of the cited references in these two publications, the authors are either explaining the mechanism of codeposition or the influence of deposition parameters on the mechanical, tribological, corrosion, and magnetic properties of obtained composite structures. Recently, Cattarin and Musiani [124] published review of the studies carried out during the last decade on the electro-synthesis of oxide-matrix composite materials for electrocatalysis of the HER [125–134] and the oxygen evolution reaction [135–145] (OER), citing 71 references. In this review [124] particular attention has been given to the oxide-matrix composite materials for electrocatalysis of the OER. The matrix materials were PbO_2, Ni, Tl_2O_3, and PbO_2/CoO_x, while as dispersed phases Co_3O_4, RuO_2 or Ru + Sn oxides, and Co_2NiO_4 were used. In comparison with the pure matrices, composite anodes exhibited much lower overpotentials and lower Tafel slopes, comparable to those of electrocatalytic materials produced by other methods.

Taking into account that the review concerning oxide-matrix composite materials for electrocatalysis of the OER has already been published [124], in the next part of this chapter deposition of composite coatings containing metals and non-metallic (oxide) particles as catalysts for the HER will be presented and discussed.

1.4.4 Ni Based Coatings with Inclusion of Nonmetallic Particles as Cathodes for the HER

Two types of Ni-based composite coatings were investigated: (i) composite coatings of large surface area containing Co_3O_4 and RuO_2 particles [146], composite coatings containing RuO_2 and IrO_2 particles [147, 148] and TiO_2 particles [134]; (ii) composite coatings containing MoO_2 particles [150–153].

1.4.4.1 Composite Coatings Containing Co_3O_4, RuO_2, IrO_2, and TiO_2 Particles

All composite coatings were deposited from suspensions stirred with a magnetic bar (at ca. RPM = 1,000) onto disc electrodes rotating at RPM = 2,500, under galvanostatic control. The Ni + RuO_2 composites were deposited from suspensions of RuO_2 in either of the following electrolytes: 0.2 M $NiCl_2$ + 2 M NH_4Cl aqueous solution, pH 4.5, at 25 °C; Watt's bath containing 1.126 M $NiSO_4{\cdot}6H_2O$ + 0.185 M + $NiCl_2{\cdot}6H_2O$ + 0.485 M H_3BO_3, pH 4, at 55 °C at low current densities (−0.02 to −0.1 A cm^{-2}). The deposits obtained from the former bath at high current densities (−0.5 to −3.0 A cm^{-2}) are called Por–Ni + RuO_2 and those obtained from the latter one are called Watt–Ni + RuO_2. Similarly, Por–Ni and Por–Ni + Co_3O_4 are plated from the former solution (at high current densities) [146]. The Tafel plots for the HER

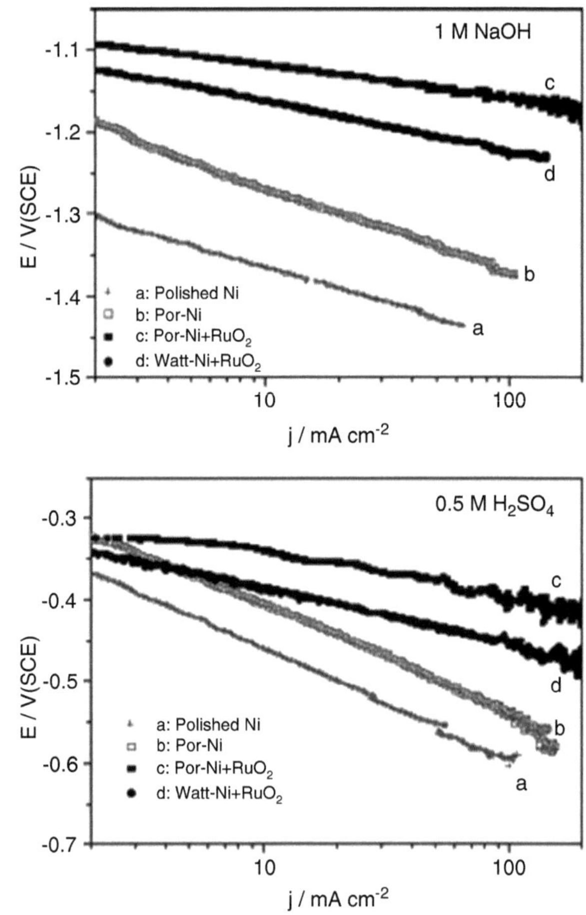

Fig. 1.49 Tafel plots of the current density potential curves for the HER measured on polished Ni (a), Por–Ni (b), Por–Ni + RuO_2 composite, $\alpha = 0.18$ (c) and Watt–Ni + RuO_2 composite, $\alpha = 0.28$ (d) in 1 M NaOH and 0.5 M H_2SO_4 (Reprinted from [146] with the permission of Elsevier)

on some of the investigated cathodes, recorded with *IR* drop correction, are presented in Fig. 1.49. As can be seen the most active electrode, in either alkaline or acidic solution, is Por–Ni + RuO_2, mainly due to high surface area in comparison with Por–Ni and Watt–Ni + RuO_2, while all composites show catalytic effect for the HER. This is confirmed by the SEM analysis of surfaces of composite coatings obtained in this work, presented in Fig. 1.50. It was also shown that composite coatings based on IrO_2 particles [148] did not possess catalytic effect for the HER.

Composite Ni–P + TiO_2 coatings were also deposited onto stationary Cu plate from a bath for deposition of Ni–P containing 99 g dm^{-3} TiO_2 powder, which was stirred with RPM = 300. It was shown that the increase in electrochemical activity of the Ni–P + TiO_2 electrode towards the HER is due to both the presence of titanium oxides and the increase in the real surface area, as compared with the Ni–P electrode [134].

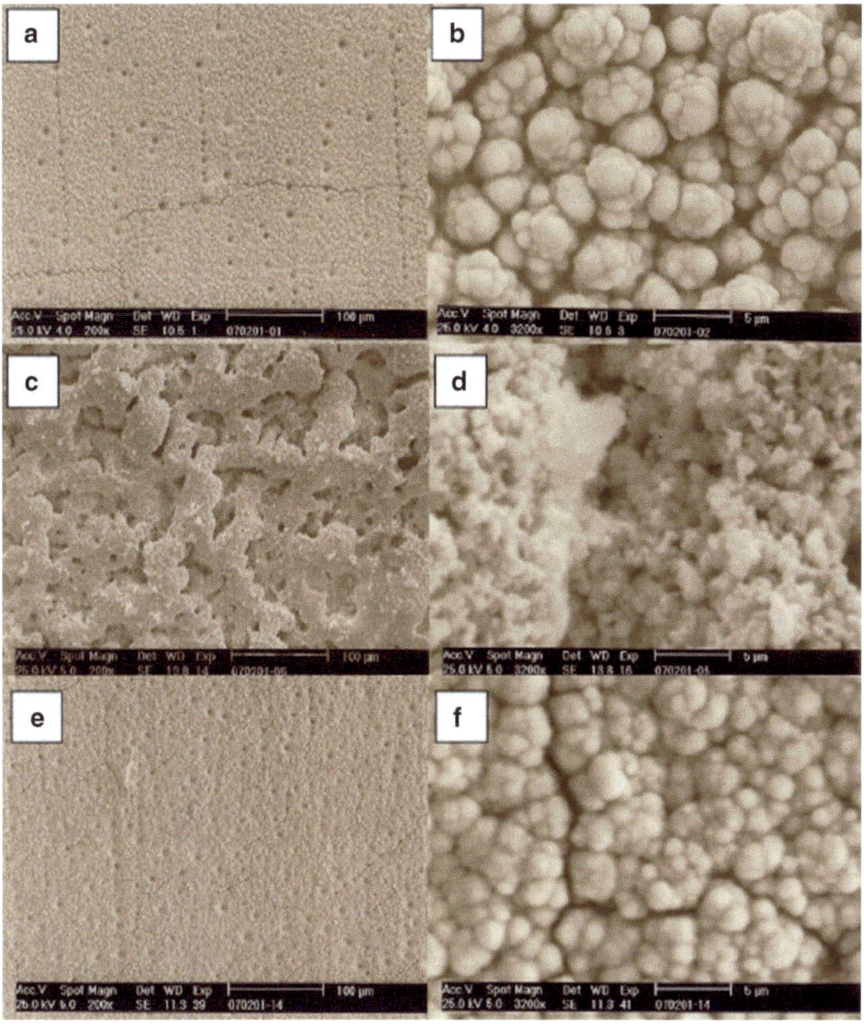

Fig. 1.50 SEM pictures of deposits obtained from 0.2 M $NiCl_2$ + 2 M NH_4Cl aqueous solution, pH 4.5, with a deposition current density −3.0 A cm^{-2}. (**a**) and (**b**) Por–Ni; (**c**) and (**d**) Por–Ni + RuO_2, from a 50 g dm^{-3} RuO_2 suspension; (**e**) and (**f**) Por–Ni + Co_3O_4, from a 50 g dm^{-3} Co_3O_4 suspension (Reprinted from [148] with the permission of Elsevier)

1.4.4.2 Composite Ni–MoO_2 Coatings as Cathodes for the HER in Industrial Electrolysis

Composite coatings containing MoO_2 [150–153] particles (average size 200 nm) were deposited at $j = -300$ mA cm^{-2} from the solution 0.2 M $NiCl_2$ + 2 M NH_4Cl (pH 3.8) containing MoO_2 powder particles, in the apparatus (pilot plant of small dimensions) schematically presented in Fig. 1.51. The volume of the electrolyte in

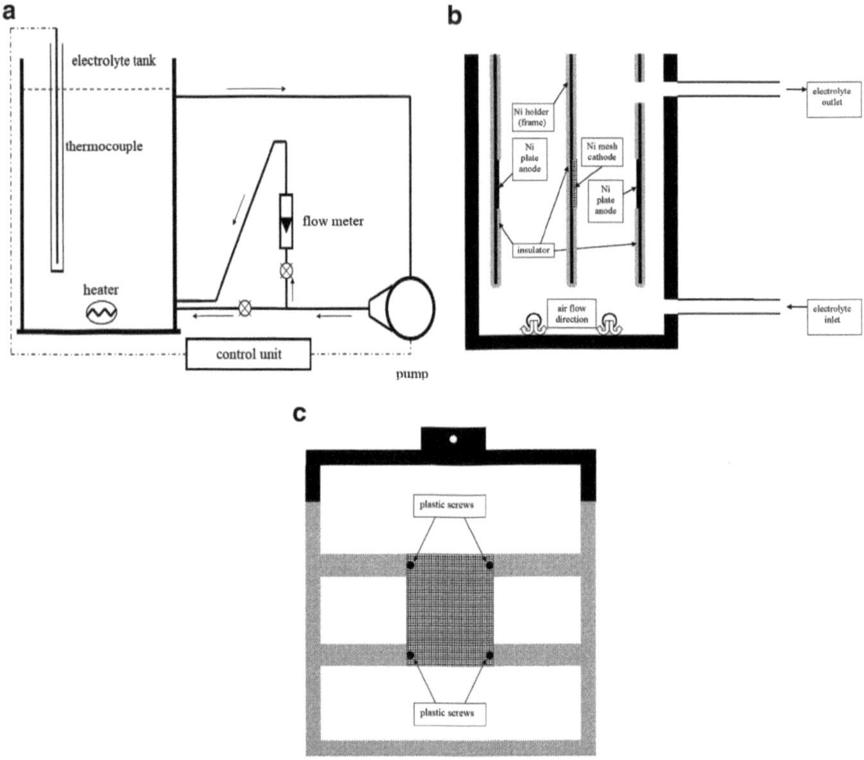

Fig. 1.51 Schematic presentation of the apparatus for deposition of Ni–MoO_2 composite coatings (Reprinted from [149] with the permission of Int. J. Hydrogen Energy)

the cell for deposition was approximately 20 dm^{-3}. The electrolyte was circulated with the pump with a maximum flow rate of 0.83 $dm^3\ s^{-1}$ (a). The flow rate in the cell was measured with a flow meter (a), being 0.33 $dm^3\ s^{-1}$. Additional mixing of the electrolyte was provided by the airflow of 0.17 $dm^3\ s^{-1}$ through two pipes with small openings facing the bottom of the cell in order to remove eventually precipitated MoO_2 particles from the bottom of the cell and to force particles to float and circulate with the electrolyte (b). The temperature of the electrolyte was kept constant by the thermocouple, heater, and the control unit (a). The Ni 40 mesh cathode, connected to a Ni holder (frame), was placed between two Ni anode plates, as is schematically presented in (c). A homemade power supply, with the ripple smaller than 1 %, was used for applying necessary current/voltage [149, 150].

Four samples were deposited under different conditions. Samples 1 and 2 were deposited from the bath with freshly added MoO_2 (after 1 h of electrolyte circulation), while deposition of samples 3 and 4 started 24 h after MoO_2 addition. It appeared that aging of the electrolyte of at least 24 h is necessary for better incorporation of MoO_2 particles in the deposit [150].

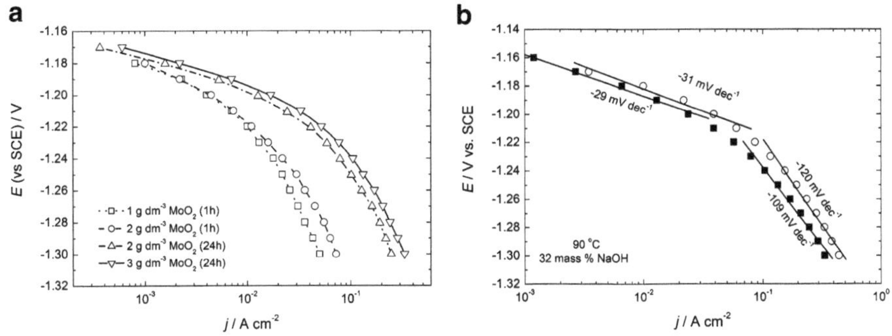

Fig. 1.52 (**a**) Polarization diagrams for the HER measured on samples 1–4. (**b**) Polarization diagrams for the HER measured on sample 4 (*open circle*) compared with that recorded for the commercial De Nora's Ni–RuO_2 cathode (*filled square*) (Reprinted from [150] with the permission of Int. J. Hydrogen Energy)

Polarization characteristics of the HER for samples 1–4, recorded in 32 mass % NaOH at 90 °C and corrected for *IR* drop, are presented in Fig. 1.52a. Significant decrease of overvoltage for the HER is recorded for samples 3 and 4. By comparing polarization diagrams for the commercial De Nora's electrode (■), and the best sample 4 (○), one can see that polarization diagrams are almost identical, Fig. 1.52b, indicating that Ni–MoO_2 composite coatings, deposited under conditions described above in the apparatus presented in Fig. 1.51, could be promising replacement for the commercial cathodes based on the Ni–RuO_2 composite coatings (DN) [150]. Taking into account that in the industrial electrolysis the current density for the HER is usually -0.3 A cm^{-2}, considering polarization curves presented in Fig. 1.52b, it could be concluded that the overvoltage for the HER onto Ni–MoO_2 composite coating is for about 20 mV lower than that on the commercial cathode.

The morphology of the coating deposited in the presence of the highest amount of MoO_2 powder particles (3 g dm^{-3}, sample 4) is presented in Fig. 1.53a. As can be seen, two regions can be detected on the deposit surface. The region rich in Ni with no cracks (low percentage of molybdenum and oxygen), usually positioned at the middle of a Ni mesh wire between the crossing of two wires, and the region rich in Mo with the presence of cracks (high percentage of molybdenum and oxygen and low percentage of nickel) are placed around the position of the crossing of two wires. It is quite difficult to uniformly incorporate MoO_2 particles in the Ni deposit on the Ni 40 mesh, since all places of the surface are not equally accessible to the MoO_2 particles and Ni ions. The position of crossing of two wires is the place where, because of the change of hydrodynamic conditions at this particular position, higher amount of MoO_2 particles is embedded in the Ni deposit, probably as a result of lower real current density for Ni deposition. The uniform composition of the Ni–MoO_2 composite coating could probably be achieved by improving hydrodynamic conditions. The cross section of the Ni-rich part of the coating is presented in Fig. 1.53b. The EDS analysis showed that white areas (b) are rich in Mo, but the

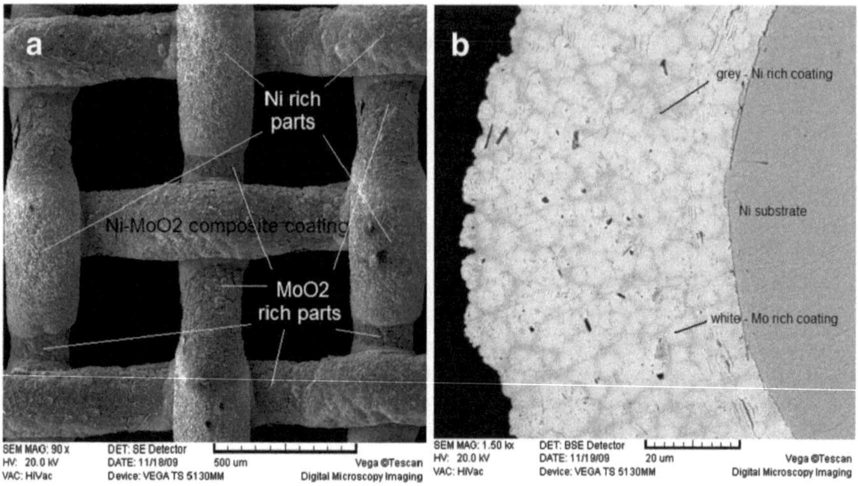

Fig. 1.53 (**a**) Surface of the sample 4, SEM. (**b**) Cross section of the Ni-rich part of the coating, backscatter SEM (Reprinted from [151] with the permission of Elsevier)

content of Ni is still high (around 79 at.%) and the content of oxygen is low. Hence, white areas correspond to agglomerates of MoO_2 surrounded with Ni and incorporated into Ni matrix (gray areas) [150].

This statement is confirmed by the XRD analysis of as-deposited sample 4 and XRD analysis after additional thermal treatment. The coating was thermally treated in air atmosphere for 5 h at two different temperatures, 300 and 600 °C. As shown in Fig. 1.54, in the XRD pattern of the as-deposited sample only Ni peaks could be detected, indicating amorphous (or nano-crystalline) MoO_2. After recrystallization at 300 and 600 °C, peaks of MoO_2 are clearly seen in the XRD pattern of sample 4 [150].

Hence, considering polarization curves presented in Fig. 1.52b it appears that the overvoltage for the HER onto Ni–MoO_2 composite coating is for about 20 mV lower than that on the commercial cathode. However, this is not the only criteria that should be satisfied for the application of such cathodes in industrial electrolysis. The "service life" test should also give satisfactory results.

Service Life Test

In the process of chlor-alkali electrolysis for chlorine production in the membrane cells [154–156], an ion-exchange membrane separates the anode and the cathode compartments and gaseous hydrogen is produced in the cathode compartment from 30 to 32 mass % caustic soda solution at a typical operating temperature of 90 °C. The efficiency of the cathodes is an important issue in this process, since the overvoltage for the HER in the cathode compartment contributes significantly to

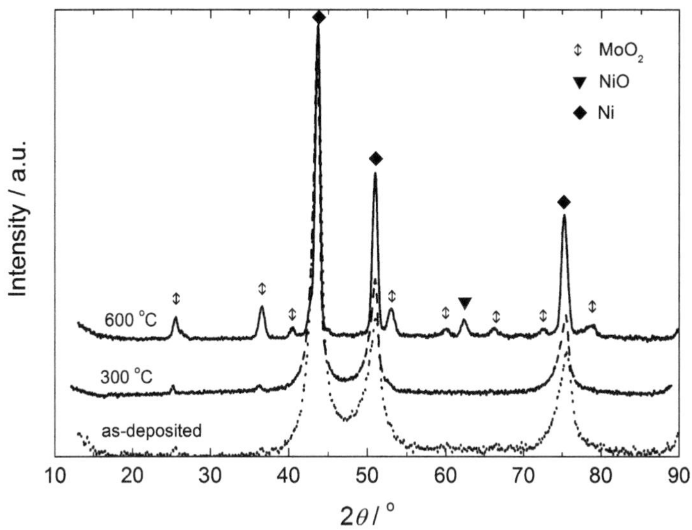

Fig. 1.54 XRD pattern of as-deposited sample 4, the same sample after annealing at 300 °C for 5 h in air (300 °C) and after additional annealing at 600 °C for 5 h in air (600 °C) (Reprinted from [150] with the permission of Int. J. Hydrogen Energy)

the overall power consumption. The efficiency of the cathodes is a result of combination of certain activity and stability at current densities (0.3–0.6 A cm^{-2}) normally used in industrial applications. The main reason for the loss of activity and stability of the cathodes during long-term operation is the so-called polarity inversion of the electrodes, which takes place during the replacement of old electrodes of an electrolyzer with new ones in the zero-gap cells. During this operation anodes and cathodes are short-circuited, causing a reverse current flow which may damage the cathodes and negatively affect their activity for the HER [156]. The manufacturers can predict how often in a certain period of time such operation should be performed and, in accordance with that, design an appropriate accelerated "service life" test for cathodes [148, 151, 153].

Such test was performed in the following way: the electrode was kept at $j = -0.3$ A cm^{-2} for 0.5 h and the corresponding potential response was recorded. After that, the electrode was cycled (five cycles) in the potential range from −1.25 V (HER) to 0.5 V (OER) with a sweep rate of 50 mV s^{-1}. In the next step the electrode was kept again at $j = -0.3$ A cm^{-2} for 500 s and the corresponding potential response was recorded. This procedure was repeated five times, until the number of cycles reached 25. The potential response measured at $j = -0.3$ A cm^{-2} for 500 s (the value recorded at the last point, 500 s) was corrected for IR_{Ω} drop using the value of ohmic resistance (R_{Ω}) obtained from EIS measurements. The results of this test for the composite Ni–MoO_2 coating (sample 4) and the commercial De Nora's electrode (DN) are presented in Fig. 1.55. As can be seen the performance of the Ni–MoO_2 coating (sample 4) after "service life" test confirmed

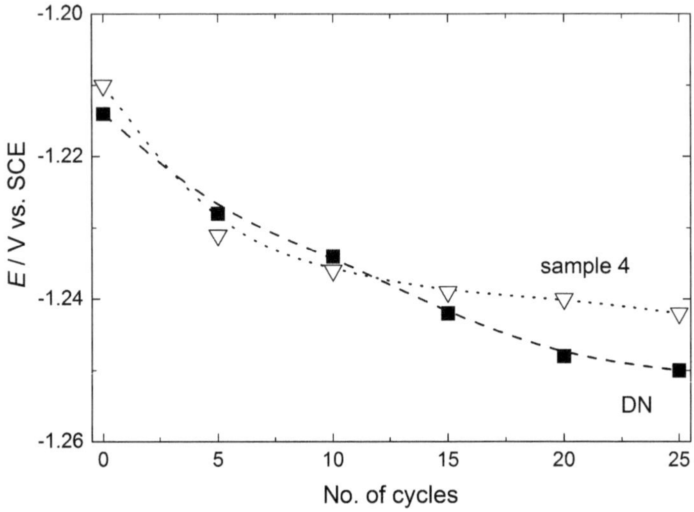

Fig. 1.55 The results of "service life" test for the commercial De Nora's electrode (DN) and composite Ni–MoO_2 coating (sample 4), potential for the HER corrected for IR_{Ω} drop at the current density $j = -0.3$ A cm^{-2}

that this electrode could be promising replacement for the commercial one, showing slightly lower overvoltage for the HER after 15 cycles.

The appearances of the coating after its cathodic polarization at -0.3 A cm^{-2} for 40 h and after the "service life" test are presented in Fig. 1.56a, b, respectively. As can be seen the appearance of the coating did not change after 40 h of the HER (Figs. 1.53a and 1.56a), while Mo-rich parts were mainly destroyed and peeled from the Ni mesh surface after the "surface life" test (Fig. 1.56b). Taking into account the presence of the large number of cracks in the Mo-rich parts of the Ni–MoO_2 coating, the loss of the coating (and consequently the loss of activity) is most likely due to erosion caused by the evolved gas. It is well known that the effective catalytic activity does not correlate always with the total surface of the catalyst, especially with the inner surface. In the case of Raney–nickel cathodes [158], utilization of the inner surface for the HER does not exceed on the average 10 % at the high current densities. This low degree of catalyst utilization is due to slow diffusive mass transport of electrogenerated dissolved hydrogen out of the pores and cracks of the catalyst. Hydrogen bubbles cannot precipitate in nanometer pores unless the concentration of the hydrogen in the electrolyte exceeds high concentration. It is clear that the higher exchange current density of the evolved gas reaction, the lower the penetration depth and catalyst utilization. Taking into account the fact that this electrode exhibits high catalytic activity for the HER, due to diffusive mass transport limitation of dissolved hydrogen, it could be expected that this reaction practically does not take place in the pores and cracks present in the coating. Consequently, the mechanical damage was not caused by the evolved hydrogen bubbles, which is confirmed in Fig. 1.56a. However, during cycling procedure

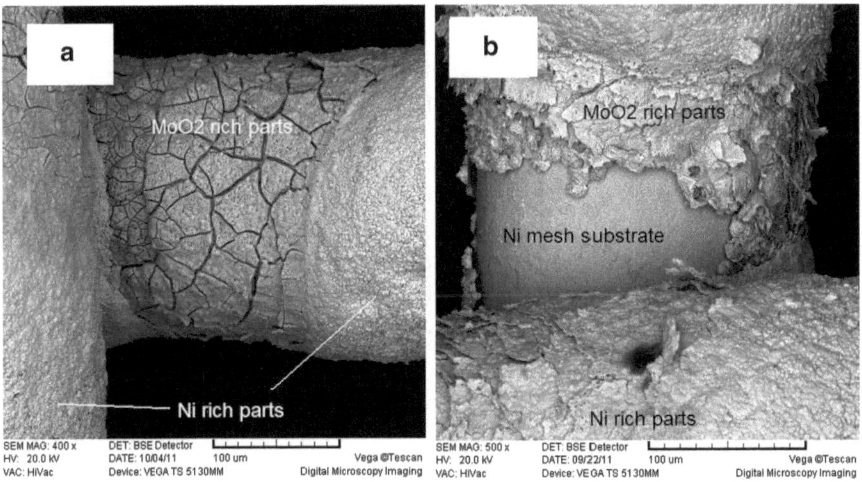

Fig. 1.56 (**a**) Surface of the sample 4 after 40 h of the HER at $j = -0.3$ A cm^{-2}, SEM. (**b**) Surface of the sample 4 after "service life" test (Reprinted from [151] with the permission of Elsevier)

(inverse polarization), the OER takes place on examined electrodes reaching very high current densities of 0.3–0.4 A cm^{-2} during anodic polarization. Unlike the HER, the OER is a highly irreversible reaction and due to its very low exchange current density it takes place at overall catalyst surface without any diffusion limitation in the pores and cracks. So, the rapid evolution of oxygen on Mo-rich parts of the Ni–MoO_2 coating generated significant pressures within the porous coating causing their mechanical damage (erosion). The EDS analysis of the surface of the Ni–MoO_2 coating showed that the percentage of Ni and Mo was practically the same before and after the "service life" test, indicating that during the replacement of old electrodes with the new ones (inverse polarization) no dissolution and further oxidation of MoO_2 could take place.

1.5 Conclusions

In this chapter basics of electrodeposition of alloys from aqueous solutions are presented, taking into account conditions that must be fulfilled for deposition of alloys, their reversible potentials in the solutions of corresponding ions, the influence of the Gibbs energy of phase formation, and stability of such alloy phases in the solutions of corresponding ions.

Different types of alloy deposition (equilibrium, irregular, regular, anomalous, and induced codeposition) are defined, being ascribed to certain systems, such as Ni–Sn, Ag–Pd, Ag–Cd, Co–Ni, and Fe–Mo, respectively.

Characterization of electrodeposited alloys by ALSV is explained in great detail for the eutectic-type alloys (systems Cu–Pb, Cd–Co, and Cd–Zn), solid solution type alloys (systems Ag–Pd and Co–Ni) and alloys containing intermediate phases and/or intermetallic compounds (systems Cd–Cu, Cd–Ni, Ag–In, and Ag–Cd). Based on the presented results of ALSV technique application to different alloy systems, a mechanism of alloy dissolution is proposed.

In the part concerning electrodeposition of composite metal coatings, two types of deposits are described: laminar metal structures and metal coatings with inclusion of nonmetallic particles. Basic equations and conditions for deposition of laminar metal structures are given, together with the explanation of the improvement of their mechanical and magnetic properties (based on the existence of the intermediate layer of high surface area) in comparison with their alloys or pure metals. In the case of metal coatings with inclusion of nonmetallic particles, particular attention is addressed to their catalytic properties for hydrogen and oxygen evolution reactions.

Acknowledgments This work was financially supported by the Ministry of Education, Science and Technological Development of the Republic of Serbia through the Project No. 172054/2011.

The authors also wish to express their gratitude to the Department of Research and Development of De Nora Industries S.p.A. for providing equipment for electrodeposition of Ni–MoO_2 coatings.

References

1. Despić AR, Jović VD (1995) Electrochemical Deposition and Dissolution of Alloys and Metal Composites – Fundamental Aspects. In: White RE, Bockris JO'M, Conway BE (eds) Modern Aspects of Electrochemistry. vol. 27, Plenum Press, New York, pp. 143–232
2. De Ruolz M (1842) C R Acad Sci 15:1140
3. Srivastava RD, Mukerjee RC (1976) J Appl Electrochem 6:321
4. Brenner A (1963) Electrodeposition of Alloys: Principle and practice. Academic Press, New York
5. Bondar VV, V. Grimina V, Pavlov VN (1980) Itogi nauki i tehniki, Elektrokhimiya. vol. 16, Izd. Viniti, Moscow
6. Spitzer F (1905) Z Electrochem 11:345
7. Schlötter M (1914) Die Elektrolytische darstellung von Legierungen aus wässserigen Lösungen (Sammlung Vieweg, Tagesfragen aus den Gebieten Naturwiss. U. der Technik, No. 19). In: Kremann R (ed), Vieweg, Braunschweig
8. Gorbunova KM, Polukarov YuM (1976) In: Tobias CW (ed) Advances in Electrochemistry and Electrochemical Engineering. vol. 5, John Willey and Sons Inc., New York
9. Fedoteev NP, Bibikor NN, Vyacheslavov PM, Grilihes CYa (1962) Elektroliticheskie Splavy, MASHGIZ, Moscow
10. Faust CL (1963) In: Lowenheim FA (ed) Modern Electroplating, Ch. 18, John Wiley and Sons Inc., New York
11. Reinders W (1902) Z Physik Chem 42:225
12. Pushin N (1907) Z Anorg Chem 56:1
13. Lačnjevac U, Jović BM, Jović VD (2012) J Electrochem Soc 159: D310

14. Handbook of Chemistry and Physics (1976–1977) 57th edition, Weast RC (ed), CRC Press Inc., Cleveland, pp. D141–146
15. Jović VD, Tošić N (1998) J Electroanal Chem 441:69
16. Daen JA (1999) Lange's Handbook of Chemistry 15th Edition, McGraw-Hill Inc., New York
17. Duffield JR, Williams DR, Kron I (1991) Polyhedron 10:377
18. Turyan IYa, Kravtsov VI, Kondratev VV (1986) Elektrokhimiya 22:1388; 1618
19. Orekhova VV, Andryuschenko FK, Sakhnenko ND (1980) Elektrokhimiya 16:1304
20. Han C, Liu Q, Ivey DG (2008) Electrochim Acta 53:8332
21. Jović VD, Lačnjevac U, Jović BM, Karanović Lj, Krstajić NV (2012) Int J Hydrogen Energy 37:17882
22. Delahay P (1954) New Instrumental Methods in Electrochemistry, Interscience, New York
23. Jović VD, Stojanović MV, Jović BM, Gajić-Krstajić Lj (1992) J Serb Chem Soc 57:951
24. Jović VD, Tošić N, Stojanović M (1997) J Electroanal Chem 420:43
25. Dahms H, Croll J (1965) J Electrochem Soc 112:771
26. Higashi K, Fukushima H, Urakawa T, Adaniya T, Matsuko K (1981) J Electrochem Soc 128:2081
27. Horkans T (1979) J Electrochem Soc 126:1861
28. Horkans T (1981) J Electrochem Soc 128:45
29. Podlaha EJ, Landolt D (1996) J Electrochem Soc 143:885
30. Podlaha EJ, Landolt D (1996) J Electrochem Soc 143:893
31. Podlaha EJ, Landolt D (1997) J Electrochem Soc 144:1672
32. Marlot A, Kern P, Landolt D (2002) Electrochim Acta 48:29
33. Zeng Y, Li Z, Ma M, Zhou S (2000) Electrochem Comm 2:36
34. Sanches LS, Domingues SH, Carubelli A, Mascaro LH, J Braz Chem Soc 14: 556
35. Jović VD, Jović BM, Lačnjevac U, Branković G, Bernik S, Rečnik A (2010) Electrochim Acta 55:4188
36. Grgur BN, Krstajić NV, Elezović N, Jović VD (2005) J Serb Chrem Soc 70:879
37. Evans UR (1928) J Inst Metals 40:99
38. Glazunov A (1953) Metallic Protective Coatings 21:1262
39. Steigerwald RF, Greene ND (1962) J Electrochem Soc 109:1026
40. Pickering HW, Wagner C (1967) J Electrochem Soc 114:698
41. Pickering HW (1968) J Electrochem Soc 115:690
42. Pickering HW, Byrne PJ (1971) J Electrochem Soc 118:209
43. Holliday JE, Pickering HW (1973) J Electrochem Soc 120:470
44. Bockris JO'M, Rubin BT, Despić A, Lovreček B (1972) Electrochim Acta 17:973
45. Mance A, Mihajlović A (1980) J Appl Electrochem 10:967
46. Lee HP, Nobe K (1984) J Electrochem Soc 131:1236
47. Petro I, Mallat T, Syabo A, Hange F (1984) J Electroanal Chem 160:289
48. Shapovalov ET, Baranova LI, Yektser O (1988) Elektrokhimicheskie metody v metallovodenii i fazovom analize, Izd. Metallurgya, Moscow
49. Stevanović J, Skibina L, Stefanović M, Despić A, Jović VD (1992) J Appl Electrochem 22:172
50. Swathirajan S (1986) J Electrochem Soc 133:671
51. Jović VD, Zejnilović RM, Despić AR, Stevanović JS (1988) J Appl Electrochem 18:511
52. Swathirajan S (1987) J Electroanal Chem 221:211
53. Andricacos PC, Tabib J, Romankiw LT (1988) J Electrochem Soc 135:1172
54. Andricacos PC, Avana C, Tabib J, Duković J, Romankiw LT (1989) J Electrochem Soc 136:1336
55. Wong KH, Andricacos PC (1990) J Electrochem Soc 137:1087
56. Horkans J, I-Chia Hsu Chang, Andricacos PC, Podlaha EJ (1991) J Electrochem Soc 138:411
57. Jović VD, Despić AR, Stevanović JS, Spaić S (1989) Electrochim Acta 34:1093
58. Jović VD, Spaić S, Despić AR, Stevanović JS, Pristavec M (1991) Mater Sci Technol 7:1021
59. Skibina L, Stevanović J, Despić AR (1991) J Electroanal Chem 310:391

60. Stevanović R, Kovrigina I, Despić AR (1991) J SerbChem Soc 56:217
61. Despić AR, Electrochemistry in Transition from the 20th to the 21st Century. In: Murphy OJ, Srinivasan S, Conway BE (1992) Plenum Press, New York, p.453.
62. Jović VD, Jović BM, Despić AR (1993) J Electroanal Chem 357:357
63. Raub E, Engel A (1950) Z Metallk 41:485
64. Jović VD, Jevtić V (1996) J Serb Chem Soc 61:479
65. Jović VD, Jevtić V (1998) Electrochim Acta 43:63
66. Fedotev NP, Vyacheslavov PM (1970) Plating 3:700
67. Hansen M, Andrenko K (1958) Constitution of binary alloys. Mc Graw-Hill Book Company, Inc., New York, Toronto, London
68. Köster W, Hern E (1952) Z Metallk 43:333
69. Lewkonja K (1980) Z Anorg Chem 59:322
70. Westgren A, Ekman W (1930) Arkiv kemi, Mineral Geol B10:1
71. Lihl F, Buhl E (1955) Z. Metallkunde 46:787
72. Jović BM, Dobrovolska Ts, Lačnjevac U, Krastev I, Jović VD (2009) Electrochim Acta 54:7565
73. Porter D, Easterling KA (1980) Phase Transformations in Metals and Alloys, Van Nostrand Reinhold, Wokingham
74. Spitzer F (1905) Z Electrochem Angew Physik Chem 23:345
75. Rotinyan AL, Molotkova EN (1959) Zh Prikl Khim 11:2502
76. Thierer Ch, (1951) Gmelin Handbuch der anorganischen Chemie, System 65A(5), Weinheim/Bergstrasse, Verlag Chemie, p.653
77. Savitski EM, Pravoverov NL (1961) Zh Neorg Khim 76:499
78. Sturzenegger B, Puippe JC (1984) Platinum Met Rev 28:117
79. Dobrovolska Ts, Jović VD, Jović BM, Krastev I (2007) J Electroanal Chem 611:232
80. Dobrovolska Ts, Krastev I, Jović BM, Jović VD, Beck G, Lačnjevac U, Zielonka A (2011) Electrochim Acta 56:4344
81. Weibke F, Eggers H (1935) Z Anorg Chem 222:145
82. Campbell AN, Wagemann R, Ferguson RB (1970) Can J Chem 48:1703
83. Dobrovolska Ts, Krastev I, Zielonka A (2005) J Appl Electrochem 35:1245
84. Dobrovolska Ts, Veleva L, Krastev I, Zielonka A (2005) J Electrochem Soc 152 :C137
85. Stevanović JS, Jović VD, Despić AR (1993) 349:365
86. Dobrovolska Ts, Krastev I, Zielonka A (2010) ECS Trans 25:1
87. Petrauskas A, Grincevičiene L, Češuniene A, Juškenas R (2005) Electrochim Acta 50:1189
88. Elkhatabi F, Benballa M, Sarret M, Müler C (1999) Electrochim Acta 44:1645
89. Stevanović J, Gojković S, Despić A, Obradović M, Nakić V (1998) Electrochim Acta 43:705
90. Stevanović J, Despić A, Poleti D (1995) J Serb Chem Soc 60:285
91. Bajat JB, Stanković S, Jokić BM, Stevanović SI (2010) Surf Coat Techn 204:2745
92. Low CTJ, Wills RGA, Walsh FC (2006) Surf Coat Techn 201:371
93. Metcalfe AG (1974) Interfaces in Metal Composites. Academic Press, New York, p. 65
94. Meyer WR, Phillips A (1938) Trans Electrochem Soc 73:377
95. Aotani K (1953) J Electrochem Soc Jpn 21:180
96. Mikhalev PF (1939) C R Acad Sci URSS 24:899
97. Cohen U, Koch FB, Sard R (1983) J Electrochem Soc 130:1987
98. Tench D, White J (1984) Metall Trans 15A:2039
99. Ogden C (1986) Plating Surf Finish 73:130
100. Verbruge MW, Tobias CW (1985) J Electrochem Soc 132:1298
101. Despić AR, Jović VD (1987) J Electrochem Soc 134:3004
102. Despić AR, Jović VD, Spaić S (1989) J Electrochem Soc 136:1651
103. Despić AR, Trišović T (1993) J Appl Electrochem 23:662
104. Power CP, Ritchie IM (1975). In: Conway BE, Bockris JO'M (eds) Modern Aspects of Electrochemistry vol. 11, Plenum Press, New York
105. Yahalom J, Zadok O (1987) J Mater Sci 22:499

106. Lashmore DS, Dariel MP (1988) J Electrochem Soc 135:1218
107. Tench DM, White JD (1990) J Electrochem Soc 137:3061
108. Tench M, White J (1991) J Electrochem Soc 138:375
109. Ebrahimi F, Liscano AJ (2001) Mat Sci Eng A301:23
110. Yang C–C, Cheh HY (1995) J Electrochem Soc 142:3034
111. Yang C–C, Cheh HY (1995) J Electrochem Soc 142:3040
112. Roy S, Landolt D (1995) J Electrochem Soc 142:3021
113. Bradley PE, Landolt D (1997) Electrochim Acta 42:993
114. Despić AR, Jović VD, Tošić N (1998) Surf Coat Techn 105:206
115. Oberle RR, Cammarata RC (1995) Scr Metall Et Mater 32:583
116. Cammarata RC, Schleisinger TE, Kim C, Qadri SB, Edelstein AS (1990) Appl Phys Lett 56:1862
117. Meneyes S, Anderson DP (1990) J Electrochem Soc 137:440
118. Simunovich D, Schleisinger M, Snyder DD (1994) J Electrochem Soc 141:L10
119. Cziraki A, Pierron–Bohnes V, Ulhaq–Bouillet C, Toth–Kadar E, Bakonyi I (1998) Thin Solid Films 318:239
120. Bakonyi I, Toth–Kadar E, Becsei T, Toth J, Tarnoczi T, Cziraki A, Geröcs I, Nabiyouni G, Schwarzacher W (1996) J Magn Magn Mater 156:347
121. Bird KD, Schleisinger M (1995) J Electrochem Soc 142:L65
122. Hart R, Alper M, Attenborough K, Schwarzacher W (1994). In: Romankiw LT, Herman DAJr (eds) Proc. 3rd Int. Symp. on Magnetic Mater. Processes and Devices, Electrodeposition Division of the Electrochem. Soc., vol. 94-96, Pennington, p. 215.
123. Hua SZ, Lashmore DS, Salamanca–Riba L, Schwarzacher W, Swartzendruber LJ, McMichael RD, Bennett LH, Hart R (1994) J Appl Phys 76:6519
124. Cattarin S, Musiani M (2007) Electrochim Acta 52:2796
125. Anani A, Mao , Srinivasan S, Appleby AJ (1991) J Appl Electrochem 21:683
126. Iwakura C, Furukawa N, Tanaka M (1992) Electrochim Acta 37:757
127. Miao HJ, Piron DL (1993) Electrochim Acta 38:1079
128. Iwakura C, Tanaka M, Nakamatsu S, Noue H, Matsuoka M, Furukawa N (1995) Electrochim Acta 40:977
129. Gierlotka G, Rowinski E, Budniok A, Lagiewka E (1997) J Appl Electrochem 27:1349
130. Assuncao NA, De Giz MJ, Tremiliosi–Filho G, Gonzalez ER (1997) J Electrochem Soc 144:2794
131. Tavares AC, Trasatti S (1998). In: S.A. Sealey SA (ed.) Modern Chlor-Alkali Technology, vol. 7, SCI, London, p. 65
132. Baruffaldi C, Cattarin S, Musiani M (2003) Electrochim Acta 48:3921
133. Panek J, Serek A, Budniok A, Rowinski E, Lagiewka E (2003) Int J Hydrogen Energy 28:169
134. Losiewicz B, Budniok A, Rowinski E, Lagiewka E, Lasia A (2003) J Appl Electrochem 34:507
135. Olesky M, Budniok A, Niedbala J, Matyja P (1994) Electrochim Acta 39:2439
136. Musiani M (1996) Chem Commun 2403
137. Musiani M, Furlanetto F, Guerriero P (1997) J Electroanal Chem 437:131
138. Musiani M, Guerriero P (1998) J Electrochem Soc 145:549
139. Musiani M, Furlanetto F, Guerriero P (1998) J Electrochem Soc 145:555
140. Musiani M, Guerriero P (1998) Electrochim Acta 44:1499
141. Bertoncello R, Furlanetto F, Guerriero P, Musiani M (1999) Electrochim Acta 44:4061
142. Musiani M, Furlanetto F, Bertoncello R (1999) J Electroanal Chem 465:160
143. Bertoncello R, Cattarin S, Frateur I, Musiani M (2000) J Electroanal Chem.492:145
144. S. Cattarin, P. Guerriero, M. Musiani, Electrochim. Acta 46 (2001)4229.
145. Huet F, Musiani M, Nogueira RP (2003) Electrochim Acta 48:3981
146. Vazquez–Gomez L, Cattarin S, Guerriero P, Musiani M (2007) Electrochim Acta 52:8055
147. Vazquez–Gomez L, Cattarin S, Guerriero P, Musiani M (2009) J Electroanal Chem 634:42

148. Antozzi AL, Bargioni C, Iacopeti L, Musiani M, Vazquez–Gomez L (2008) Electrochim Acta 53:7410
149. Jović VD, Lačnjevac U, Jović BM, Gajić-Krstajić Lj, Krstajić NV, (2013) J Serb Chem Soc 78:689
150. Krstajić NV, Lačnjevac U, Jović BM, Mora S, Jović VD (2011) Int J Hydrogen Energy 36:6450
151. Jović VD, Lačnjevac U, Jović BM, Krstajić NV (2012) Electrochim Acta 63:124
152. Lačnjevac UČ, Jović BM, Jović VD, Krstajić NV (1012) J Electroanal Chem 677: 31
153. Jović BM, Lačnjevac U, Jović VD, Gajić-Krstajić Lj, Krstajić NV (2012) J Serb Chem Soc 77:211
154. Nagamura M, Ukihashi H, Shiragami O (1983). In: Jackson C (ed) Modern Chlor-Alkali Technology, vol. 2, Ellis Horwood, Chichester, p. 61
155. Grove DE (1986) In: Wall K (ed) Modern Chlor-Alkali Technology, vol. 3, Ellis Horwood, Chichester, p. 250
156. Trasatti S (1992). Electrocatalysis of hydrogen evolution: Progress in cathode activation. In: Gerischer H, Tobias CW (eds), Advances in Electrochemical Science and Engineering, Weinheim, Wiley-VCH Verlag GmbH, p. 1
157. Iwakura C, Tanaka M, Nakamatsu S, Noue H, Matsuoka M, Furukawa N (1995) Electrochim Acta 40:977
158. Rausch S, Wendt H (1995) J Appl Electrochem 22:1025

Chapter 2
A New Approach to the Understanding of the Mechanism of Lead Electrodeposition

Nebojša D. Nikolić and Konstantin I. Popov

2.1 Introduction

The most important applications of lead include the production of high purity active materials for acid battery [1], for semiconductors [2, 3], and for the fabrication of electrochromic devices [4]. In the form of powder, lead is widely used in industries of gas and oil exploration, radiological medical protective clothing, as an industrial X-ray shield, golf club manufacture, and antifriction products [5]. The electrodeposition technique is a very suitable way to obtain lead in the form suitable for the application in the above-mentioned technologies. For example, the advantage of use of electrodeposition technique in the production of lead in the powder form lies in the fact that lead powder is produced at low overpotentials and hence with small spent of energy. The open porous structures of lead with the extremely high surface area (the honeycomb-like ones), which are ideally situated for electrodes in electrochemical devices such as fuel cells, batteries, and sensors, are also possible to get by the electrodeposition techniques [6].

Aside from the practical significance, the processes of lead electrodeposition also have the high scientific significance in the examination of the fast electrochemical processes. In the Winand's classification of metals [7], Pb belongs to the group of normal metals together with Ag, Cd, Zn, and Sn. These metals are characterized by low melting points and high exchange current densities, j_0.

N.D. Nikolić (✉)
ICTM-Institute of Electrochemistry, University of Belgrade, Njegoševa 12, 11001 Belgrade, Serbia
e-mail: nnikolic@tmf.bg.ac.rs

K.I. Popov
ICTM-Institute of Electrochemistry, University of Belgrade, Njegoševa 12, 11001 Belgrade, Serbia

Faculty of Technology and Metallurgy, University of Belgrade, Karnegijeva 4, 11001 Belgrade, Serbia
e-mail: kosta@tmf.bg.ac.rs

S.S. Djokić (ed.), *Electrodeposition and Surface Finishing: Fundamentals and Applications*, Modern Aspects of Electrochemistry 57,
DOI 10.1007/978-1-4939-0289-7_2, © Springer Science+Business Media New York 2014

For lead electrodeposition, both alkaline [8, 9] and acidic [4, 10–22] electrolytes are widely used. The most often employed acidic electrolytes are those based on chloride [10, 11], bromide [12], iodide [12], nitrate [4, 13, 14], fluorborate [15], acetate [16], fluorosilicate [17–20], methanesulfonate [21, 22], etc.

Depending on the type and composition of the electrolyte and electrodeposition conditions, various powders, e.g. dendritic [16], honeycomb-like structures [6], etc. can be obtained. On the other hand, smooth, adherent, and compact deposits can be formed in the presence of additives such as phenol, ethanol, gelatin [23], sodium lignin sulfonate, chestnut extract, and animal glue [17–20]. Furthermore, the additions of sorbitol [24] and glycerol [8] to the electroplating solutions have some effect on the morphology of the electrodeposited lead.

It is believed that the electrodeposition processes of normal metals are diffusion controlled. However, it was shown recently [25–29] that the linear dependence of the current density on the overpotential during silver electrodeposition from nitrate solution could be ascribed to an ohmic controlled electrodeposition process. Regarding the affiliation to the same group of metals, the well-defined linear dependence of the current density on overpotential during lead electrodeposition from a concentrated nitrate solution was also observed [30, 31]. The new investigations of the mechanism of lead electrodeposition showed that electrodeposition of this metal can follow the mixed ohmic-diffusion control [32]. For that reason, the aim of this chapter was a comprehensive survey of all aspects related with the processes of lead electrodeposition.

2.2 Mechanism of Lead Electrodeposition: The Mixed Ohmic-Diffusion and the Full Ohmic Control

2.2.1 Polarization Characteristics: Experimental Evidence

Figure 2.1 shows the polarization curves for lead electrodeposition from solutions containing 0.050, 0.10, 0.20 (Fig. 2.1a), and 0.30 and 0.45 M $Pb(NO_3)_2$ (Fig. 2.1b) in 2.0 M $NaNO_3$. The following experimental procedure, usual for the recording of the polarization curves of fast electrochemical processes, was applied to obtain a reproductive shape of the polarization curves for lead [32]: the values of the current density obtained after stabilization of their values at every 5 mV were used for constructing the polarization curves. The time required for stabilization of the current density values was 3 s at all overpotentials. After the determined values of overpotential (the inflection point), the current density increased dramatically and then the values of the current density at the moment selected values of the overpotential were attained were employed for the further recording of the polarization curves.

The common characteristic of all the polarization curves is the existence of an inflection point at an overpotential of 55 mV, followed by a sharp increase in the

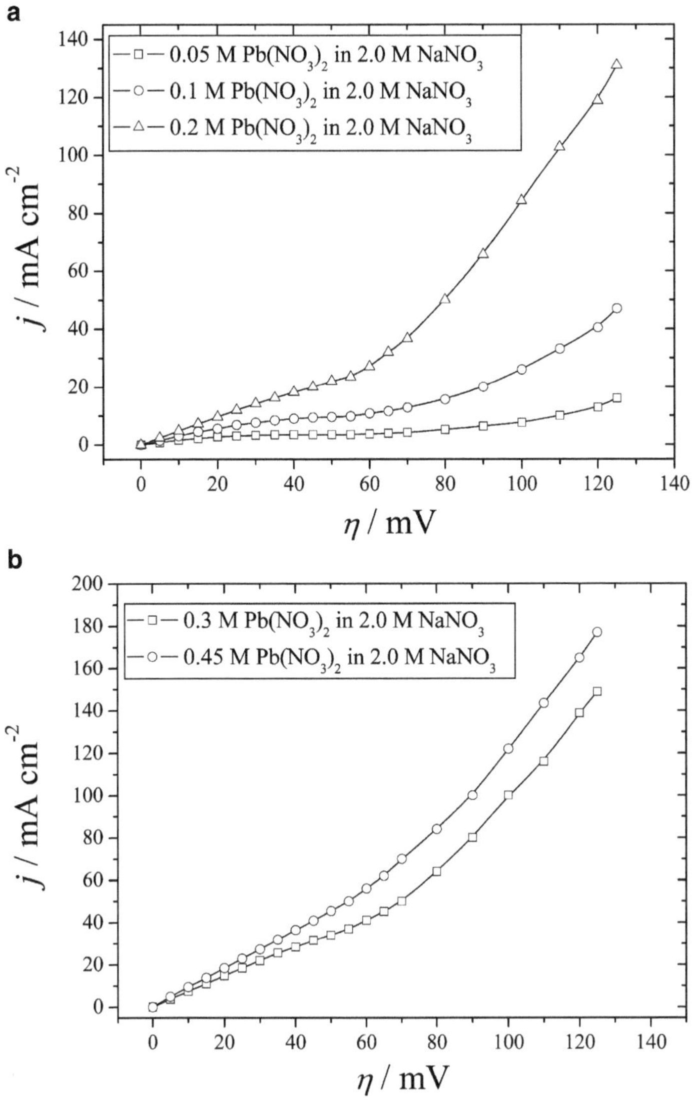

Fig. 2.1 Polarization curves for lead electrodeposition from: (**a**) 0.050, 0.10, 0.20 and (**b**) 0.30 and 0.45 M $Pb(NO_3)_2$ in 2.0 M $NaNO_3$ (Reprinted from [32] with permission from Elsevier.)

current density on further increasing the overpotential. A careful analysis of the parts of polarization curves up to an overpotential of 55 mV (Fig. 2.2) showed that the polarization curves recorded from solutions containing 0.050, 0.10, 0.20, and 0.30 M $Pb(NO_3)_2$ in 2.0 M $NaNO_3$ consisted of two parts: (a) linear dependencies of the current density on overpotential, which were ascribed to ohmic controlled electrodeposition and (b) the parts corresponding to diffusion controlled

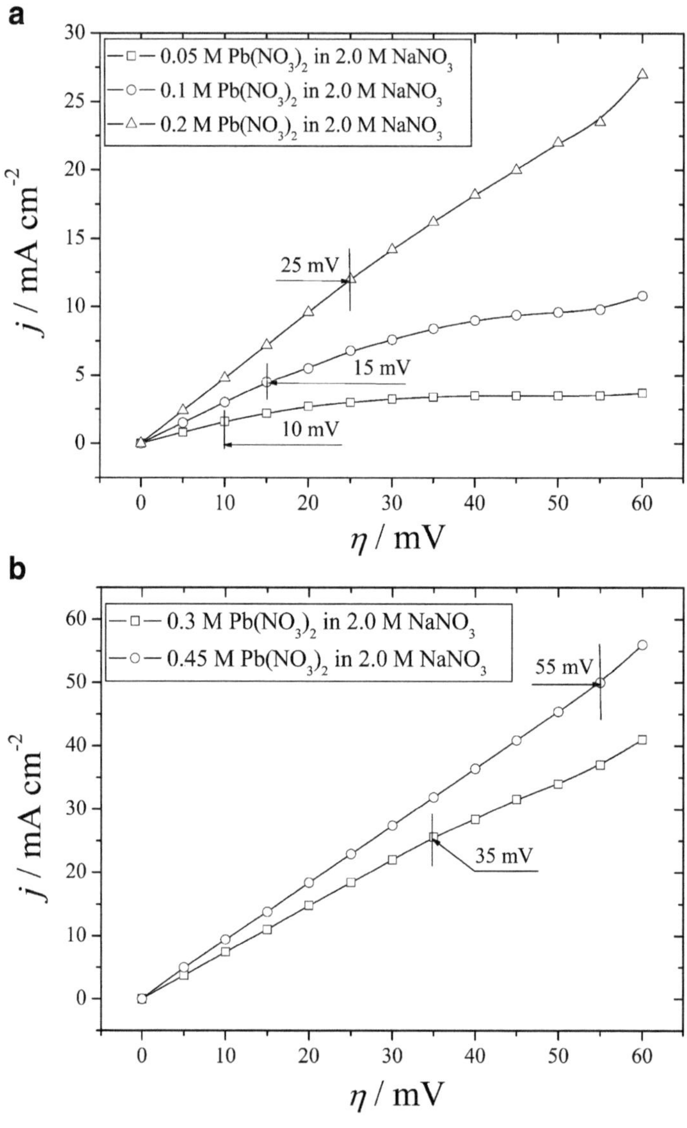

Fig. 2.2 The parts of the polarization curves up to an overpotential of 55 mV: (**a**) 0.050, 0.10, 0.20 and (**b**) 0.30 and 0.45 M $Pb(NO_3)_2$ in 2.0 M $NaNO_3$ (Reprinted from [32] with permission from Elsevier.)

electrodeposition. A well-defined plateau of the limiting diffusion current density could be noticed on the polarization curve recorded from the solution of the lowest analyzed Pb(II) concentration (0.050 M $Pb(NO_3)_2$ in 2.0 M $NaNO_3$). For this solution, the linear dependence of the current density on overpotential (ohmic control of electrodeposition) was in the interval of overpotentials from 0 to 10 mV.

The length of this linear dependence increased with increasing concentration of Pb (II) ions. Simultaneously, the region of diffusion control of the electrodeposition decreased, while the slope of the inclined limiting diffusion current density plateau increased. The values of the overpotentials corresponding to the end of the ohmic control of the electrodeposition process are denoted in Fig. 2.2 by a vertical line. Nevertheless, these curves can be denoted as typical representatives of mixed ohmic-diffusion control of the electrodeposition process.

The plateaus of the limiting diffusion current density vanish with further increase of the Pb(II) ions concentration, becoming an inflection point at an overpotential of 55 mV (0.45 M $Pb(NO_3)_2$ in 2.0 M $NaNO_3$—Fig. 2.1b). A linear dependence of the current density on overpotential was observed up to the inflection point. This polarization curve represents a typical example of complete ohmic control of the electrodeposition process. A similar shape of the polarization curve was previously recorded from 0.50 M $Pb(NO_3)_2$ in 2.0 M $NaNO_3$ [30, 31], as well as in the case of electrodeposition of some other metals from this group, such as silver [25–29]. A sharp increase in the current density with increasing overpotential after the inflection point was also observed.

2.2.2 General Discussion of the Mechanism of Lead Electrodeposition

Electrodeposition of lead occurred under conditions of natural convection, which results in a linear dependence of the logarithm of the limiting diffusion current density on the logarithm of the concentration of the depositing ions.

Namely, the mass and heat transfer induced by the density gradient in liquids in the gravitational field is natural or free convection [25, 33, 34]. Levich [33] gave an approximate analytical solution for the estimation the diffusion flux of the component that reacts and disappears on the surface of a vertical plate in a liquid. The electrodeposition of metals is a fair example of this kind of reaction [25].

According to Levich, in non-stirred electrolytes under conditions of natural convection, the limiting diffusion current density, j_L, varies with concentration of solution, c, as

$$j_L \approx c^n, \tag{2.1}$$

where:

$$1 \leq n \leq 1.25. \tag{2.2}$$

When the temperatures of the thermostated solutions (with a free surface) and of the surrounding air are the same, $n = 1.25$ and this value corresponds to proper isothermal conditions [34]. The values $n \leq 1.25$ correspond to non-isothermal conditions, i.e., when there are differences between these temperatures.

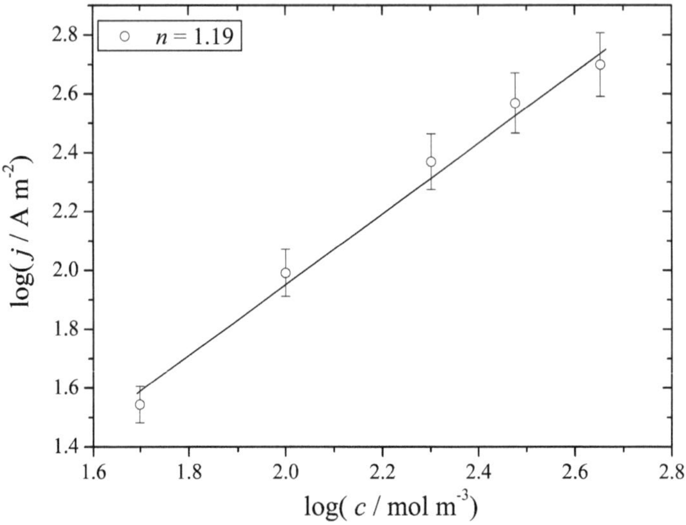

Fig. 2.3 The logarithmic dependence of the limiting diffusion current density, j_L, on the concentration of solution, c. The values of the limiting diffusion current density correspond to the inflection points on the polarization curves ($\eta = 55$ mV) (Reprinted from [32] with permission from Elsevier.)

The logarithmic dependence of j_L on c is shown in Fig. 2.3, and the slope n was found to be 1.19. It is necessary to stress that the current densities corresponding to the inflection points on the polarization curves, i.e., an overpotential of 55 mV, were taken as the limiting ones. The excellent agreement between theoretical prediction and the experimentally obtained value confirmed that the electrodeposition of lead occurs under conditions of natural convection.

The polarization curves (Fig. 2.2) standardized to the values of the limiting diffusion current density are given in Fig. 2.4. An analysis of Fig. 2.4 clearly indicated that the ratio of the ohmic control to the overall control of electrodeposition process increased with increasing concentration of Pb(II) ions.

There is no evidence about the true reaction mechanism for lead electrodeposition, but two reaction mechanisms are possible:

(a)

$$Pb^{2+} + 2e^- \rightarrow Pb \tag{I}$$

with $b_a = 60$ mV/dec and $b_c = 60$ mV/dec (the case of cadmium [35]), and

(b)

$$\begin{aligned} &Pb^{2+} + e^- \rightarrow Pb^+ \\ &Pb^+ + e^- \rightarrow Pb \end{aligned} \tag{II}$$

with $b_a = 40$ mV/dec and $b_c = 120$ mV/dec (the case of copper [36]).

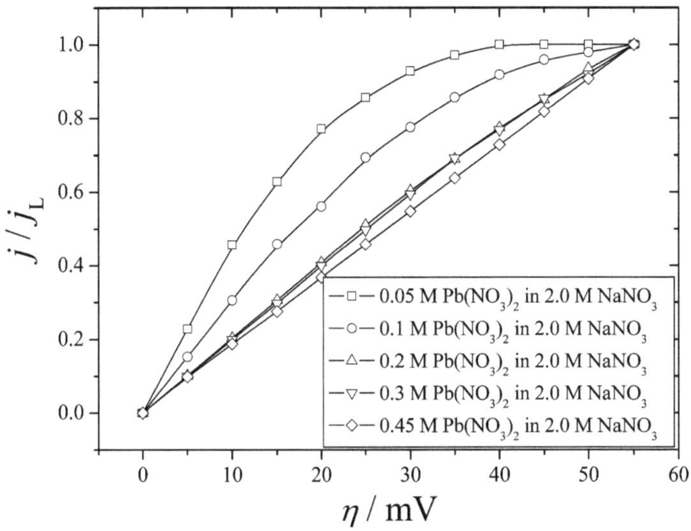

Fig. 2.4 The polarization curves standardized to the values of the limiting diffusion current density (Reprinted from [32] with permission from Elsevier.)

Although the reaction mechanism (I) is more probable than mechanism (II), we are of the opinion that reaction mechanism (II) should not be completely disregarded [32].

For the very fast electrodeposition reactions, the following form of the polarization curve equation is valid [27, 28]:

$$j = \frac{j_0(f_c - f_a)}{1 + \frac{j_0}{j_L}(f_c - f_a)}. \tag{2.3}$$

In Eq. (2.3), $f_c = 10^{\eta/b_c}$, $f_a = 10^{-\eta/b_a}$, j is the current density, and η is the overpotential (the cathodic current density, j and η are taken as positive values).

Figure 2.5 shows the dependencies of j/j_L on η for the different j_0/j_L ratios [(a)—$j_0/j_L = 100$, (b)—$j_0/j_L = 75$, (c)—$j_0/j_L = 50$, (d)—$j_0/j_L = 25$, and (e)—$j_0/j_L = 10$] obtained by the simulation of Eq. (2.3) for the two possible mechanisms of lead deposition: (a) (I)—Fig. 2.5a and (b) (II)—Fig. 2.5b. Both dependencies are practically the same when $j_0/j_L \geq 10$, indicating that there is no effect of the different Tafel slopes on the shape of the polarization curves. This is better illustrated in Fig. 2.6, which shows a comparative survey of the dependencies of j/j_L on η for the j_0/j_L ratios of 100, 50, and 10 for the both reaction mechanisms. From Fig. 2.5, it can also be seen that for $j_0/j_L \geq 50$, the systems enter full diffusion control at $\eta > 10$ mV, being under complete ohmic control up to $j \approx 0.9 j_L$. Moreover, differences in the polarization characteristics of lead between these two reaction mechanisms are not expected due to the absence of activation control of the electrodeposition. The dependencies shown in Figs. 2.5 and 2.6 were calculated without inclusion of an ohmic potential

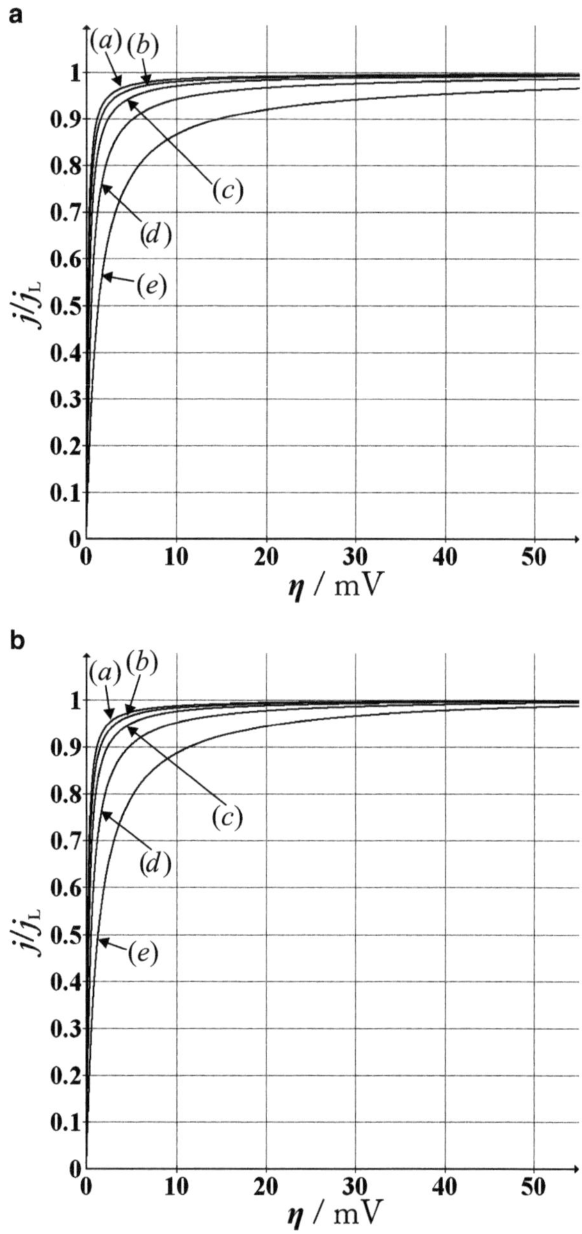

Fig. 2.5 The dependencies of j/j_L on η for the different j_0/j_L ratios obtained by the simulation of Eq. (2.3) for the possible lead reaction mechanisms: (**a**) (I) and (**b**) (II). The markings in the figure: (*a*) $j_0/j_L = 100$, (*b*) $j_0/j_L = 75$, (*c*) $j_0/j_L = 50$, (*d*) $j_0/j_L = 25$, and (*e*) $j_0/j_L = 10$ (Reprinted from [32] with permission from Elsevier.)

drop in the overpotential. On the other hand, an ohmic potential drop is included in the experimentally obtained polarization curves (Fig. 2.4). This means that a comparison of the calculated and experimentally obtained polarization curves is impossible without elimination of the ohmic potential drop from the experimentally obtained polarization curves.

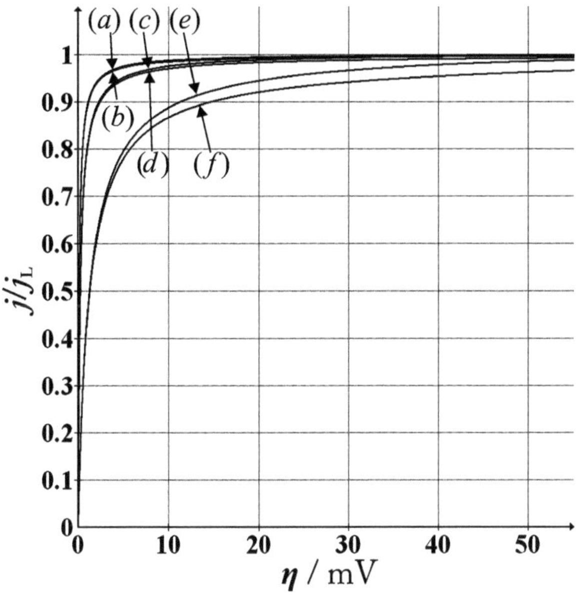

Fig. 2.6 The comparative survey of the dependencies of j/j_L on η for the different j_0/j_L ratios obtained for the both groups of Tafel slopes. The markings in the figure: (*a*), (*c*), and (*e*)—Reaction mechanism: (I), and (*b*), (*d*), and (*f*)—Reaction mechanism: (II), (*a*) and (*b*) $j_0/j_L = 100$; (*c*) and (*d*) $j_0/j_L = 50$; (*e*) and (*f*) $j_0/j_L = 10$ (Reprinted from [32] with permission from Elsevier.)

The recently proposed procedure [25] was used for the elimination of the ohmic potential drop from the experimentally obtained polarization curves. This procedure can be presented briefly as follows: at a specified current density, the difference between the experimentally measured overpotential and the one corresponding to full ohmic control of the electrodeposition is the diffusion overpotential, η_{diff}. Naturally, the above consideration is only valid before the abrupt increase in the current density with the increasing overpotential after the inflection point. This way of correction for the ohmic potential drop is illustrated in Fig. 2.7a for the part of the polarization curve up to the inflection point recorded from the solution containing 0.050 M $Pb(NO_3)_2$ in 2.0 M $NaNO_3$. In this way, corrected polarization curves, shown in Fig. 2.7b, suitable for discussion of the initiation of dendritic growth were obtained. It is necessary to remember that the corrected values of the overpotentials represent only diffusion ones, η_{diff}.

Hence, it could be concluded from Fig. 2.7b that in mixed ohmic-diffusion control of the electrodeposition (the systems with $j_0 \gg j_L$), increasing concentration of Pb(II) ions causes a decrease in both the critical overpotential for dendritic growth initiation, η_i, and the critical overpotential for instantaneous dendritic growth, η_c. A qualitative comparison with the simulated dependencies shown in Fig. 2.5 was thus enabled by standardization of the current density values from Fig. 2.7b to the corresponding limiting diffusion current density values.

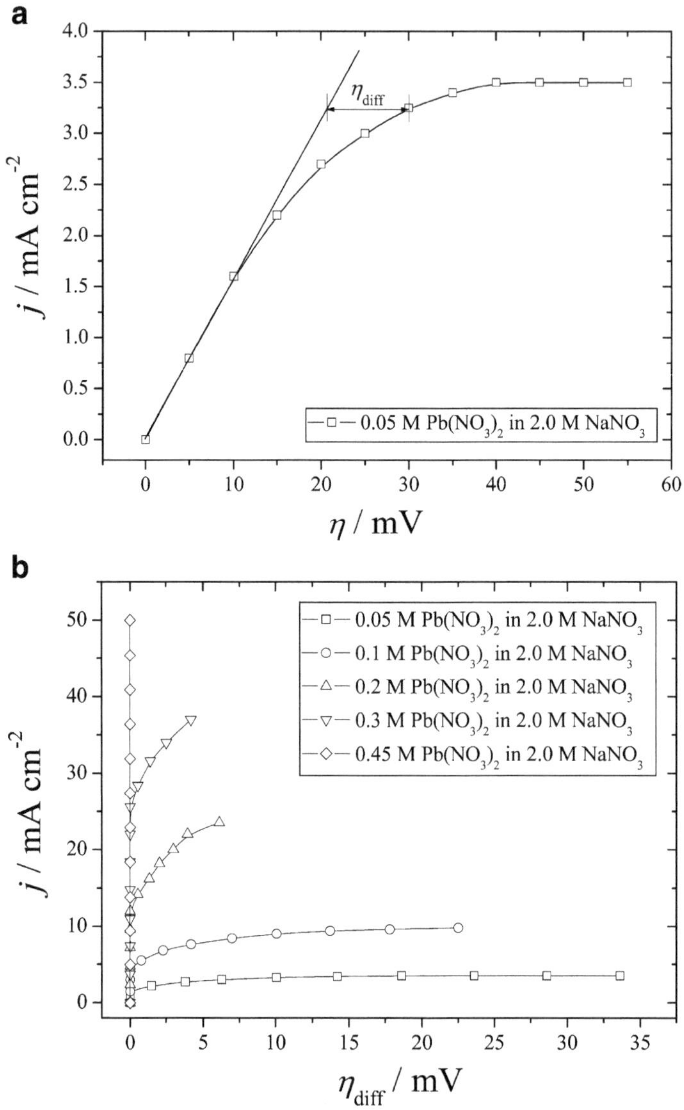

Fig. 2.7 (**a**) The way of the correction of the ohmic potential drop in the experimentally obtained polarization curves and (**b**) the corrected parts of polarization curves up to the inflection point j on η_{diff} (Reprinted from [32] with permission from Elsevier.)

The standardized dependencies $j/j_{\mathrm{L}} - \eta_{\mathrm{diff}}$ are shown in Fig. 2.8. Based on the dependencies shown in Figs. 2.5 and 2.8, it is very clear that increasing the concentration of Pb(II) ions leads to a strong increase in the j_0/j_{L} ratio.

This situation is completely different from the one observed in the activation–diffusion controlled electrodeposition (the systems characterized by $j_0 < j_{\mathrm{L}}$; e.g., copper). In the activation–diffusion control of electrodeposition process, both η_{i} and

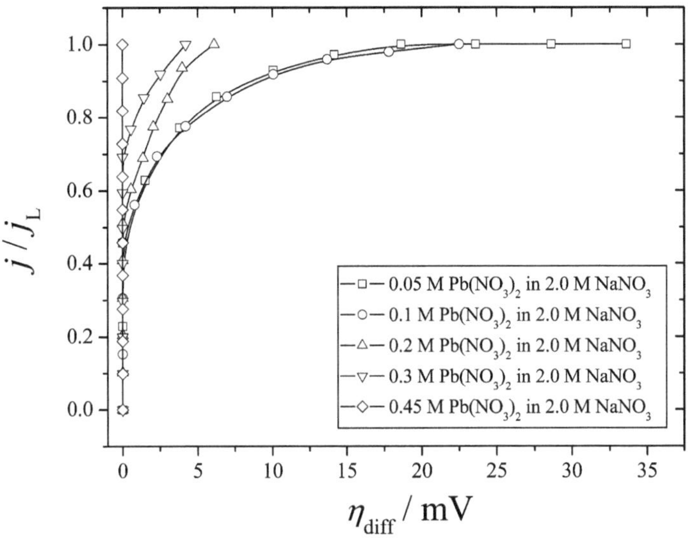

Fig. 2.8 The standardized dependencies of j/j_L on η_{diff} (Reprinted from [32] with permission from Elsevier.)

η_c increased with increasing concentration of the depositing ions, indicating a decrease of the j_0/j_L ratios with the increasing concentration of metal ions.

This difference can be discussed in the following way: the critical overpotential for dendritic growth initiation in the systems where $j_L > j_0$ is given by Eq. (2.4):

$$\eta_i = \frac{b_c}{2.3} \ln \frac{j_L}{j_0} \tag{2.4}$$

and the critical overpotential for the instantaneous dendritic growth is given by Eq. (2.5):

$$\eta_c = \frac{b_c}{2.3} \ln \left[\frac{j_L}{j_0} \left(\frac{\delta}{h} \right)^{\gamma} \right], \tag{2.5}$$

where δ is the diffusion layer thickness, and h is the height of protrusion from which the growth of dendrite commences and $\gamma = \frac{\mathrm{dlog} j_0}{\mathrm{dlog} c}$ [28].

According to Newman [37];

$$j_0 \approx c^{\gamma}, \tag{2.6}$$

where $\gamma = 0.5$ [38] for

$$\mathrm{Me}^{2+} + 2\mathrm{e}^- \rightarrow \mathrm{Me}$$

and $\gamma = 0.75$ [37] for

$$Me^{2+} + e^- \rightarrow Me^+$$
$$Me^+ + e^- \rightarrow Me$$

reaction mechanisms.

Taking into account Eqs. (2.1) and (2.6), one obtains

$$\frac{j_L}{j_0} \approx c^{n-\gamma}, \tag{2.7}$$

where $n - \gamma > 0$ in both cases.

From Eqs. (2.4), (2.5), and (2.7), it could be confirmed that in the activation–diffusion controlled electrodeposition, both η_i and η_c increase with increasing j_L/j_0 ratios (or with the decreasing j_0/j_L ratios). This means that increasing the concentration of the depositing ions leads to an increase in both η_i and η_c, as already observed in the case of copper electrodeposition [39].

On the other hand, if $j_L/j_0 \ll 1$ (very fast electrodeposition processes), Eqs. (2.4) and (2.5) become

$$\eta_i = \frac{RT}{nF}\frac{j_L}{j_0} \tag{2.8}$$

and

$$\eta_c = \frac{RT}{nF}\frac{j_L}{j_0}\left(\frac{\delta}{h}\right)^\gamma, \tag{2.9}$$

where nF is the number Faradays per mole of consumed ions, R is gas constant, and T is temperature. These relations show that dendritic growth is to be expected at very low overpotentials.

The difference between η_c and η_i can be written as

$$\eta_c - \eta_i = \frac{RT}{nF}\frac{j_L}{j_0}\left[\left(\frac{\delta}{h}\right)^\gamma - 1\right] \tag{2.10}$$

from which it can be seen that this difference decreases with the decreasing j_L/j_0 value (or increasing j_0/j_L value). Hence, the difference between η_c and η_i decreases with increasing concentration of Pb(II) ions and becomes negligible at the sufficiently large concentrations of Pb(II) ions. In an ohmic-diffusion controlled electrodeposition process, the maximum current density corresponding to the ohmic controlled electrodeposition increases with increasing j_0/j_L ratio. This causes, after elimination of the ohmic potential drop, a decrease in both the overpotential at which electrodeposition system enters diffusion control, η_i and the overpotential at which full diffusion control is attained, η_c, as well as of the difference between them, with the increasing concentration of Pb(II) ions.

In addition, a semiquantitative analysis of Eq. (2.10) confirms very well the presented results. As already mentioned, for fast electrochemical processes $j_0 \rightarrow \infty$ and $j_0 \gg j_L$ (or $j_L/j_0 \ll 1$). The j_L/j_0 ratio decreases with increasing concentration of Pb(II) ions and at the sufficiently high concentration of Pb(II) ions, this ratio will tend to zero ($j_L/j_0 \rightarrow 0$). Then, the right side of Eq. (2.10) will also tend to zero and hence $\eta_c - \eta_i \rightarrow 0$ (or $\eta_c \approx \eta_i$), which corresponds to full ohmic control of the electrodeposition process.

On the other hand, in the case of the electrodeposition under mixed activation–diffusion control ($j_0 < j_L$), the difference between η_c and η_i (Eqs. (2.4) and (2.5), respectively) could be presented by Eq. (2.11):

$$\eta_c - \eta_i = \frac{\gamma b_c}{2.3} \ln \frac{\delta}{h}. \tag{2.11}$$

It is very clear from Eq. (2.11) that this difference does not depend on the j_L/j_0 ratio and hence, on the concentration of the depositing ions. Furthermore, this difference will always have a finite value which will never tend to zero. This is one of the substantial differences between this electrodeposition system (activation–diffusion control of the electrodeposition; $j_0 < j_L$) and the one characterized by an extremely large exchange current density (ohmic-diffusion control of the electrodeposition; $j_0 \gg j_L$) from the point of view of the formation and growth of dendrites.

2.3 Correlation Between Polarization Characteristics and Morphology of Lead Deposits

2.3.1 Morphology of Lead Deposits Obtained from the Nitrate Electrolytes with the Different Concentration of Pb(II) Ions

One of the best ways to analyze an electrodeposition system is to establish a correlation between the polarization characteristics and the morphology of the deposits. All electrodepositions which results are presented in this chapter were performed in an open cell at room temperature from the corresponding solutions [32]. Lead was electrodeposited on vertical cylindrical copper electrodes, while the reference and counter electrodes were of pure lead.

The solution containing 0.30 M $Pb(NO_3)_2$ in 2.0 M $NaNO_3$ was taken as the basic solution for the determination of this correlation.

Figure 2.9 shows the morphologies of lead deposits electrodeposited from 0.30 M $Pb(NO_3)_2$ in 2.0 M $NaNO_3$ at overpotentials corresponding to ohmic

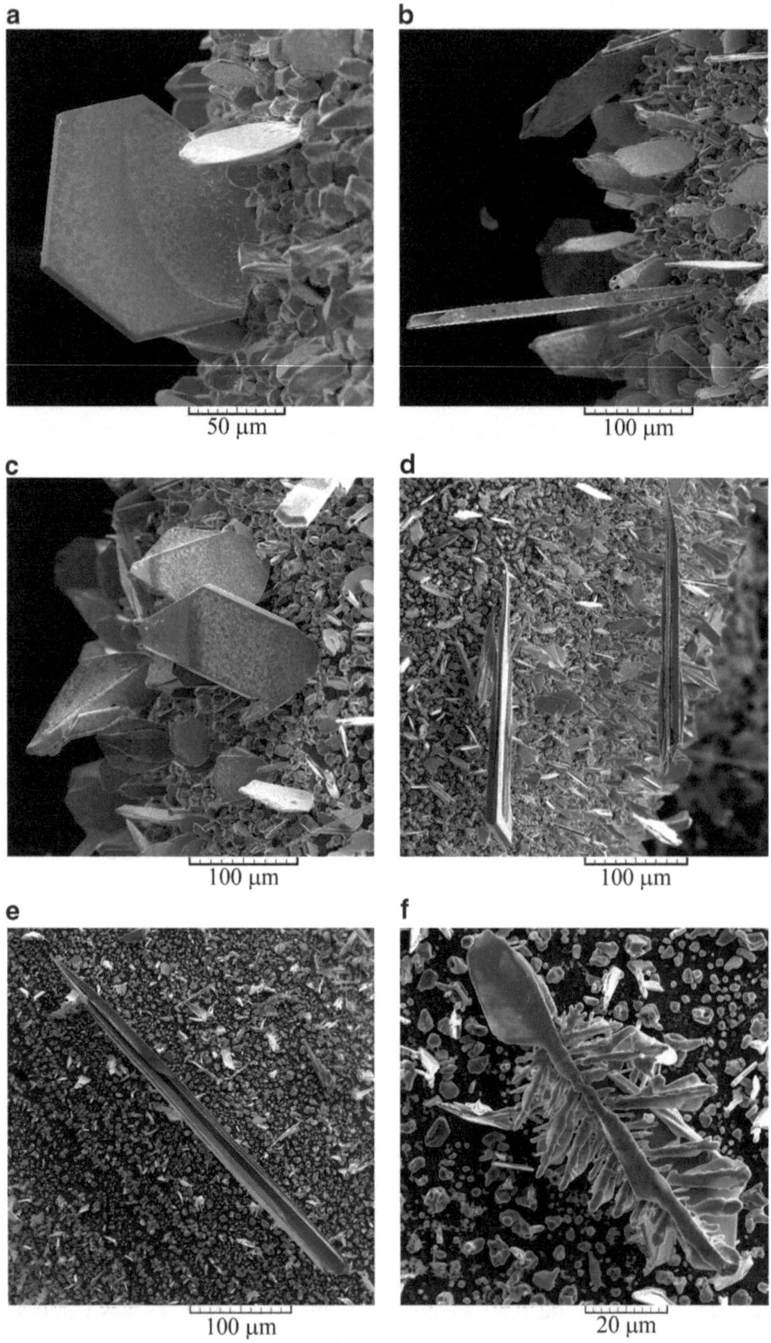

Fig. 2.9 Morphologies of lead deposits obtained from 0.30 M $Pb(NO_3)_2$ in 2.0 M $NaNO_3$ at overpotentials, η of: (**a**) 30 mV, (**b**) and (**c**) 55 mV, (**d**) 80 mV, (**e**) and (**f**) 120 mV (Reprinted from [32] with permission from Elsevier.)

controlled electrodeposition ($\eta = 30$ mV; Fig. 2.9a), the end of the limiting diffusion current density plateau ($\eta = 55$ mV; Fig. 2.9b, c), and during the fast increase in the current density with increasing overpotential ($\eta = 80$ mV; Fig. 2.9d, and $\eta = 120$ mV; Fig. 2.9e, f).

Individual regular crystals were formed at an overpotential of 30 mV, i.e., when the electrodeposition was ohmic control (Fig. 2.9a). Apart from individual regular lead crystals, crystals of irregular shape and needle-like dendrites were formed during electrodeposition at an overpotential corresponding to the end of the limiting diffusion current density plateau (Fig. 2.9b, c). Needle-like dendrites were dominantly formed at an overpotential of 80 mV (Fig. 2.9d), which was 25 mV greater than the overpotential corresponding to the inflection point. These types of dendrites were also formed at an overpotential of 120 mV (Fig. 2.9e) but, in addition, fern-like dendrites were formed as well (Fig. 2.9f).

In order to gain a comprehensive insight into the effect of the concentration of Pb (II) ions on the morphology of the lead deposits, the obtained surface morphologies were compared with those obtained from 0.050 to 0.45 M $Pb(NO_3)_2$ in 2.0 M $NaNO_3$ at some of the above-employed overpotentials.

The morphologies of the lead deposits electrodeposited from 0.050 M $Pb(NO_3)_2$ in 2.0 M $NaNO_3$ at overpotentials of 30, 55, and 80 mV are shown in Figs. 2.10 and 2.11. Crystals of irregular shape and dendritic forms were formed at $\eta = 30$ mV (Fig. 2.10a–c). The lead crystals formed from this solution were narrower and more elongated than those formed by electrodeposition from 0.30 M $Pb(NO_3)_2$ in 2.0 M $NaNO_3$, and could be considered as precursors of dendrites [30, 31]. During lead electrodeposition from 0.050 M $Pb(NO_3)_2$ in 2.0 M $NaNO_3$, the formation of dendritic forms and precursors of dendrites at $\eta = 30$ mV is understandable because this overpotential belongs to the plateau of the limiting diffusion current density (Figs. 2.1a and 2.2a). Aside from irregular crystals, needle-like and fern-like dendrites were formed at the overpotential corresponding to the end of the limiting diffusion current density plateau ($\eta = 55$ mV; Fig. 2.10d, e). It is necessary to note that the fern-like dendrite shown in Fig. 2.10e developed from an individual lead crystal. A similar development of a fern-like dendrite was also observed during electrodeposition from 0.30 M $Pb(NO_3)_2$ in 2.0 M $NaNO_3$ at $\eta = 120$ mV (Fig. 2.9f).

The morphology of the electrodeposited lead obtained from 0.050 M $Pb(NO_3)_2$ in 2.0 M $NaNO_3$ at an overpotential of 80 mV is shown in Fig. 2.11. The dendrites formed under these electrodeposition conditions were completely different from those already observed and analyzed. These dendrites consisted of a beam of saw-like forms. A similar surface morphology was also obtained by lead electrodeposition from this solution at overpotentials larger than 80 mV.

Lead deposits obtained from 0.45 M $Pb(NO_3)_2$ in 2.0 M $NaNO_3$ at overpotentials of 55, 80, and 120 mV are shown in Fig. 2.12. Individual regular crystals were formed at an overpotential of 55 mV, corresponding to the end of ohmic control of electrodeposition (Fig. 2.12a). The numerous elongated crystals (precursors of dendrites) were formed by electrodeposition at 80 mV (Fig. 2.12b). Finally, a

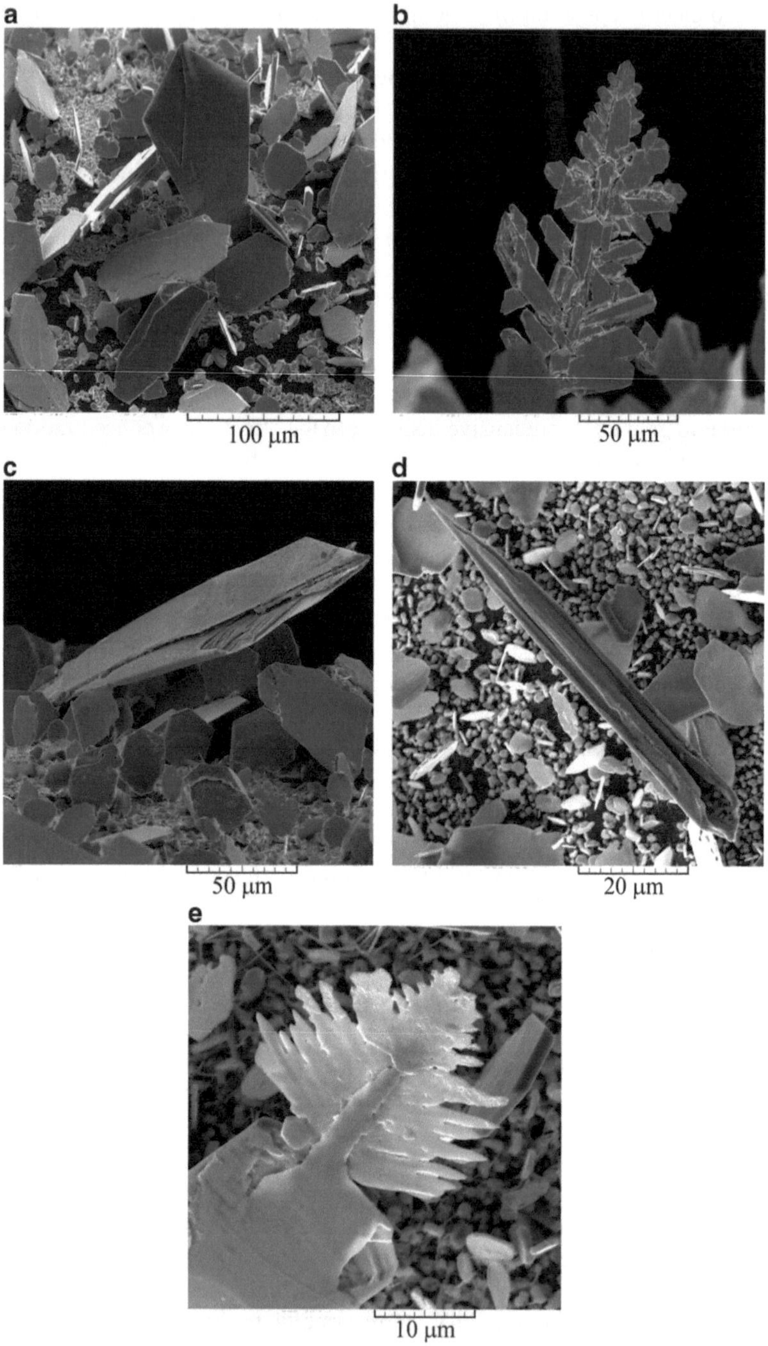

Fig. 2.10 Morphologies of lead deposits obtained from 0.050 M $Pb(NO_3)_2$ in 2.0 M $NaNO_3$ at overpotentials, η of: (**a**), (**b**), and (**c**) 30 mV, (**d**) and (**e**) 55 mV (Reprinted from [32] with permission from Elsevier.)

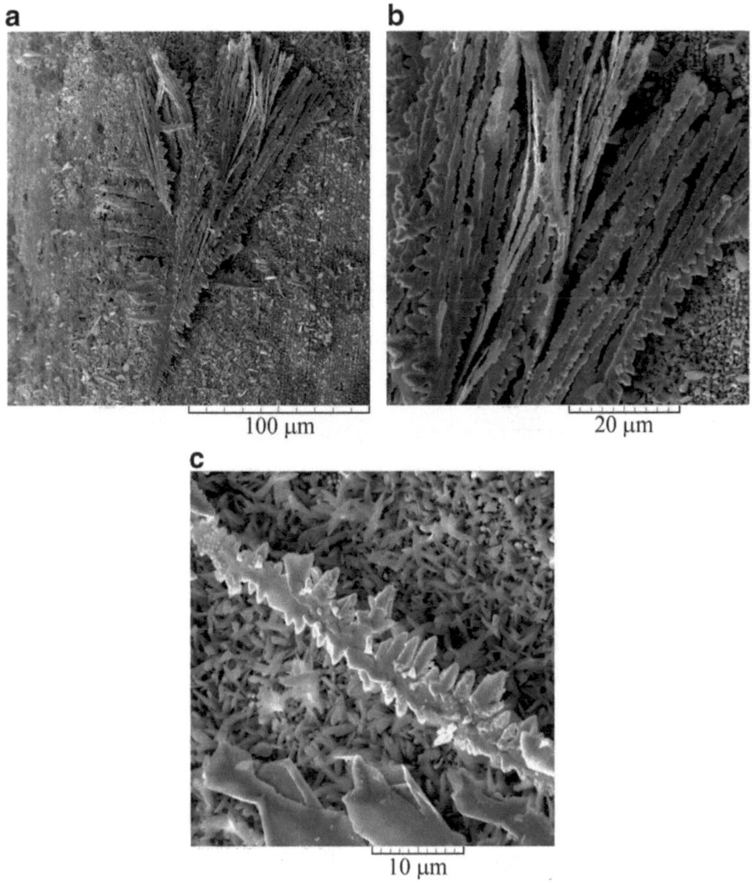

Fig. 2.11 Morphologies of lead deposits obtained from 0.050 M $Pb(NO_3)_2$ in 2.0 M $NaNO_3$ at an overpotential, η, of 80 mV (Reprinted from [32] with permission from Elsevier.)

mixture of needle-like and fern-like dendrites was formed by lead electrodeposition at an overpotential of 120 mV (Fig. 2.12c, d, respectively).

An analysis of the polarization characteristics and morphologies of lead deposits showed that in the mixed ohmic-diffusion control of the electrodeposition process, dendritic growth is initiated at overpotentials belonging to the diffusion part of the polarization curve. Therefore, as in the case of the diffusion controlled electrodeposition, the overpotential at which the system enters diffusion control can be denoted as the critical overpotential for dendritic growth initiation, η_i, while the overpotential at which diffusion control becomes complete can be denoted as the critical overpotential for instantaneous dendritic growth, η_c. Morphological analysis of electrodeposited lead showed that the critical overpotential for instantaneous dendritic growth, η_c, corresponds to the inflection point on the polarization curve. The difference between η_i and η_c decreases with increasing concentration of Pb

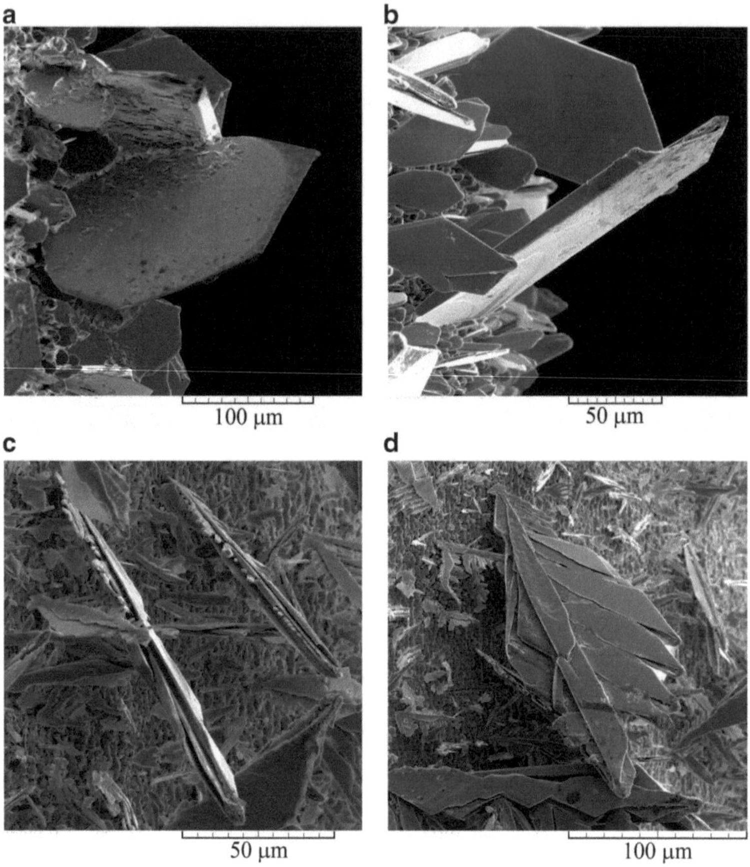

Fig. 2.12 Morphologies of lead deposits obtained from 0.45 M $Pb(NO_3)_2$ in 2.0 M $NaNO_3$ at overpotentials, η, of: (**a**) 55 mV, (**b**) 80 mV, and (**c**) and (**d**) 120 mV (Reprinted from [32] with permission from Elsevier.)

(II) ions and finally they become the same when complete ohmic control of the electrodeposition process is attained. In this way, it is experimentally proven that the inflection point on the polarization curve for the case of full ohmic control of the electrodeposition process corresponds to the condition $\eta_i = \eta_c$.

Increasing the concentration of Pb(II) ions led to a shift of the formation of characteristic morphological forms towards higher electrodeposition overpotentials. As already observed, fern-like dendrites were formed from 0.050 M Pb $(NO_3)_2$ in 2.0 M $NaNO_3$ at $\eta = 55$ mV, while this shape of dendrites was formed from 0.30 M $Pb(NO_3)_2$ in 2.0 M $NaNO_3$ at $\eta = 120$ mV. On the other hand, regular crystals were formed from 0.45 M $Pb(NO_3)_2$ in 2.0 M $NaNO_3$ at $\eta = 55$ mV, while this morphological form was obtained at $\eta = 30$ mV from 0.30 M $Pb(NO_3)_2$ in 2.0 M $NaNO_3$. The same trend in the development of surface morphology was also observed during copper electrodeposition from solutions of different $CuSO_4$ concentrations [39].

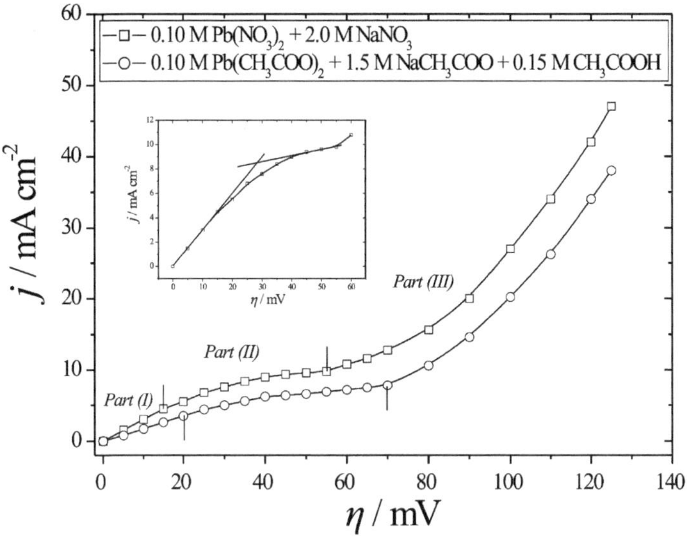

Fig. 2.13 The polarization curves for lead electrodeposition from 0.10 M $Pb(CH_3COO)_2$ + 1.5 M $NaCH_3COO$ + 0.15 M CH_3COOH and 0.10 M $Pb(NO_3)_2$ + 2.0 M $NaNO_3$ (Reprinted from [43] with permission from Elsevier.)

2.3.2 *Comparison of Lead Electrodeposition Processes from the Basic (Nitrate) and Complex (Acetate) Electrolytes*

The complex electrolytes represent the special type of electrolytes, and they can be from the group of both acid and alkaline ones. In the metal electrodeposition processes, the effect of complex formation is limited not only on the polarization behavior of the metal [31, 40], but to the morphology of metal deposits, as observed in the production of silver [31, 40, 41] and bismuth [42] powdered deposits.

2.3.2.1 Comparative Analysis of the Polarization Curves

The polarization curves for lead electrodeposition from the basic (nitrate) and complex (acetate) electrolytes are shown in Fig. 2.13. The polarization curves were recorded on cylindrical copper electrodes using experimental procedure described in Sect. 2.2.1. Similar to the polarization curve recorded from the nitrate electrolyte, the polarization curve recorded from the acetate electrolyte also consisted of three parts. These three parts are clearly denoted in Fig. 2.13.

The characteristic of the first part (*Part (I)*) is the linear dependence of the current density on overpotential and, as already mentioned, this linear part corresponds to the ohmic controlled electrodeposition process. With the nitrate ions used, the interval of overpotentials corresponding to the ohmic control is between

0 and 15 mV. When the acetate ion was used, the end of the ohmic control of the electrodeposition corresponded to an overpotential of 20 mV.

The second part of the polarization curves (*Part (II)*) corresponds to the diffusion controlled electrodeposition process. The ranges of overpotentials belonging to the diffusion control of the electrodeposition are from 15 to 55 mV for the basic electrolyte and from 20 to 70 mV for the complex electrolyte. The beginning of the plateau of the limiting diffusion current density is determined as intersect of straight lines joining current densities in the ohmic and diffusion control of electrodeposition, as inserted in Fig. 2.13. The plateaus of the limiting diffusion current density, determined in this way, were between 28.5 and 55 mV for the nitrate electrolyte and between 33 and 70 mV for the acetate electrolyte.

The inflection point at the polarization curves denoted the ends of the plateau of the limiting diffusion current density. The fast increase of the current density after the inflection point is the characteristic of the third part of the polarization curves (*Part (III)*). The inflection point corresponds to an overpotential of 55 mV for the nitrate and 70 mV for the acetate electrolytes.

The electrical conductivities of the electrolytes were 4.22 mS for the nitrate electrolyte and 7.06 mS for the acetate electrolyte. As seen from Fig. 2.13, this noticeable increase of the electrical conductivity of the acetate electrolyte of about 40 % showed the strong effect on the polarization characteristics of lead electrodeposition system.

The following differences in the polarization curves obtained from the basic (nitrate) and complex (acetate) electrolytes were observed [43]: (a) the intervals of overpotentials corresponding to the both ohmic and diffusion controls of the electrodeposition were larger for the complex than for the basic electrolytes, and (b) the plateau of the limiting diffusion current density was wider for the complex than for the basic electrolytes. The smaller value of the limiting diffusion current density for the complex electrolyte (for about 27 %) is probably due to the decrease of diffusion coefficient caused by the process of complexing of Pb (II) ions.

2.3.2.2 Comparative Analysis of Morphologies of Lead Deposits

In the next step, the polarization characteristics were correlated with morphologies of lead deposits obtained by the electrodeposition processes at overpotentials corresponding to the different positions at the polarization curves. In this way, the very important information related with the type of used electrolytes was obtained.

Figure 2.14 shows morphologies of lead deposits obtained at an overpotential of 10 mV belonging to the ohmic controlled electrodeposition for the basic (Fig. 2.14a) and complex (Fig. 2.14b) electrolytes. In both cases, the individual regular crystals were formed. It is necessary to note that the formed regular crystals were very similar to each others.

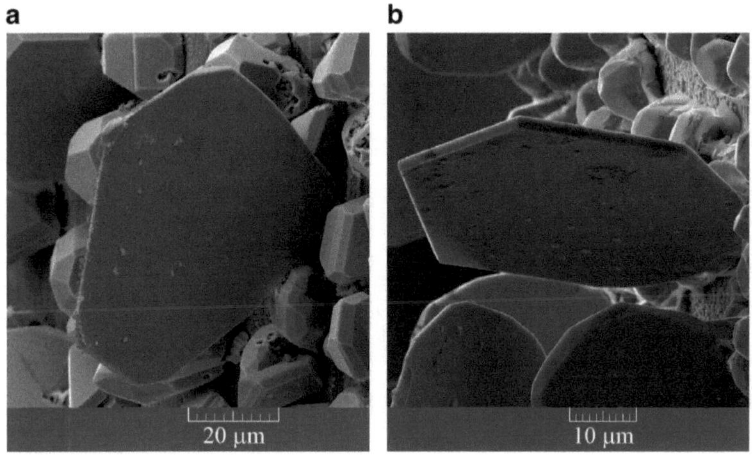

Fig. 2.14 Morphologies of electrodeposited lead obtained in the ohmic controlled electrodeposition at $\eta = 10$ mV from: (**a**) 0.10 M $Pb(NO_3)_2$ + 2.0 M $NaNO_3$ and (**b**) 0.10 M Pb $(CH_3COO)_2$ + 1.5 M $NaCH_3COO$ + 0.15 M CH_3COOH (Reprinted from [43] with permission from Elsevier.)

Morphologies of electrodeposited lead obtained at the plateaus of the limiting diffusion current density ($\eta = 50$ mV) from the both solutions are shown in Fig. 2.15. The regular crystals, different irregular forms denoted as precursors of dendrites (Fig. 2.15a), and the needle-like dendrites (Fig. 2.15b) were formed by the electrodeposition from the nitrate electrolyte. The regular crystals and needle-like dendrites were also formed by the electrodeposition from the acetate solution (Fig. 2.15c). However, aside from these morphological forms, the use of the acetate electrolyte led to the formation of the fern-like dendrites during the electrodeposition process at $\eta = 50$ mV (Fig. 2.15d).

The number of the formed fern-like dendrites increased with the increase of overpotential of electrodeposition inside the plateau of the limiting diffusion current density (Fig. 2.16a–c). From Fig. 2.16 which shows the fern-like dendrites electrodeposited at an overpotential of 60 mV from the acetate electrolyte can be clearly seen that these fern-like dendrites were more branchy structures than those formed at 50 mV. The regular crystals, the precursors of dendrites, and the needle-like dendrites can be also obtained by the electrodeposition at this overpotential (Fig. 2.16d).

The typical surface morphologies of lead obtained by the electrodeposition processes at an overpotential of 100 mV which belong to the third part of the polarization curves are shown in Figs. 2.17 (the nitrate electrolyte) and 2.18 (the acetate electrolyte). Although dendrites are formed by the electrodeposition processes from the both solutions, the difference between them was very evident.

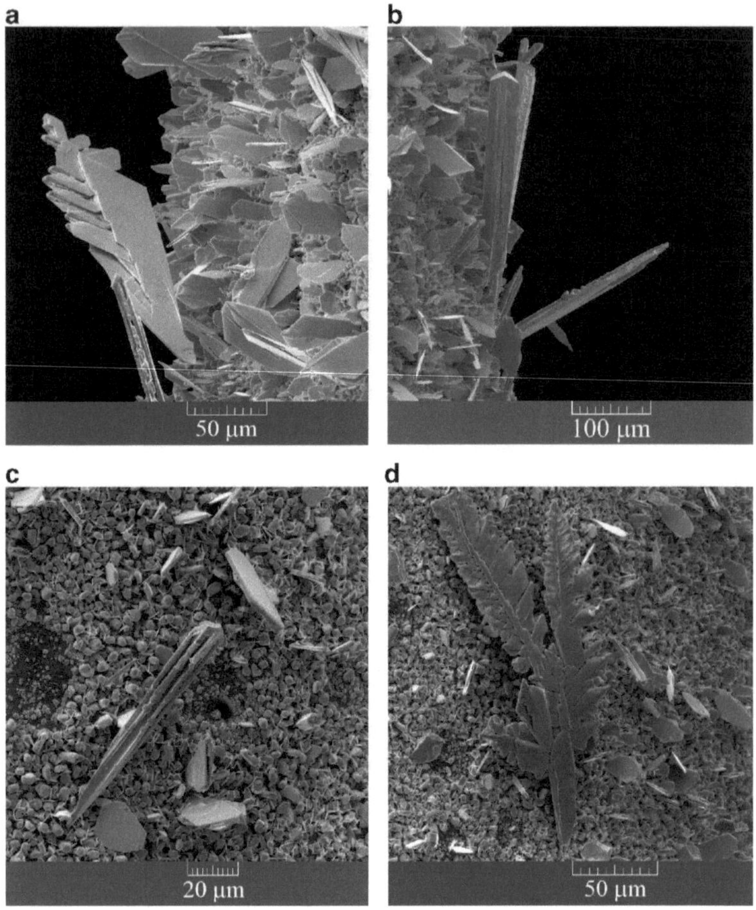

Fig. 2.15 Morphologies of electrodeposited lead obtained at the plateaus of the limiting diffusion current density ($\eta = 50$ mV) from: (**a**) and (**b**) 0.10 M $Pb(NO_3)_2$ + 2.0 M $NaNO_3$, (**c**) and (**d**) 0.10 M $Pb(CH_3COO)_2$ + 1.5 M $NaCH_3COO$ + 0.15 M CH_3COOH (Reprinted from [43] with permission from Elsevier.)

The mixture of needle-like and the fern-like dendrites was obtained at an overpotential of 100 mV when the nitrate solution was used (Fig. 2.17). However, these fern-like dendrites were completely different of those obtained from the acetate electrolyte at $\eta = 60$ mV (Fig. 2.16), as well as of those formed from the same electrolyte at $\eta = 100$ mV (Fig. 2.18). As shown in Fig. 2.18, the very branchy fern-like dendrites, similar to those obtained at $\eta = 60$ mV, were obtained by the electrodeposition from the acetate solution at an overpotential of 100 mV.

Fig. 2.16 (**a**)–(**c**) The fern-like dendrites and (**d**) the precursors of dendrites and the needle-like dendrites electrodeposited at an overpotential of 60 mV from 0.10 M $Pb(CH_3COO)_2$ + 1.5 M $NaCH_3COO$ + 0.15 M CH_3COOH (Reprinted from [43] with permission from Elsevier.)

2.3.3 General Discussion of the Correlation Between the Polarization Characteristics and Morphologies of Lead Deposits Obtained from the Different Electrolytes

For the purpose of easier analysis of lead electrodeposition processes from the basic and complex electrolytes, the comparative survey of the polarization characteristics and the typical morphologies of lead deposits is shown in Fig. 2.19.

Electrodeposition of lead from the both electrolytes occurs in the conditions of the mixed ohmic-diffusion control of the electrodeposition process. The shape of

Fig. 2.17 Lead dendrites formed at an overpotential of 100 mV from 0.10 M $Pb(NO_3)_2$ + 2.0 M $NaNO_3$ (Reprinted from [43] with permission from Elsevier.)

regular crystals formed in the ohmic control of the electrodeposition (Fig. 2.14) did not depend on the type of used electrolytes. On the other hand, the shape of dendrites was strongly affected by the type of electrolytes. Lead dendrites obtained from the nitrate solution were less branchy and completely different form than those obtained from the acetate solution. The common characteristic of all lead morphological forms is their 2D (two-dimensional) form. The shape of the dendrites formed from the nitrate solution was similar to silver dendrites [31, 40], while the dendrites formed from the acetate solution were very similar to cadmium dendrites [44]. The needle-like dendrites like those obtained from the both nitrate and acetate solutions were also observed during tin electrodeposition [45]. This similarity in the shape of dendrites is understandable because all the above-mentioned metals belong to the same group of metals ("normal metals") in Winand's classification of metals [7].

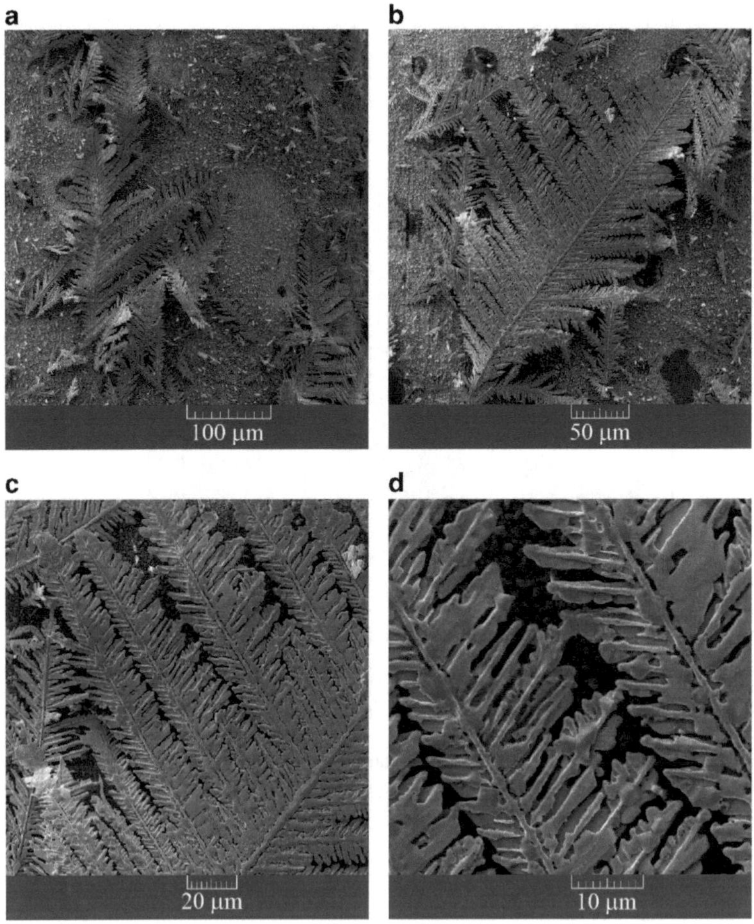

Fig. 2.18 Lead dendrites formed at an overpotential of 100 mV from 0.10 M $Pb(CH_3COO)_2$ + 1.5 M $NaCH_3COO$ + 0.15 M CH_3COOH (Reprinted from [43] with permission from Elsevier.)

It is very clear that the classical Wranglen's definition of a dendrite can be applied to all the above observed dendritic forms. According to Wranglen [46], a dendrite consists of a stalk and of branches (primary, secondary, etc.) and resembles a tree. The flat and fern-like dendrites are referred as two-dimensional (2D) dendrites. The dendrite consisting only of the stalk and primary branches are referred as primary (P) dendrite. If the primary branches in turn develop secondary branches, the dendrite is called secondary (S). It is very clear from the above consideration that the needle-like dendrites are composed only of stalk (Figs. 2.9d, e, 2.10d, 2.12c, 2.15b, c, 2.16d, and 2.17a, d). From Figs. 2.9f, 2.10e, 2.12d, and 2.17b, c, it can be seen that the primary type of dendrites was formed by the electrodeposition from the basic nitrate solution. The primary branches are poorly developed, and this type of primary branches can be denoted as

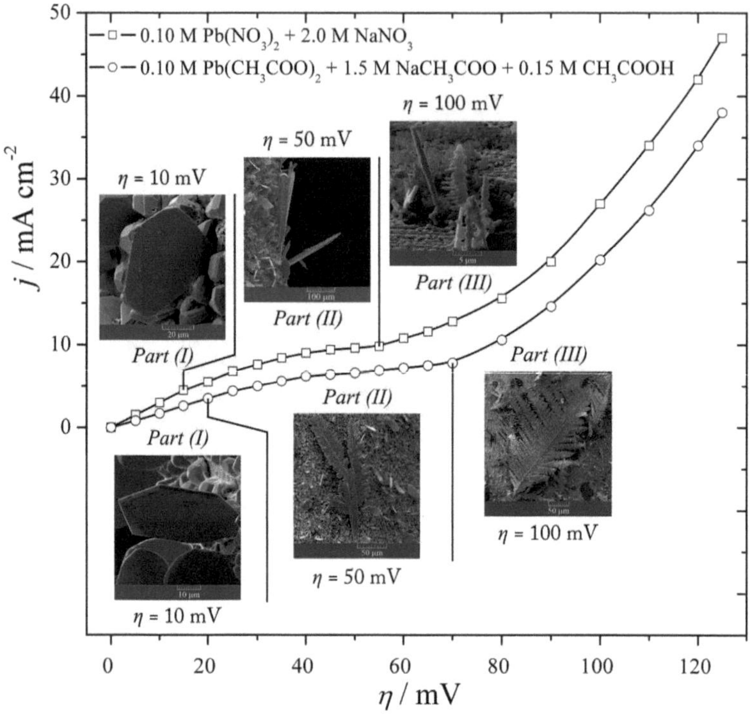

Fig. 2.19 The comparative survey of the polarization curves and the typical morphologies of lead deposits electrodeposited from 0.10 M $Pb(NO_3)_2$ + 2.0 M $NaNO_3$ and 0.10 M $Pb(CH_3COO)_2$ + 1.5 M $NaCH_3COO$ + 0.15 M CH_3COOH (Reprinted from [43] with permission from Elsevier.)

"tooth of saw" [32, 43]. On the other hand, the secondary type of dendrites was formed by the electrodeposition from the complex acetate solution (Fig. 2.18). In this type of dendrite, the primary branches were very well developed, and the secondary type of branches was developed from these primary branches.

From the electrochemical point of view, a dendrite is defined as an electrode surface protrusion that grows under activation control, while electrodeposition to the macroelectrode is predominantly under diffusion control [27, 28, 44, 47–49]. Dendritic growth is initiated at some overpotential belonging to the diffusion part at the polarization curve. The simultaneous formation of the various irregular forms from regular crystals and precursors of dendrites to the both needle-like and fern-like dendrites clearly confirms the fact that the critical overpotential for dendritic growth initiation, η_i, belongs to the diffusion part of the polarization curve, and that it denotes overpotential at which the system enters diffusion control of the electrodeposition. The outer limit of the diffusion layer of the macroelectrode was not disrupted during formation of these irregular morphological forms. The critical overpotential for instantaneous dendritic growth, η_c, is overpotential at which diffusion control becomes complete, and this overpotential corresponds to the inflection point at the polarization curve [32]. The formation of only dendritic forms at an overpotential of 100 mV just confirmed this assumption.

Finally, after the inflection point, the electrodeposition systems remain diffusion controlled ones, and the rapid increase of the current density after the inflection point is a result of the fast growth of dendrites and strong increase of the surface area. Then, using the electrochemical definition of a dendrite, it follows that this sudden and rapid increase of the current density with the increasing overpotential after the inflection point [*(Part (III)* in the polarization curves] can be mainly ascribed to the activation controlled electrodeposition at the tips of the formed dendrites. It is very clear that the tips of both primary and secondary branches contribute to the overall control of electrodeposition process causing the disruption of the outer limit of the diffusion layer of the macroelectrode.

In summary, the following differences in the polarization and morphological characteristics between the complex (acetate) and basic (nitrate) electrolytes were observed:

(a) The increase of the plateau of the limiting diffusion current density and the shift of the inflection point to the higher overpotential of electrodeposition
(b) Formation of more branchy dendrites from the complex electrolyte (the secondary (S) type) in relation to those formed from the basic (nitrate) electrolyte (the primary (P) type).

These differences can be ascribed to the lower exchange current density, j_0, and hence, to the lower rate of the electrochemical process for the complex acetate electrolyte than for the basic nitrate electrolyte. The proof that the processes of complex formation lowers the exchange current density can be found in the comparison with silver electrodeposition processes from the nitrate and ammonium electrolytes [31, 40]. Electrodeposition of silver from the nitrate electrolytes, similar to lead electrodeposition, belongs to the fast electrochemical processes characterized by $j_0 \to \infty$ [7, 27, 28, 30, 47]. On the other hand, ammonium ions makes complex with Ag(I) ions approaching the polarization and morphological characteristics of silver to those of copper [50–56]. In this way, it is enabled the determination of j_0 from the intercept of anodic and cathodic Tafel slopes [57]. For metals like copper and silver (the ammonium complex), the determined exchange current density values, j_0, were smaller than the limiting diffusion current density, j_L ($j_0 < j_L$), and these metals, together with Au, are classified in the group of intermediate metals (Winand's classification of metals) [7]. This considerably decrease of the exchange current density by the complexing process was manifested by the increase of the plateau of the limiting diffusion current density (from 75–175 mV for the basic nitrate electrolyte to 250–700 mV for the ammonium complex electrolyte [31, 40, 58]), and by the strong ramification of silver dendrites and by the change of their shape of those like lead electrodeposited from the nitrate electrolyte (Fig. 2.17) to those similar to copper [50–56, 59, 60].

This analysis enables to make an analogy with lead electrodeposition processes from the basic and complex electrolytes. It is necessary because there is no a precise and reference method for the determination of the exchange current density for the majority of normal metals, such as Pb, Ag, Sn, and Zn, characterized by the extremely large exchange current density. For these metals, the exchange current density values are referred as $j_0 \rightarrow \infty$. For that reason, it is necessary to find some auxiliary criteria for the comparison of electrodeposition systems characteristic by $j_0 \rightarrow \infty$ with point of view of the different j_0 values (i.e., rate of electrochemical process). It is very clear from the above consideration that the method based on the comparison of polarization and morphological characteristics of the different systems can be a very suitable way for this analysis. In this way, it can be concluded that the increase of the plateau of the limiting diffusion current density and the formation of more branchy lead dendrites from the acetate electrolyte than those formed from the nitrate one can be definately ascribed to the lower exchange current density for the acetate complex electrolyte than for the basic nitrate electrolyte.

The another proof that the lead electrochemical system based on the acetate ions has lower exchange current density than the one based on the nitrate ions can be derived on the basis of similarity with the shape of silver and cadmium dendrites. The silver dendrites electrodeposited from the nitrate solution [31, 40] were similar to lead dendrites obtained from the nitrate solution (Figs. 2.9f, 2.10e, 2.12d, and 2.17). The exchange current density for silver (from the silver nitrate solution) was estimated to be between 1 and 7 A/cm^2 [38]. On the other hand, the cadmium dendrites [44] were very similar to the lead dendrites obtained from the acetate solution (Fig. 2.18). The exchange current density for the cadmium system was estimated to be 1.5×10^{-3} A/cm^2 [47, 61], which was for about one thousand and more smaller than the one for silver.

Also, it can be supposed that the degree of the change of the both polarization and morphological characteristics is closely related with the stability of formed complexes. Namely, Pb(II) ions form relatively weak complexes with acetate ions, and the relatively small increase of the plateau of the limiting diffusion current density was observed in the relation to the basic electrolyte. The stability constant for these complexes are $\log K_1 = 2.33$ and $\log \beta_2 = 3.60$. Contrary, the ammonium ions make the strong complex with Ag(I) ions, which is followed by the considerable increase of the plateau of the limiting diffusion current density. The stability constant for these complexes are $\log K_1 = 3.24$ and $\log \beta_2 = 7.24$. Hence, the stronger complex is formed, the stronger effect on the polarization characteristics and surface morphology of powder particles is observed.

The final morphology of electrodeposited lead was determined by overpotential and the composition of solution. However, the complete insight in the mechanism of lead electrodeposition processes and formation of different morphological forms from regular crystals to dendrites cannot be obtained without knowledge of the phenomena related with the process of nucleation. Mostany et al. [13] investigated the nucleation of lead onto vitreous carbon electrodes from the solutions of the different $Pb(NO_3)_2$ concentrations in 1.0 M KNO_3 and they found that the number density of active sites for nucleation increases with overpotential. Also, the number density of active sites is approximately constant at a given overpotential and it is independent of concentration of Pb(II) ions. The type of nucleation changed from progressive nucleation in the most dilute solution to instantaneous nucleation in more concentrated one. Considering the fact that the number of active sites was found to be independent of concentration, the transition from a progressive nucleation regime in dilute solutions to instantaneous at higher concentrations indicates that the nucleation rates increased with concentration [13]. The type of nucleation is related with the initial stages of electrodeposition (less than 0.5 s), while the final surface morphology is not only determined by overpotential and the composition of solution but also to the time of electrodeposition and the type of working electrode. Anyway, the number of formed regular crystals, dendrites, and the other irregular forms depend on the number density of active sites indicating on a close correlation between the nucleation process and the final morphology of electrodeposited lead.

2.4 Correlation Between the Morphology of Lead Deposits and the Crystallographic Structure

Electrodepostion of lead from 0.45 M $Pb(NO_3)_2$ in 2.0 M $NaNO_3$ is the typical representative of the ohmic controlled electrodeposition with the inflection point at an overpotential of 55 mV (Fig. 2.1b).

The regular individual crystals of lead formed in the ohmic control at an overpotential of 20 mV are shown in Fig. 2.20 [62].

Electrodeposition process at an overpotential of 80 mV led to the formation largely irregular individual crystals (precursors of dendrites) as shown in Fig. 2.21. The increase of overpotential of electrodeposition from 80 to 140 mV led to the formation of different dendritic forms (Fig. 2.22). The two types of dendrites were formed by the electrodeposition at 140 mV: needle-like and fern-like dendrites.

Anyway, the two limiting morphological forms were obtained by the processes of lead electrodeposition from the concentrated nitrate solution—the individual

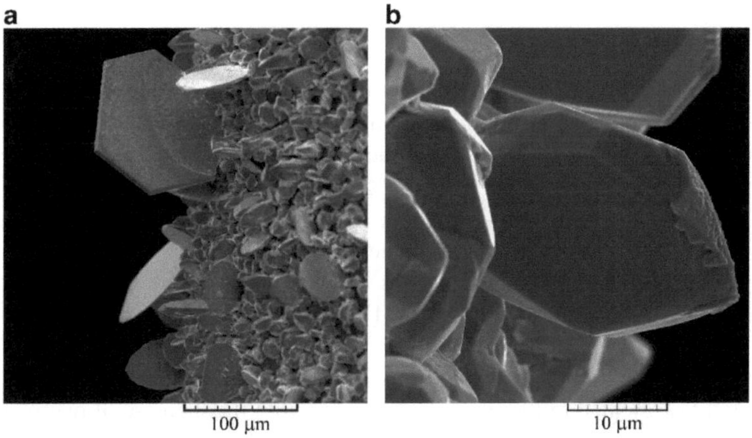

Fig. 2.20 Morphologies of lead deposits obtained from 0.45 M $Pb(NO_3)_2$ in 2.0 M $NaNO_3$ at an overpotential, η, of 20 mV

Fig. 2.21 Morphologies of lead deposits obtained from 0.45 M $Pb(NO_3)_2$ in 2.0 M $NaNO_3$ at an overpotential, η, of 80 mV (Reproduced from [62] by permission of The Royal Society of Chemistry.)

crystals in the ohmic controlled electrodeposition (Fig. 2.20) and the dendrites in the activation controlled electrodeposition at their tips (Fig. 2.22).

Powder particles for the X-ray diffraction (XRD) analysis were obtained by tapping the lead deposits shown in Figs. 2.20, 2.21, and 2.22, and the corresponding XRD patterns are shown in Fig. 2.23. The crystallites of Pb obtained at an overpotential of 20 mV were predominately oriented in the (111) plane. The ratio of crystallites oriented in the other planes was negligible. Aside from the dominant orientation of Pb crystallites in the (111) plane, the presence of Pb crystallites oriented in the (200), (220), (311), and (331) planes was observed in the XRD patterns of powder particles obtained at overpotentials of 80 and 140 mV. The ratio of crystallites oriented in these planes increased with the increase of overpotential of electrodeposition from 80 to 140 mV. Anyway, the strong preferred orientation in the (111) plane was observed in all electrodeposited forms.

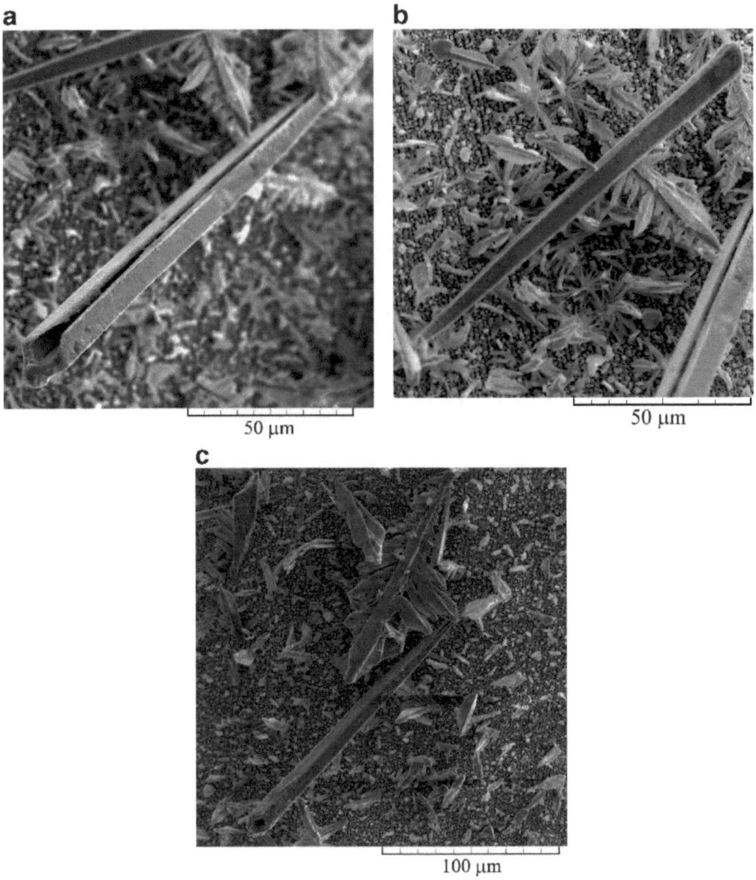

Fig. 2.22 Morphologies of lead deposits obtained from 0.45 M $Pb(NO_3)_2$ in 2.0 M $NaNO_3$ at an overpotential, η, of 140 mV

Analysis of Figs. 2.20, 2.21, 2.22, and 2.23 enables to correlate morphologies of lead deposits with its crystallographic structure. Then, formation of the different morphological forms from the regular crystals to dendrites of the different shape with the strong (111) preferred orientation can be discussed by the following consideration of general characteristics of growth layers in electrodeposition processes: lead crystallizes in a face-centered cubic (FCC-type) lattice, and Fig. 2.24 shows a crystal with a simple cubic lattice and (100), (111), and (110) faces [63]. The planes of the growth layers are the closely packed atom plane of the crystal lattice [46]. The growth layers start from certain active centers or growth centers on the crystal face. The beginning of growth layers is mainly determined by a current density or overpotential. At low current densities and overpotentials, the growth layers are invariably originated in the interior of the crystal faces, and a new layer is not started until the former has reached the edges of the crystal. At higher

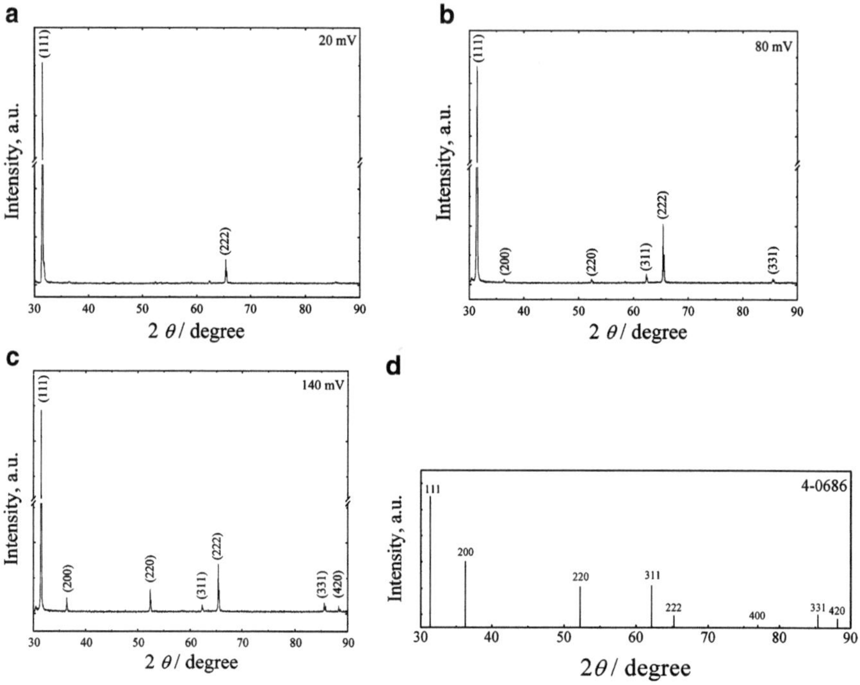

Fig. 2.23 The XRD patterns of the powder particles obtained by tapping the lead deposits electrodeposited at overpotentials of: (**a**) 20 mV, (**b**) 80 mV, and (**c**) 140 mV, and (**d**) for standard for Pb powder

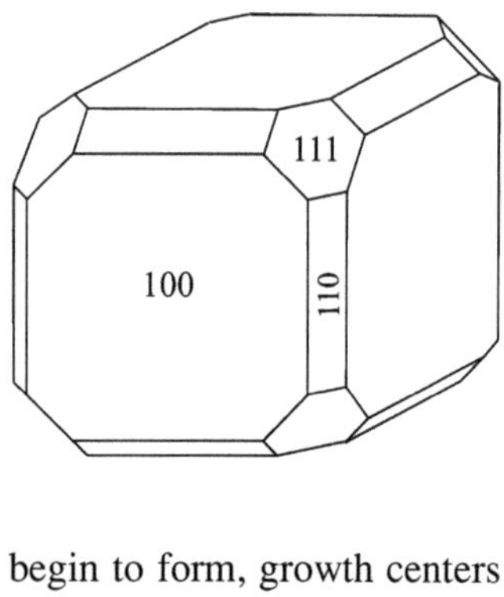

Fig. 2.24 A crystal with a simple cubic lattice and (100), (111), and (110) faces (According to [63]) (Reproduced from [62] by permission of The Royal Society of Chemistry.)

current densities and overpotentials when dendrites begin to form, growth centers are also observed at the edges and corners. In this case, new layers are developed before the forerunners have reached their final size so that at certain times a great many layers may be flowing simultaneously.

Hence, the origin of the growth layers is of both growth centers present in the interior of the crystal faces (center type; these growth layers are formed at low current densities and overpotentials) and growth centers present on the edges and

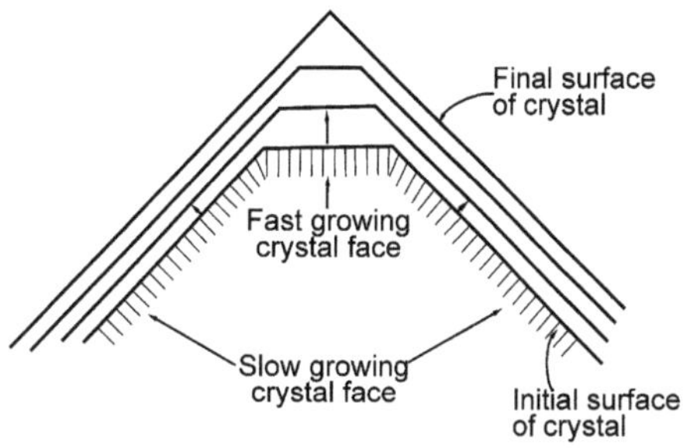

Fig. 2.25 The growth of a crystal illustrating how a fast-growing face grow out of existence, while the slow-growing crystal faces remain (According to [64]) (Reprinted from [65] with permission from Serbian Chemical Society.)

corners (edge and corner type; these growth layers are formed at high current densities and overpotentials).

At constant overpotential, the rates of electrodeposition for a FCC lattice increase in the order of (110) > (100) > (111) [64]. This different deposition rate onto different faces had an important consequence; fast-growing faces tend to grow out of existence and disappear, and slow-growing faces tend to survive [64], as illustrated in Fig. 2.25.

The individual regular crystals shown in Fig. 2.20 are formed at the low overpotential ($\eta = 20$ mV), and hence, the growth centers present in the interior of the crystal faces (center type) are only responsible for their formation. During the electrodeposition process, the layers flowed rapidly to the edges of the crystal. Several growth centers were present on a surface and the growth layers originating from them grew together completely without discontinuities. As result of this, well-developed crystal plates with flat faces and sharp edges and corners, containing angles of 60° and 120°, are obtained [46]. The typical hexagonal crystal plates (the parts in the circles) recorded from the side of regular crystal formed at an overpotential of 20 mV are shown in Fig. 2.26a. On the basis of the analysis of Fig. 2.24 and the XRD analysis of Pb particles electrodeposited at 20 mV (Pb crystallites only oriented in the (111) plane), it is clear that these hexagonal crystal faces correspond to the (111) plane. The regular crystals are only constructed from them, and hence, they represent single crystals of the (111) orientation. Simultaneously, the powder obtained by tapping the deposit electrodeposited at an overpotential of 20 mV is a collection of Pb single crystals which is confirmed by the XRD analysis. Summary, Pb crystallites oriented in the (111) plane are formed at the low overpotential which indicates the affiliation of this plane to the group of slow-growing faces.

On the other hand, irregular crystals, needle-like and fern-like dendrites, are formed by the electrodeposition processes at the high overpotentials, and the both types of growth centers are responsible for their formation and growth. The growth layers of center type appear on the sides of the needle-like form because the current density and overpotential are lower on the sides than at the tip of a dendrite [46]. An inspection of a space lattice model showed that a morphological form may very well be bounded by only (111) faces, as shown in Fig. 2.26b for a side of one irregular crystal. In this way, the existence of the strong (111) preferred orientation

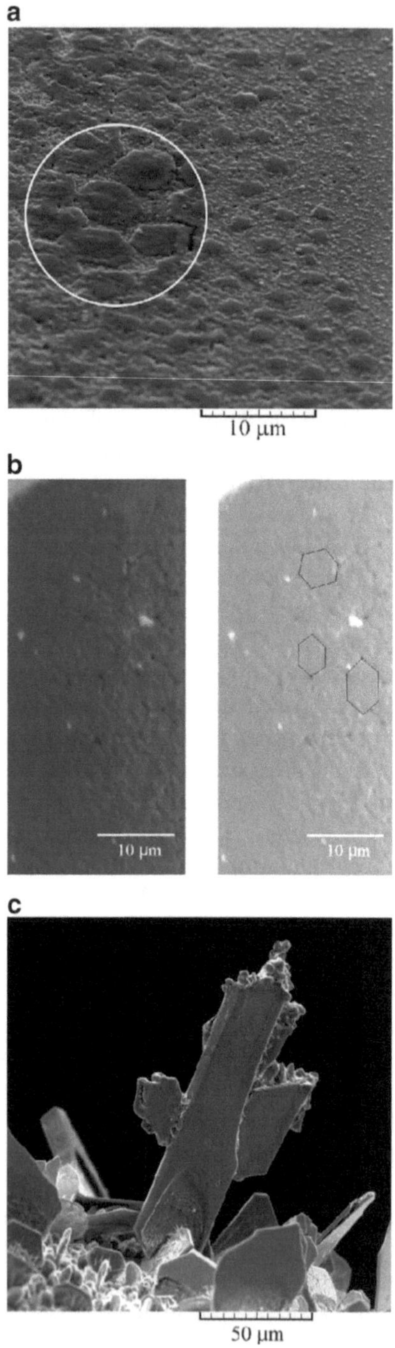

Fig. 2.26 The hexagonal growth layers formed at overpotentials of: (**a**) 20 mV and (**b**) 80 mV, and (**c**) a rounded tip of the irregular form electrodeposited at 80 mV (Reproduced from [62] by permission of The Royal Society of Chemistry.)

of powder particles primarily consisting of irregular crystals (precursors of dendrites) and dendrites is very clear (Figs. 2.21, 2.22, and 2.23). At the higher overpotentials, the growth layers flowed slowly and often stop before they have reached the edges of the crystal face [46]. New layers form which stop still earlier. The growth layers spread from many centers and they merge incompletely making rounded edges and corners, as shown in Fig. 2.26c for the tip of the dendrite "in the construction." Hence, it is obvious that the origin of the crystallites oriented in the (200), (220), (311), and (331) planes is of growth centers at the edges and corners, and these planes can be denoted by fast-growing faces [65]. The larger the overpotential, the larger the ratio of the lead crystallites oriented in the (200), (220), (311), and (331) planes is obtained (Fig. 2.23).

Then, the strong (111) preferred orientation of lead deposits can be summarized as follows: overpotentials and current densities responsible for the crystal growth based on growth centers of "edge and corner type" are higher than those responsible for crystal growth based on growth centers of "center type." It means that the current densities are higher at the tips of the growing forms than at their sides. In the growth process, slow-growing (111) face will survive constructing all elements of dendrites (both stalk and branches), and hence, causing the predominant orientation of Pb crystallites in this plane, as confirmed in Fig. 2.23. Simultaneously, the fast-growing (200), (220), (311) and (331) planes will disappear which explain considerably smaller the presence of Pb crystallites in these planes than in the (111) plane.

As already mentioned, the two groups of dendrites are formed by the electrodeposition at 140 mV: the needle-like and fern-like ones (Fig. 2.22). The needle-like dendrites grow along the cube face diagonal [110] of the FCC lattice [46]. The fern-like dendrites are constructed from the stalk and primary branches (primary (P) type) with an approximate angle of 60° between the stalk and the branches. Then, it is obvious that the dendrites branch along the face diagonals [110] of the unit cube which is the most favored from an energetic point of view [46]. Due to these characteristics, the fern-like lead dendrites are denoted 2D[110]60° in Wranglen's classification of a dendrite.

Also, the effect of the current density distribution [66] should be not excluded in the consideration of growth of dendrites because the current lines are primarily concentrated at the tips of these growing needle-like forms.

As already mentioned, morphology of lead dendrites strongly depended on the type of electrolyte (see Sect. 2.3.2.2). In Figs. 2.27 and 2.28 are shown the typical powder particles obtained by the tapping lead deposits electrodeposited at an overpotential of 100 mV from the nitrate (0.10 M $Pb(NO_3)_2$ + 2.0 M $NaNO_3$; Fig. 2.27) and the complex (0.10 M $Pb(CH_3COO)_2$ + 1.5 M $NaCH_3COO$ + 0.15 M CH_3COOH; Fig. 2.28) electrolytes.

From Figs. 2.27 and 2.28, it is very clear that there is no difference between the lead surface morphologies electrodeposited at the electrode surface (Figs. 2.17 and 2.18) and powder particles obtained after their removal from the electrode surface (Figs. 2.27 and 2.28, respectively) [43].

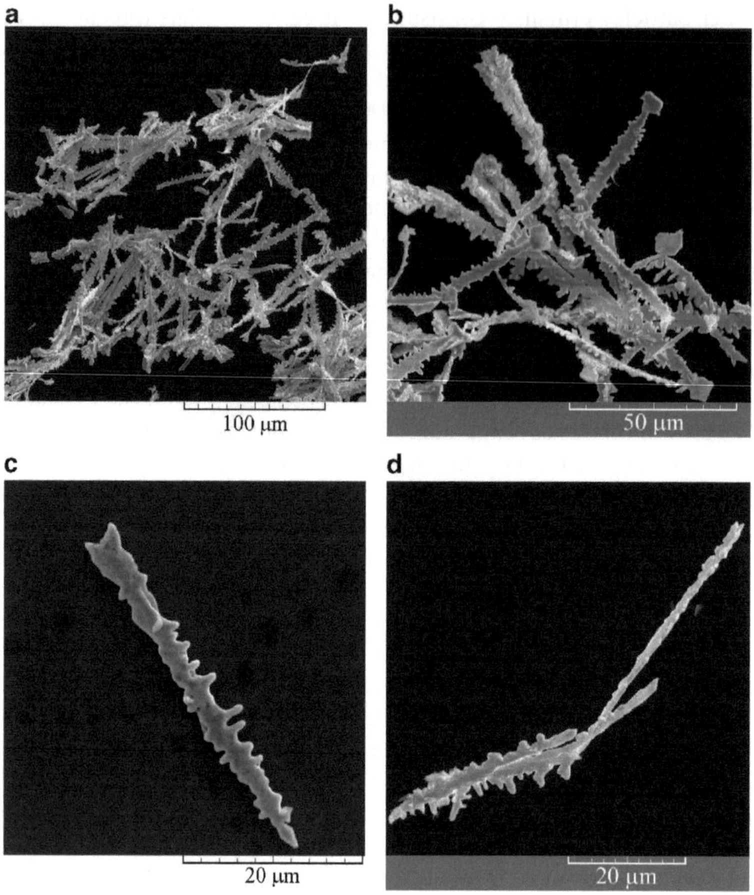

Fig. 2.27 Powder particles obtained by tapping the lead powdered deposit formed by electrodeposition at an overpotential of 100 mV from 0.10 M $Pb(NO_3)_2$ + 2.0 M $NaNO_3$ (Reprinted from [43] with permission from Elsevier and [65] with permission from Serbian Chemical Society.)

Powder particles shown in Figs. 2.27 and 2.28 are used for the XRD analysis, and the corresponding XRD patterns are shown in Fig. 2.29 [65]. It is very clear from Fig. 2.29 that the shown XRD patterns were very similar to each other. Pb crystallites were predominately oriented in the (111) plane. Aside from Pb crystallites oriented in this plane, the presence of crystallites oriented in the (200), (220), (311), and (331) planes was also observed. Anyway, unlike of the completely different morphology of powder particles, the crystallographic structure was practically the same in the both cases (the strong (111) preferred orientation).

The different shapes of Pb deposits, such as octahedral, nonregular crystals, zigzag nanowires, and nanoclusters, were synthesized by the electrodeposition from complex acetate electrolytes in the galvanostatic regime on the different substrates, like Cu, Ti, and Al [67]. The common characteristic of all these morphological

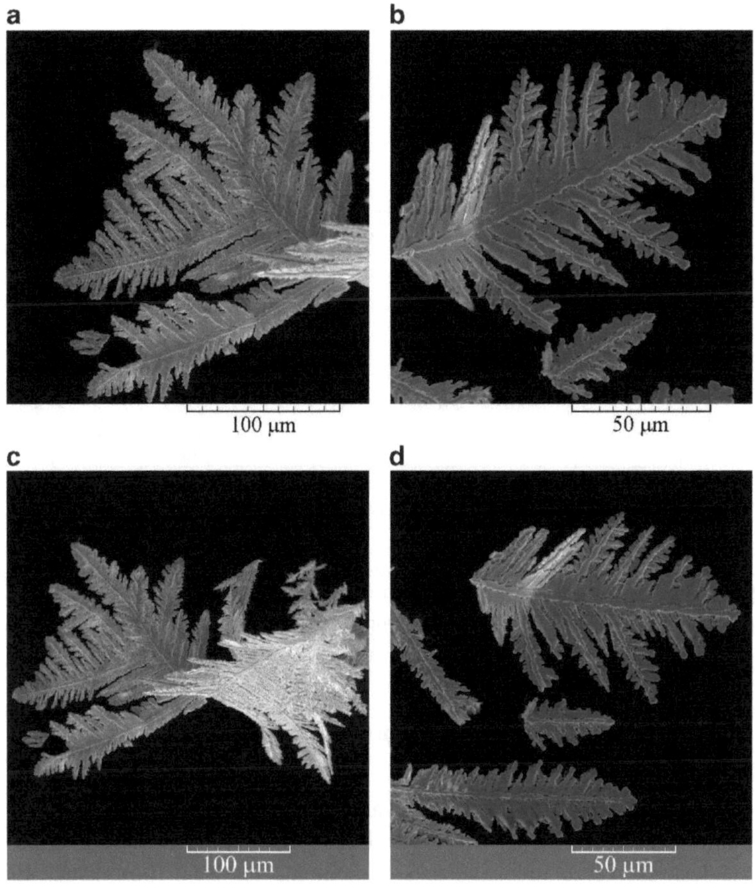

Fig. 2.28 Powder particles obtained by tapping the lead powdered deposit formed by electrodeposition at an overpotential of 100 mV from 0.10 M $Pb(CH_3COO)_2$ + 1.5 M $NaCH_3COO$ + 0.15 M CH_3COOH (Reprinted from [43] with permission from Elsevier and [65] with permission from Serbian Chemical Society.)

forms is the dominant orientation of Pb crystallites in the (111) plane. Electron diffraction (ED) study also confirmed a growth direction in single zigzag lead nanowire along the [1 1 0] axis [67].

It can be concluded from the hitherto consideration that the final morphology of electrodeposited metal (in this case, Pb) is determined by the current density or overpotential applied, composition of solution, the type of working electrode, temperature, etc., while the interior structure of these surface morphologies is only determined by the characteristics and nature of metal, and it does not depend on the electrodeposition conditions. In our case, morphologies of lead deposits were substantially different, but the strong (111) preferred orientation was always obtained.

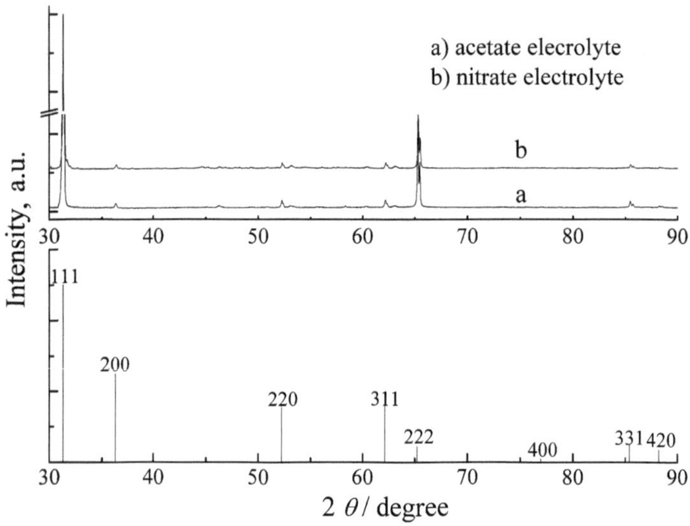

Fig. 2.29 The XRD patterns of the lead powder particles electrodeposited from electrolytes 0.10 M $Pb(CH_3COO)_2$ + 1.5 M $NaCH_3COO$ + 0.15 M CH_3COOH and 0.10 M $Pb(NO_3)_2$ + 2.0 M $NaNO_3$ (Reprinted from [65] with permission from Serbian Chemical Society.)

2.5 Electrodeposition of Pb from Alkaline Electrolytes

Unlike of most acid electrolytes which are toxic, alkaline electrolytes are more appropriate from the environmental standpoint [68]. Additionally, alkaline solutions are less corrosive to plants and equipment. The two most usual alkaline electrolytes used for the deposition of lead are the pyrophosphate and plumbite (based on hydroxide ions) processes. From alkaline electrolytes, a solution containing 0.10 M $Pb(NO_3)_2$ in 2.0 M NaOH is analyzed because it gives the highest rate of deposition which is very suitable for industrial purposes [8].

2.5.1 Polarization Characteristics

Polarization curve for lead electrodeposition from 0.10 M $Pb(NO_3)_2$ in 2.0 M NaOH is shown in Fig. 2.30. In relation to acid electrolytes (Fig. 2.13), the beginning of the lead electrodeposition process is shifted to the higher overpotential for about 5 mV indicating a sensitivity of the reaction of Pb electrodeposition on the type of electrolyte. Also, Pb electrodeposition reaction is sensitive to the substrate type. For example, Pb electrodeposition on SS316 stainless steel occurs at a slightly higher overpotential than at a platinum one [9]. Similar to the polarization curves for Pb electrodeposition from nitrate and acetate electrolytes, this polarization curve

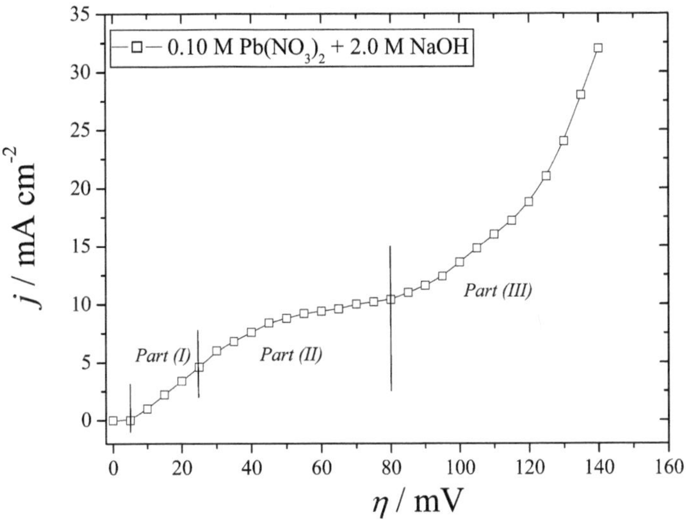

Fig. 2.30 The polarization curve for lead electrodeposition from 0.10 M $Pb(NO_3)_2$ + 2.0 M NaOH

consists of three parts: (a) the ohmic part—*Part (I)*; in the range of overpotentials between 5 and 25 mV, (b) the diffusion part—*Part (II)*; in the range of overpotentials between 25 and 80 mV, and (c) the part characterized by the fast increase of the current with increasing the overpotential after the inflection point ($\eta = 80$ mV)—*Part (III)* [69]. The plateau of the limiting diffusion current density determined in a way described in Sect. 2.3.2.1 was between 39.5 and 80 mV.

2.5.2 Morphological and Crystallographic Analysis of Pb Powdered Deposits

Morphologies of lead deposits obtained by electrodeposition in the potentiostatic mode at overpotentials corresponding to different positions at the polarization curve are shown in Figs. 2.31, 2.32, 2.33, and 2.34.

In Fig. 2.31 are shown the regular crystals obtained in the ohmic control at an overpotential of 20 mV. As expected, the fern-like dendrites were dominant morphological forms obtained during the diffusion controlled electrodeposition. The typical dendrites obtained at an overpotential of 50 mV are shown in Fig. 2.32. Aside from the dendrites, the small irregular crystals (precursors of dendrites) were formed by the electrodeposition at 50 mV (part in the circle in Fig. 2.32a). The more branchy dendrites and the considerably larger their number was formed during electrodeposition at an overpotential of 80 mV, which corresponds to the end of the

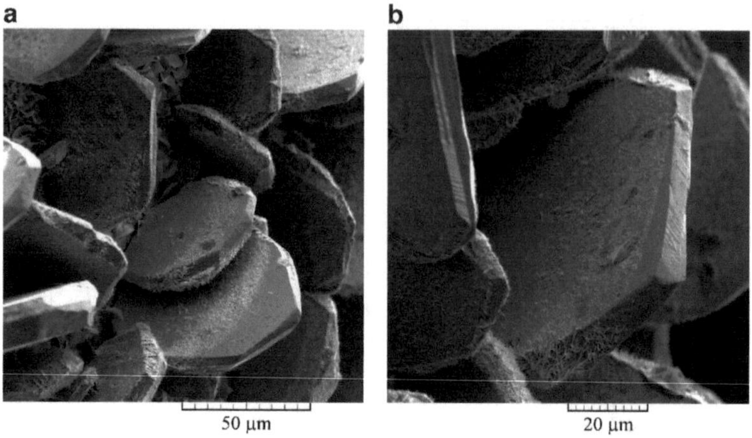

Fig. 2.31 Morphologies of electrodeposited lead obtained in the ohmic controlled electrodeposition at $\eta = 20$ mV from 0.10 M $Pb(NO_3)_2$ + 2.0 M NaOH

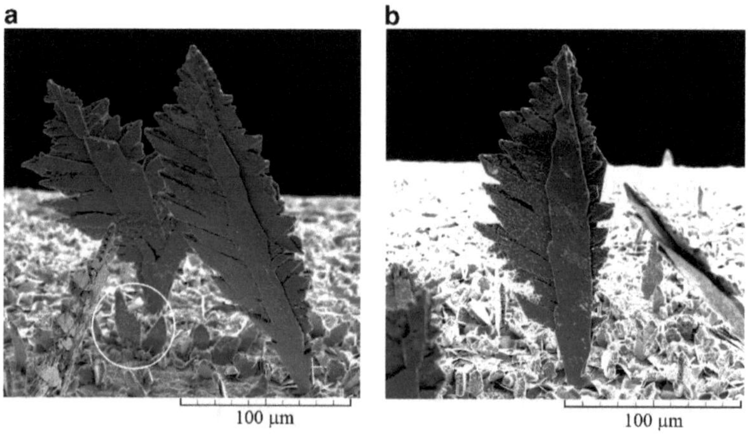

Fig. 2.32 Morphologies of electrodeposited lead obtained at the plateau of the limiting diffusion current density ($\eta = 50$ mV) from 0.10 M $Pb(NO_3)_2$ + 2.0 M NaOH

plateau of the limiting diffusion current density (Fig. 2.33). It is necessary to note that the dendrites were only morphological forms formed during the electrodeposition at this overpotential. The very branchy dendrites, as the only type of the surface morphology, were also obtained at 100 mV (Fig. 2.34) in the zone of the fast growth of the current density with the overpotential.

From Fig. 2.32, it can be clearly seen that the primary (P) type of a dendrite is formed at 50 mV. Increasing overpotential causes branching of dendrites, and the mixture of primary (P) and secondary (S) dendrites was predominately formed at 80 mV (Fig. 2.33). Finally, the dendrites of S type are

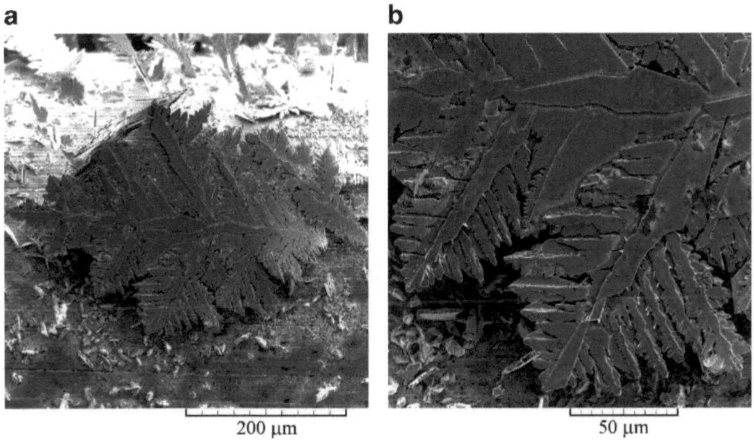

Fig. 2.33 Morphologies of electrodeposited lead obtained at the end of the plateau of the limiting diffusion current density ($\eta = 80$ mV) from 0.10 M $Pb(NO_3)_2$ + 2.0 M NaOH

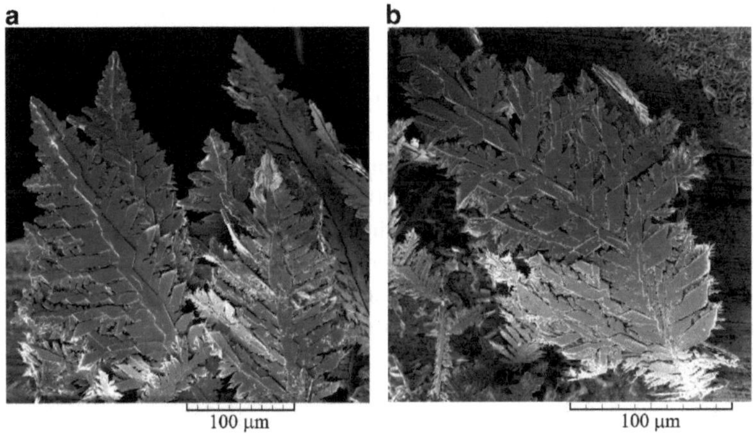

Fig. 2.34 Morphologies of electrodeposited lead obtained in the zone of the fast growth of the current density with the overpotential ($\eta = 100$ mV) from 0.10 M $Pb(NO_3)_2$ + 2.0 M NaOH

predominately formed at 100 mV (Fig. 2.34). Although the (P) and (S) dendrites are the dominant shapes of dendrites formed from this electrolyte, the presence of the tertiary branches developed from some secondary branches is also visible (Fig. 2.33b; the enlarged detail from Fig. 2.33a). The two-dimensional (2D) shape of the fern-like dendrites is clearly visible from Figs. 2.32, 2.33, and 2.34.

It is necessary to note that the shape of regular (single) crystals did not depend on the type of electrolyte. Contrary, the surface morphology of dendrites strongly depended on the type of electrolyte; the dendrites electrodeposited from this

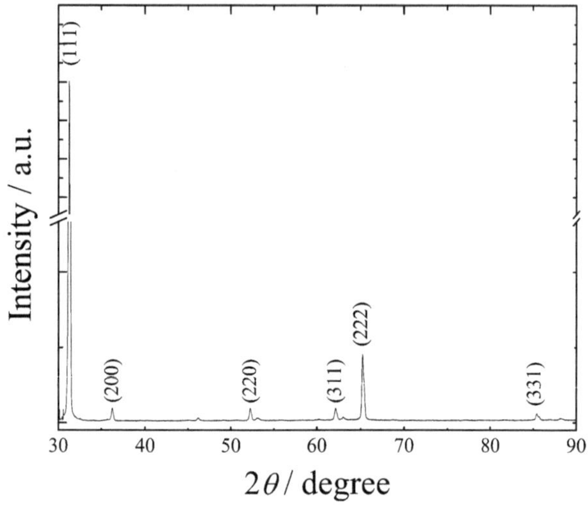

Fig. 2.35 The XRD pattern of the powder particles obtained by the tapping lead deposit electrodeposited at an overpotential of 100 mV from 0.10 M Pb $(NO_3)_2$ + 2.0 M NaOH

alkaline electrolyte are more similar to those obtained from the acetate electrolyte than those obtained from the nitrate ones.

The crystallographic characteristics of Pb crystals were examined by the XRD analysis of powder particles obtained by tapping the lead deposit electrodeposited at an overpotential of 100 mV, and the corresponding XRD pattern is shown in Fig. 2.35. Lead crystallites were predominately oriented in the (111) plane. The ratio of crystallites oriented in the (200), (220), (311), and (331) planes was considerably smaller than those oriented in the (111) plane. The size of crystallites is determined by Williamson–Hall method and was estimated to be 36.5 nm. The quantitative analysis of this Pb powder showed its 100 % purity indicating the significance of the electrolysis processes in the production of powdered forms of high purity.

It is confirmed that the crystallographic (interior) structure of Pb crystals is determined by the nature of lead, and it is not associated with the type of used electrolyte, and hence, with the surface morphology.

2.5.3 Final Discussion of the Presented Results

The crystallographic aspects of formation of powder particles cannot be analyzed separately than electrochemical ones. Schematic diagram illustrating the overpotential dependent formation and growth of fern-like dendrites is given in Fig. 2.36. At the low overpotential (I. in Fig. 2.36), current lines are equally distributed, and regular crystals like those shown in Fig. 2.31 are formed. Crystal growth based only on the growth centers present in the interior of the crystal faces is responsible for formation of the regular crystals, which is confirmed by the XRD

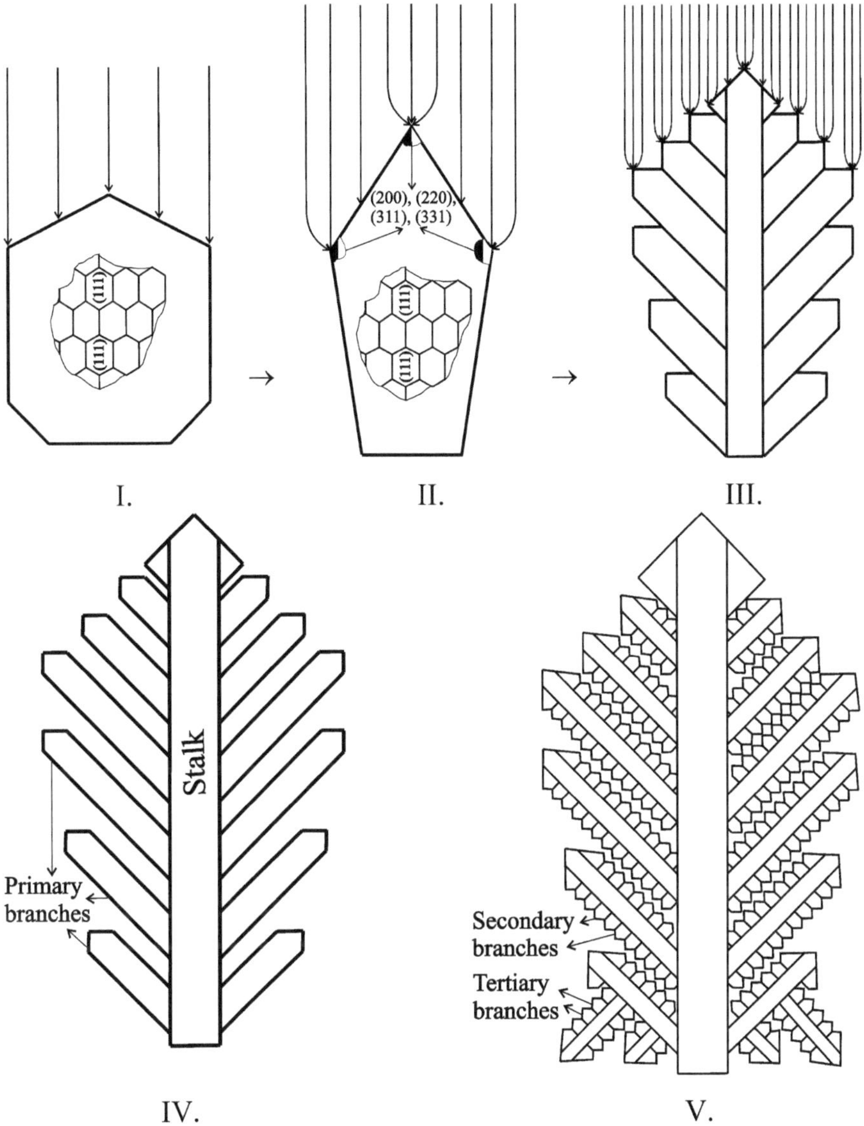

Fig. 2.36 Schematic diagram illustrating the overpotential dependent growth of fern-like dendrites

analysis of this surface morphology type. As already mentioned, these regular crystals represent of single crystals of the (111) orientation [62]. Increasing overpotential leads to a concentration of current lines in the first place at the tips and corners of growing forms causing faster growth on them than at the sides of crystals and in the vicinity of the electrode surface (the current density distribution

Table 2.1 The ranges of overpotentials belonging to the ohmic and diffusion parts of the polarization curves, as well as the length of the plateaus of the limiting diffusion current density for the given electrolytes

Solution for lead electrodeposition	The ohmic part of the polarization curve (mV)	The diffusion part of the polarization curve (mV)	The plateau of the limiting diffusion current density (mV)
0.10 M $Pb(NO_3)_2$ + 2.0 M $NaNO_3$	0–15	15–55	28.5–55
0.10 M $Pb(CH_3COO)_2$ + 1.5 M $NaCH_3COO$ + 0.15 M CH_3COOH	0–20	20–70	33–70
0.10 M $Pb(NO_3)_2$ + 2.0 M NaOH	5–25	25–80	39.5–80

effect) [66]. Hence, the current densities are higher at the tips and corners of the growing forms (the growth based on edge and corner growth centers) than at their sides (the growth based on growth centers of the center type). As a result of this, irregular crystals or precursors of dendrites are formed (II. in Fig. 2.36). The origin of Pb crystallites oriented in the (200), (220), (311), and (331) planes is just of crystal growth on the tips and corners of growing crystals. Simultaneously, the sides of the crystals are constructed from Pb crystallites oriented in the (111) plane [62]. The typical irregular crystals (precursors of dendrites) obtained from this alkaline electrolyte are shown in the circle in Fig. 2.32a. The current density, and hence, concentration of current lines at the growing crystals increases with increasing the overpotential leading to the appearance of dendrites at the higher overpotentials (III. in Fig. 2.36). Please note that for reasons of simplification, the only one part of current lines is illustrated in Fig. 2.36. During the growth of dendrites, the current densities increase exponentially. The larger the applied overpotential, the faster exponential increase during electrodeposition is observed [30]. Due to the increase of the number of growth centers at which a concentration of current lines takes place, branching of dendrites is observed resulting in the development of primary, secondary, and tertiary branches in them (IV. and V. in Fig. 2.36 and Figs. 2.32, 2.33, and 2.34).

For the alkaline electrolyte containing 0.10 M $Pb(NO_3)_2$ in 2.0 M NaOH, both ohmic and diffusion parts at the polarization curve were shifted to higher overpotentials in relation to those observed for the acid electrolytes based on nitrate and acetate ions (for the same concentration of 0.10 M Pb(II) ions). In Table 2.1 are given the ranges of overpotentials belonging to the ohmic and diffusion control, as well as the lengths of the plateau of the limiting diffusion current density for the nitrate, acetate, and alkaline electrolytes. The increasing trend in row nitrate < acetate < hydroxide ions is clearly visible from Table 2.1.

From the morphological point of view, morphologies of Pb dendrites (Figs. 2.32, 2.33, and 2.34) are more similar to those obtained from the acetate than from the nitrate electrolytes. The ramified dendrites belonging to the secondary (S) type are formed from both acetate and alkaline electrolytes. Unlike of this type of a dendrite, the needle-like and primary (P) dendrites were only formed from the nitrate electrolyte [30, 32, 43, 62, 65].

The common characteristic of the acetate and alkaline electrolytes is their affiliation to the group of complex electrolytes. In the case of the acetate electrolyte, the increase of the plateau of the limiting diffusion current density and ramification of dendrites in relation to the nitrate electrolyte was ascribed to the process of the complex formation of Pb(II) ions with the acetate ions [43]. One of the consequences of the process of complex formation was a lowering of the exchange current density, j_0, for electrodeposition process, and hence, the decrease of a rate of electrochemical process. For the alkaline electrolytes based on hydroxide ions, the reduction of lead occurs entirely through the complex species and not through Pb(II) ions [8]. Lead makes different complexes with hydroxide ions, such as $[PbOH]^+$, $[Pb(OH)_2]$, $[Pb(OH)_3]^-$, $[Pb_2(OH)]^{3+}$, $[Pb_3(OH)_4]^{2+}$, $[Pb_4(OH)_4]^{4+}$, and $[Pb_6(OH)_8]^{4+}$ [70, 71]. In the very alkaline electrolyte solution, pH over 12 (as in this case), $[Pb(OH)_3]^-$ complex becomes a major species in the solution [72–74], and probably the reduction of lead occurs entirely through this complex.

Hence, as a result of the processes of the complex formation, the increase of both the plateau of the limiting diffusion current density and ramification of dendrites is observed. It is very clear that this is very good auxiliary diagnostic criteria for the comparison of electrodeposition systems characterized by the large exchange current density ($j_0 \to \infty$) with point of view of the different j_0 values (i.e., rate of electrochemical process) for which there is no precise and reference method for the determination of j_0 values.

The degree of the change of the both polarization and morphological characteristics can be correlated with the stability of formed complexes. Namely, Pb(II) ions form relatively weak complexes with acetate ions. The stability constants for the acetate complexes are $\log K_1 = 2.33$ and $\log \beta_2 = 3.60$. On the other hand, Pb makes stronger complex with hydroxide ions than with the acetate ions. For example, the stability constant of $[Pb(OH)_3]^-$ complex, which is primary species in very alkaline solutions (2 M NaOH), is $\log \beta_3 = 13.3$ [70, 74]. This constant is about 5×10^9 times larger than the constant for the acetate complex. Anyway, the stronger complex is formed, the stronger effect on the polarization characteristics and surface morphology of electrodeposited metal is observed. In this case, this is manifested through the increase of the plateau of the limiting diffusion current density and formation of larger number of dendrites of S type, as well as by the appearance of tertiary branches in the dendrites of S type from the alkaline electrolyte in relation to the acetate one.

2.6 Conclusions

A detailed mechanism of lead electrodeposition is presented in this chapter. Unlike of long-time belief that electrodeposition of metals characterized by high exchange current densities is the diffusion controlled process in the whole range of overpotentials, it is shown that lead electrodeposition occurs in the conditions of the mixed ohmic-diffusion, or even, the full ohmic control. With the increasing concentration of Pb(II) ions in the solution, the ratio of the ohmic control to the total control of electrodeposition process increased. At the higher concentrations of Pb (II) ions, the electrodeposition process became completely ohmic controlled.

The polarization characteristics of lead were correlated with morphology of electrodeposited lead. The regular crystals were obtained in the ohmic control. The XRD analysis showed that they represent single crystals of the (111) preferred orientation. The shape of single crystals did not depend on the type of used electrolyte.

The shape of dendrites formed in the diffusion control strongly depends on the type of electrolyte. Dendrites formed from the complex electrolytes (acetate and alkaline ones) are more branchy than those obtained from the basic (nitrate) electrolyte. The dendrites obtained from the basic electrolyte largely belonged to the primary (P) type. Contrary, dendrites obtained from the complex electrolytes were of the secondary (S) type. Unlike of the strong effect of the type of electrolyte on morphology of dendrites, all dendritic forms were predominately oriented in the (111) plane.

For all examined electrolytes, the fast increase of the current density with increasing the overpotential after the inflection point was ascribed to the activation control at the tips of formed dendrites.

Also, the polarization characteristics and the morphology of lead electrodeposited from the complex electrolytes were correlated with the stability constant of lead complex.

Acknowledgments The authors are grateful to Dr. Goran Branković for SEM analysis of investigated systems, to Dr. Vesna Maksimović for the XRD analysis of the powder particles, as well as to Prof. Dr. Predrag Živković for the digital simulations.

The work was supported by the Ministry of Education, Science and Technological Development of the Republic of Serbia under the research project: "Electrochemical synthesis and characterization of nanostructured functional materials for application in new technologies" (No. 172046).

References

1. Pavlov D (1993) J Power Sources 42:345
2. Rashkova B, Guel B, Potzschke RT, Staikov G, Lorenz WJ (1998) Electrochim Acta 43:3021
3. Ehlers C, Konig U, Staikov G, Schultze JW (2002) Electrochim Acta 47:379
4. Avellaneda CO, Napolitano MA, Kaibara EK, Bulhoes LOS (2005) Electrochim Acta 50:1317
5. http://www.nuclead.com/leadpowderapps.html.
6. Cherevko S, Xing X, Chung C-H (2011) Appl Surf Sci 257:8054
7. Winand R (1994) Electrochim Acta 39:1091
8. Carlos IA, Malaquias MA, Oizumi MM, Matsuo TT (2001) J Power Sources 92:56
9. Wong SM, Abrantes LM (2005) Electrochim Acta 51:619
10. Doulakas L, Novy K, Stucki S, Comninellis Ch (2000) Electrochim Acta 46:349

11. Scharifker B, Hills G (1983) Electrochim Acta 28:879
12. Mostany J, Parra J, Scharifker BR (1986) J Appl Electrochem 16:333
13. Mostany J, Mozota J, Scharifker BR (1984) J Electroanal Chem 177:25
14. Popov KI, Krstajić NV, Pantelić RM, Popov SR (1985) Surf Technol 26:177
15. Exposito E, Gonzalez-Garcıa J, Bonete P, Montiel V, Aldaz A (2000) J Power Sources 87:137
16. Popov KI, Stojilković ER, Radmilović V, Pavlović MG (1997) Powder Technol 93:55
17. Ghergari L, Oniciu L, Muresan L, Pantea A, Topan VA, Ghertoiu D (1991) J Electroanal Chem 313:303
18. Muresan L, Oniciu L, Froment M, Maurin G (1992) Electrochim Acta 37:2249
19. Muresan L, Oniciu L, Wiart R (1993) J Appl Electrochem 23:66
20. Muresan L, Oniciu L, Wiart R (1994) J Appl Electrochem 24:332
21. Hazza A, Pletcher D, Wills R (2004) Phys Chem Chem Phys 6:1773
22. Pletcher D, Wills R (2004) Phys Chem Chem Phys 6:1779
23. Ghali E, Girgis M (1985) Metall Mater Trans B 16:489
24. Carlos IA, Siqueira JLP, Finazzi GA, de Almeida MRH (2003) J Power Sources 117:179
25. Popov KI, Živković PM, Grgur BN (2007) Electrochim Acta 52:4696
26. Popov KI, Živković PM, Krstić SB, Nikolić ND (2009) Electrochim Acta 54:2924
27. Popov KI, Živković PM, Nikolić ND (2010) The effect of morphology of activated electrodes on their electrochemical activity. In: Djokić SS (ed) Electrodeposition: Theory and Practice, Modern Aspects of Electrochemistry, vol 48. Springer, pp 163–213
28. Popov KI, Nikolić ND (2012) General Theory of Disperse Metal Electrodeposits Formation. In: Djokić SS (ed) Electrochemical Production of Metal Powders, Modern Aspects of Electrochemistry, vol. 54. Springer, pp 1–62
29. Popov KI, Nikolić ND, Živković PM, Branković G (2010) Electrochim Acta 55:1919
30. Nikolić ND, Branković G, Lačnjevac U (2012) J Solid State Electrochem 16:2121
31. Jović VD, Nikolić ND, Lačnjevac UČ, Jović BM, Popov KI (2012) Morphology of Different Electrodeposited Pure Metal Powders. In: Djokić SS (ed) Electrochemical Production of Metal Powders, Modern Aspects of Electrochemistry, Vol. 54. Springer, pp 63–123
32. Nikolić ND, Popov KI, Živković PM, Branković G (2013) J Electroanal Chem 691:66
33. Levich VG (1962) Physicochemical Hydrodynamics. Prentice – Hall, Inc: Englewood Cliffs, Ch. 2, NJ
34. Blagojević NS, Maksimović MD, Popov KI (1978) Chem Eng J 16:35
35. Lorenz W (1954) Z Electrochem 58:912 (in German)
36. Mattsson BE, Bockris JO'M (1959) Trans Faraday Soc 55:1586
37. Newman JS (1973) Electrochemical Systems. Prentice-Hall, Inc. Engelwood Cliffs, N. J.
38. Fetter K (1967) Electrochemical Kinetics. Moscow, Khimiya, (in Russian)
39. Nikolić ND, Popov KI, Pavlović LjJ, Pavlović MG (2007) Sensors 7:1
40. Djokić SS, Nikolić ND, Živković PM, Popov KI, Djokić NS (2011) ECS Trans 33:7
41. Mandke MV, Han S-H, Pathan HM (2012) CrystEngComm 14:86
42. Ni Y, Zhang Y, Zhang L, Hong J (2011) CrystEngComm 13:794
43. Nikolić ND, Vaštag DjDj, Živković PM, Jokić B, Branković G (2013) Adv Powder Technol 24:674
44. Popov KI, Krstajić NV, Čekerevac MI (1996) The mechanism of formation of coarse and disperse electrodeposits. In: White RE, Conway BE, Bockris JO'M (eds) Modern Aspects of Electrochemistry, vol 30. Plenum Press, New York, pp 261–312
45. Popov KI, Pavlović MG, Jovićević JN (1989) Hydrometallurgy 23:127
46. Wranglen G (1960) Electrochim Acta 2:130
47. Popov KI, Djokić SS, Grgur BN (2002) Fundamental aspects of electrometallurgy. Kluwer Academic/Plenum Publishers, New York
48. Despić AR, Popov KI (1972) Transport controlled deposition and dissolution of metals. In: Conway BE, Bockris JO'M (eds) Modern Aspects of Electrochemistry, vol 7. Plenum Press, New York, pp 199–313

49. Diggle JW, Despić AR, Bockris JO'M (1969) The mechanism of the dendritic electrocrystallization of zinc. J Electrochem Soc 116:1503
50. Nikolić ND, Popov KI, Pavlović LjJ, Pavlović MG (2006) J Electroanal Chem 588:88
51. Nikolić ND, Pavlović LjJ, Pavlović MG, Popov KI (2008) Powder Technol 185:195
52. Nikolić ND, Popov KI (2010) Hydrogen co-deposition effects on the structure of electrodeposited copper. In: Djokić SS (ed) Electrodeposition: Theory and Practice, Modern Aspects of Electrochemistry, vol 48. Springer, pp 1–70
53. Nikolić ND (2012) Porous Copper Electrodes Formed by the Constant and the Periodically Changing Regimes of Electrolysis. In: Djokić SS (ed) Electrochemical Production of Metal Powders, Modern Aspects of Electrochemistry, Vol. 54, Springer, pp 187–249
54. Nikolić ND, Branković G, Pavlović MG (2012) Powder Technol 22:271
55. Orhan G, Hapci G (2010) Powder Technol 201:57
56. Orhan G, Gezgin GG (2012) J Serb Chem Soc 77:651
57. Živković PM, Nikolić ND, Gvozdenović M, Popov KI (2009) J Serb Chem Soc 74:291
58. Popov KI, Krstajić NV, Popov SR (1983) Surf Technol 20:203
59. Nikolić ND, Popov KI (2012) Electrodeposition of Copper Powders and Their Properties. In: Djokić SS (ed) Electrochemical Production of Metal Powders, Modern Aspects of Electrochemistry, Vol. 54, Springer, pp 125–185
60. Nikolić ND, Pavlović LjJ, Krstić SB, Pavlović MG, Popov KI (2008) Chem Eng Sci 63:2824
61. Popov KI, Grgur BN, Stojilković ER, Pavlović MG, Nikolić ND (1997) J Serb Chem Soc 62:433
62. Nikolić ND, Maksimović VM, Branković G, (2013) RSC Adv 3:7466
63. Milchev A (2002) Electrocrystallization, Fundamentals of Nucleation and Growth. Kluwer Academic Publishers, Boston/Dordrecht/London, p. 24
64. Bockris JO'M, Reddy AKN, Gamboa-Aldeco M (2000) Modern Electrochemistry 2A, Fundamentals of Electrodics, Kluwer Academic / Plenum Publishers, New York, p. 1333
65. Nikolić ND, Maksimović VM, Branković G, Živković PM, Pavlović MG (2013) J Serb Chem Soc 78:1387
66. Popov KI, Živković PM, Nikolić ND (2011) J Serb Chem Soc 76:805
67. Yao C-Z, Liu M, Zhang P, He X-H, Li G-R, Zhao W-X, Liu P, Tong Y-X (2008) Electrochim Acta 54:247
68. Jordan M (2010) Electrodeposition of lead and lead alloys, in: M. Schlesinger, M. Paunovic (Eds.), Modern Electroplating, Fifth Edition, John Wiley & Sons, Inc., p. 251
69. Nikolić ND, Vaštag DjDj, Maksimović VM, Branković G, Trans Nonferrous Met Soc China, in press
70. NIST Standard Reference Database 46, NIST Critically Selected Stability Constants of Metal Complexes Database, version 3.0, compiled by R.M. Smith and A.E. Martell, U.S. Department of Commerce, National Institute of Standards and Technology, Gaithersburg, MD 20899, USA
71. Kratgen J (1978) Atlas of Metal-Ligand Equilibria in Aqueous Solution. Series in Analitical Chemistry, Ellis Horwood, Chichester. UK
72. Cukrowska E, Cukrowski I (1998) Talanta 47:1175
73. Pereira M, Mantas PQ (1998) J Eur Ceram Soc 18:565
74. Wang Y, Chai L, Chang H, Peng X, Shu Y (2009) Trans Nonferrous Met Soc China 19:458.

Chapter 3
Electrophoretic Deposition of Ceramic Coatings on Metal Surfaces

Vesna B. Mišković-Stanković

3.1 Introduction

Electrophoretic deposition (EPD) is a special colloidal processing technique that uses the electrophoresis mechanism for the movement of charged particles suspended in a solution under an electric field and to deposit them in an ordered manner on a substrate to develop thin and thick films, coatings, and free-standing bodies. EPD emerges as a method of choice due to formation of uniform thin or multilayer films of controlled thickness and morphology, even on substrates of complex shape [1–13]. The thickness and structure of deposited films can be easily controlled by changing the deposition parameters, i.e., applied voltage, deposition time, and particles concentration in suspension, while mechanical characteristics are good [14]. The interest in EPD is based not only on its high versatility to be used with different materials but also because EPD is a cost-effective technique usually requiring simple processing equipment and infrastructure [1]. Electrophoretic deposition is gaining increasing attention both in science and industry, due to novel applications in the processing of advanced ceramic materials.

This chapter represents an overview of mechanism of electrodeposition of alumina, boehmite, monetite, brushite, hydroxyapatite, hydroxyapatite/lignin, silver/hydroxyapatite, and silver/hydroxyapatite/lignin coatings on metal surfaces from aqueous suspensions using sol–gel routes. The determination of the optimal deposition parameters, i.e., applied voltage, deposition time, bath temperature, and particles concentration in suspension is also discussed. The characteristics of deposited ceramic coatings determined by different techniques are described in detail. It was shown that electrodeposition is a powerful method to design the thickness, porosity,

V.B. Mišković-Stanković (✉)
Faculty of Technology and Metallurgy, University of Belgrade, Karnegijeva 4,
11120 Belgrade, Serbia
e-mail: vesna@tmf.bg.ac.rs

S.S. Djokić (ed.), *Electrodeposition and Surface Finishing: Fundamentals and Applications*, Modern Aspects of Electrochemistry 57,
DOI 10.1007/978-1-4939-0289-7_3,

thermal stability, structure, and morphology of different ceramic coatings on metal surfaces aimed for different applications.

Moreover, the electrochemical synthesis of nanosized monetite and hydroxyapatite powders, aimed for electrophoretic deposition of monetite and hydroxyapatite coatings, is reviewed. It was demonstrated that using electrochemical synthesis under galvanostatic conditions, the nanosized monetite and hydroxyapatite powders were obtained, and the influence of applied current density and pH value on the phase composition, crystallite size, morphology, and thermal characteristics of obtained powders were proved.

3.2 Electrophoretic Deposition of Ceramic Coatings

The majority of the work on electrophoretic deposition has been done in non-aqueous medium [15–18]. The reason was that nonaqueous medium had much better performances in respect to aqueous medium (e.g., low viscosity, good chemical stability). On the other hand, one of the major problems associated with use of organic solution is environmental problem. That was the reason why electrophoretic deposition is performed in aqueous solution whenever is possible [11, 19–21]. The use of sol–gel routes in ceramic processing has many advantages, such as greater purity, higher homogeneity, and ultrafine grain size distribution, in comparison to conventional powder-based processing techniques. Moreover, as the sol contains very sinter-active ceramic particles on a nanometer scale, the sintering temperatures can be lowered by several hundred degrees [22].

The necessary condition that provides successful electrophoretic deposition is a stable suspension/sol, where the particles have a high zeta potential while the ionic conductivity of the suspension is kept at a low value [23]. Factors influencing the EPD process are electrical conditions (voltage and time) and parameters related to suspension (particle charging, solid loading, dispersants, suspension viscosity, particle size distribution). According to proposed mechanism [24–27], the deposition process occurs in several steps. Charged particles attract oppositely charged ions (e.g., counterions) around the particles. In the case of cataphoretic deposition positively charged particles migrate toward the cathode. The rate of migration that the particles can achieve, v, depends on applied electric field, E, suspension viscosity, η, particle radius, r, and particle charge, z, and is given by equation:

$$v = zE/6\pi\eta r \tag{3.1}$$

indicating the linear dependence of migration rate upon applied electric field. Because the particles are close enough to the cathode, attractive forces dominate and coagulation/deposition occurs.

In the case of cataphoretic deposition of water-born ceramic coatings on a metal electrode, the primary process is the OH^- ions generation and hydrogen evolution on the cathode by H_2O discharge:

$$2H_2O + 2e^- \rightarrow H_2 + 2OH^- \quad (pH\ 7 - 14) \tag{3.2}$$

followed by electrocoagulation of the ceramic particles at the cathode surface by neutralization of positively charged groups with electrochemically generated OH^- ions. The deposition of a coating will occur when the hydroxyl ion concentration (pH of system) achieves a critical value:

$$R^+ + OH^- \rightarrow \text{film} + H_2O \tag{3.3}$$

At the same time, the hydrogen evolves on the cathode (Eq. (3.2)) and oxygen on the anode:

$$2H_2O \rightarrow 4H^+ + O_2 + 4e^- \tag{3.4}$$

Evolved hydrogen on the cathode goes out through the coating, leaving the vacancies inside the deposited film and causing its porous structure.

Electrophoretic deposition is usually applied in order to obtain ceramic, composite ceramic/ceramic, and polymer/ceramic coatings for various applications: biomedical implants [1–5, 28–51], photocatalysts [52–55] often used for wastewater treatment [52], solid oxide fuel cells (SOFC) [56–61], thermal protective coatings for high temperature devices, such as steam turbines, gas turbines and combustion chambers [62–65], zeolites [66], and solar cells [67–71].

3.3 Alumina Coatings

In order to obtain the alumina coatings of low porosity on steel, the alumina aqueous suspension of high zeta potential and low viscosity has to be prepared, because both parameters induce the larger amount of particles which reach the electrode surface, causing the thicker alumina films. The statistical analysis of the results obtained from optical microscopy coupled by quantitative image analysis gives information about morphology of alumina films (pore number, percentage of film surface covered by pores, and mean pore diameter) and enables the determination of the optimal deposition parameters, i.e., applied voltage, deposition time, and concentration of alumina powder in suspension, to obtain the alumina coatings of the highest thickness and the lowest porosity [11]. The solid phase content of alumina powder in suspension was 10, 20, and 30 wt%. The alumina coatings were electrodeposited on steel from the aqueous suspensions of alumina powder at different values of constant voltage between 30 and 90 V, during different deposition time between 1 and 10 min and over temperature range of 20–40 °C It was shown that increase in applied deposition voltage over 100 V decreases the alumina film thickness and increases the film porosity due to larger amount of evolved hydrogen on cathode (Eq. (3.2)) leaving more vacancies in the film and causing more porous structure. The effect of applied deposition voltage and concentration

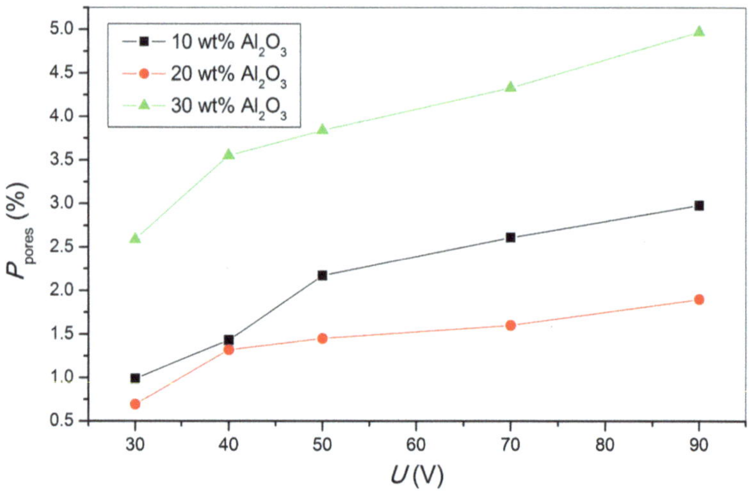

Fig. 3.1 The percentage of alumina film surface covered by pores vs. applied deposition voltage for different concentration of alumina powder in suspension (deposition time 10 min) (reprinted from [11] with permission from Elsevier)

of alumina powder in aqueous suspension on the percentage of alumina film surface covered by pores is shown in Fig. 3.1.

It can be observed that the films obtained from 20 wt% of alumina powder in suspension have the lowest porosity for all applied deposition voltages. This can be explained by the optimal ratio of number of suspended particles and number of water molecules, which enables the best particle dispersion and consequently, the best particle packing on the cathode during deposition process. The films obtained from 10 wt% and 30 wt% suspension have greater porosity due to larger amount of evolved hydrogen (in the case of 10 wt% suspension) or due to agglomerates detected (in the case of 30 wt% suspension). On the other hand, the increase in applied voltage increases the percentage of film surface covered by pores due to larger amount of evolved hydrogen on the cathode at higher voltages. The minimum values of the alumina film surface covered by pores (Fig. 3.1), as well as pore number and mean pore diameter [11], were obtained for deposition voltage of 30 V and 20 wt% of alumina powder in aqueous suspension.

The effect of deposition time on alumina film thickness and morphology was investigated for coatings electrodeposited from 20 wt% suspension at 30, 40, and 50 V [27]. At lower applied voltages (30 and 40 V), the increase in deposition time increases the alumina film thickness (Fig. 3.2) due to much more particles that reach the cathode and discharge on it.

For longer deposition time, at higher applied voltages (50 V), the film thickness achieves a maximum and then decreases due to larger amount of evolved hydrogen, which accumulates on the cathode and causes more vacancies in the deposited film. This explains the greater values of mean pore diameter (Fig. 3.3), as well as percentage of alumina film surface covered by pores and mean pore area [27], for alumina coatings electrodeposited at higher voltages and for longer deposition time.

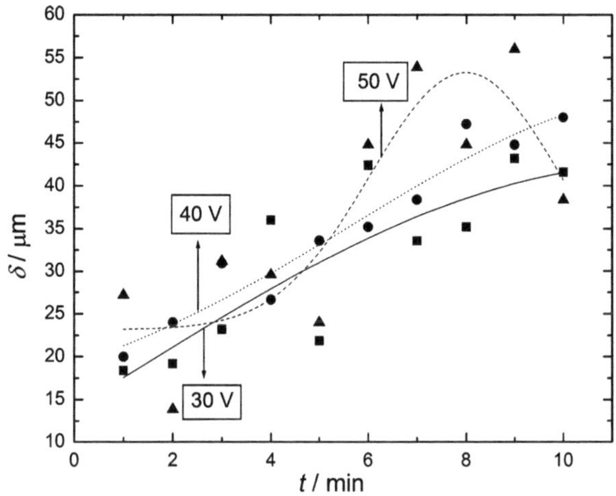

Fig. 3.2 Dependence of alumina coating thickness on deposition time for 20 wt % alumina suspension at different applied voltages (reprinted from [27] with permission from Expert Fachmedien GmbH)

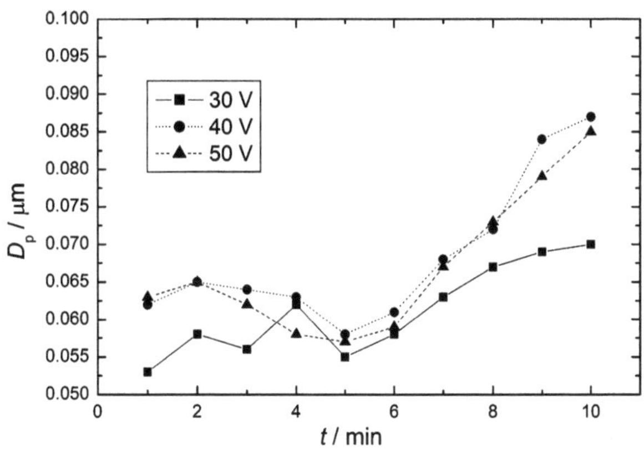

Fig. 3.3 The mean pore diameter vs. deposition time for 20 wt% alumina suspension at different applied voltages (reprinted from [27] with permission from Expert Fachmedien GmbH)

Since the increase in deposition time does not increase the film porosity significantly at the lowest deposition voltage of 30 V (Fig. 3.3), and on the other hand increases the film thickness (Fig. 3.2), it can be concluded that optimal deposition time is 10 min for applied voltage of 30 V and 20 wt% suspension.

The temperature dependence of pore number (Fig. 3.4), mean pore area, and percentage of film surface covered by pores show the minimum at 30 °C for all deposition times, while the mean pore diameter doesn't depend on bath temperature and deposition time [26].

This behavior can be explained by the proposed mechanism of alumina deposition process, which occurs in several steps, as mentioned earlier. The increase in bath temperature decreases the viscosity of water suspension, and consequently increases the rate of particles migration according to Eq. (3.1). On the other hand, the rate of the hydrogen evolution reaction (Eq. (3.2)) as an undesirable process increases with increasing bath temperature. At 30 °C these two processes, i.e., particle migration and coagulation and hydrogen evolution, are in equilibrium,

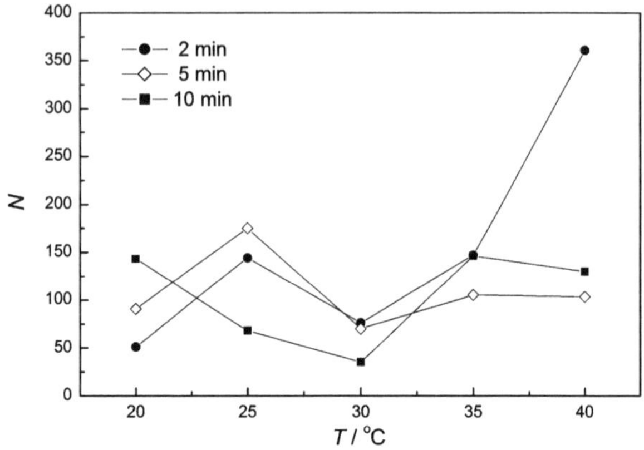

Fig. 3.4 Dependence of pore number on bath temperature for 20 wt% alumina suspension and different deposition time (reprinted from [26] with permission from Expert Fachmedien GmbH)

which results in the minimum porosity of deposited films, e.g., minimum values of pore number (Fig. 3.4), mean pore area, and percentage of film surface covered by pores [26]. At temperatures lower than 30 °C, although the rate of hydrogen reaction is low, the particles migration is too slow, which causes the bad packing of alumina particles and greater porosity of deposited films. On the other hand, at temperatures higher than 30 °C the particles' migration is fast, but the hydrogen evolution is too fast also. As a result, the greater amount of evolved hydrogen goes out through the film, leaving more vacancies inside the films and causing the more porous structures. The morphology of alumina coating obtained under optimal deposition conditions, i.e., applied voltage 30 V, 20 wt% alumina powder suspension, bath temperature 30 °C, and deposition time 10 min, is shown in Fig. 3.5.

3.4 Boehmite Coatings

The boehmite coatings were electrodeposited on titanium from boehmite aqueous sol at different values of constant voltage and deposition time with the aim to determine the optimal deposition parameters for boehmite coatings of the highest thickness and the lowest porosity at lower sintering temperature [6].

3.4.1 Synthesis and Characterization of Boehmite Powder

Boehmite sol was prepared by the peptization of freshly precipitated $Al(OH)_3$, obtained by the addition of NH_4OH to $AlCl_3 \cdot 6H_2O$ aqueous solution, with molar $n(HNO_3)/n(Al(OH)_3)$ ratio of 0.1. Boehmite powder was obtained by drying the sol at 90 °C for 48 h [6, 12]. The XRD pattern of the boehmite powder is represented in

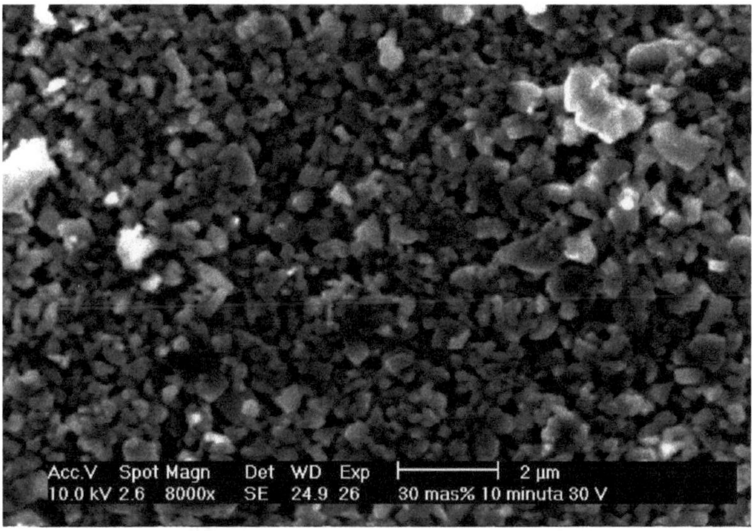

Fig. 3.5 The microphotograph of alumina coating obtained from 20 wt% of alumina powder in suspension at 30 V and 30 °C (deposition time 10 min) (reprinted from [26] with permission from Expert Fachmedien GmbH)

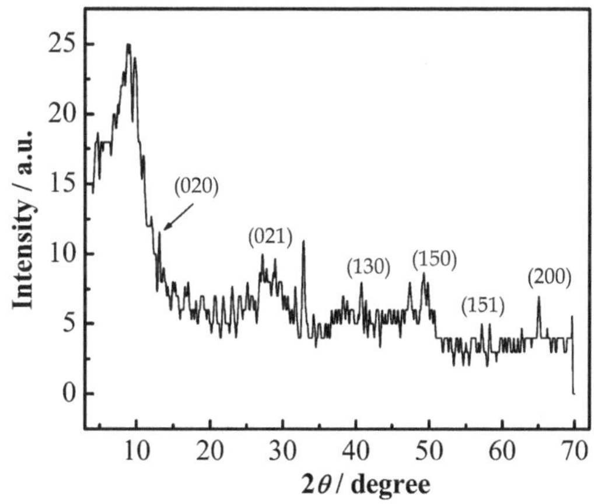

Fig. 3.6 X-ray diffraction pattern of the boehmite powder (reprint from [6] with permission from Elsevier)

Fig. 3.6, where all the peaks correspond to the pseudoboehmite phase, except the one at the angle 10°2θ, which corresponds to scarboit ($Al_2(OH)_6 \cdot 6H_2O$). This peak probably originates from $Al(OH)_3$, which remains as a residue after peptization with HNO_3.

From the peak position for the (020) plane in Fig. 3.6, the d-spacing was calculated to be 0.674 nm, which corresponds to the pseudoboehmite phase. Pseudoboehmite has lower crystallinity compared to boehmite [72]. The IR spectrum of the boehmite powder is represented in Fig. 3.7, where characteristic bonds

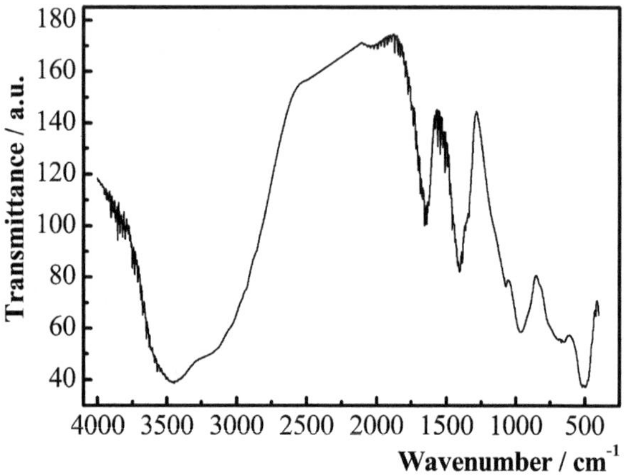

Fig. 3.7 IR spectrum of the boehmite powder (reprint from [6] with permission from Elsevier)

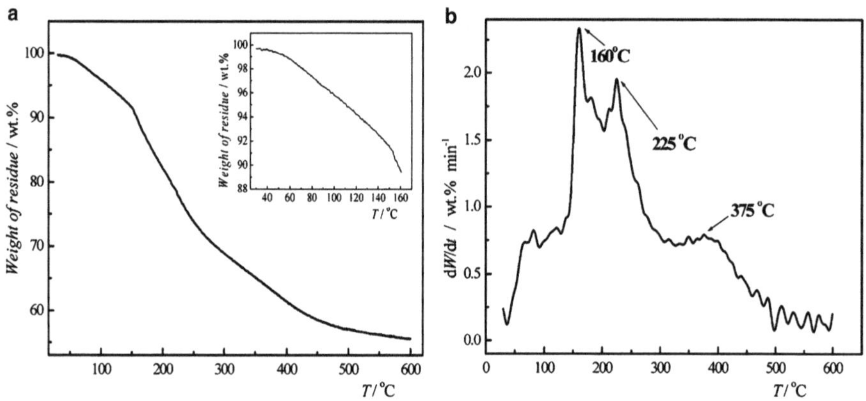

Fig. 3.8 TG curve (**a**) and differential TG (DTG) curve (**b**) of the boehmite powder, inset: TG curve between 23 and 160 °C (reprint from [6] with permission from Elsevier)

for boehmite and pseudoboehmite may be seen. The peak at 430 cm^{-1} originates from octahedral-coordinated Al ions; the peak at 1,060–1,080 cm^{-1} corresponds to the bending vibration of Al–OH, the band at 1,400 cm^{-1} to the NO_3^- group from HNO_3 added for peptization and the peak at 1,650 cm^{-1} to water of hydration. The band at 3,200–3,700 cm^{-1} corresponds to the stretching vibration of the OH group connected to the Al cation, while the peak at 3,460 cm^{-1} corresponds to the symmetric and asymmetric bond stretching vibration of the H_2O molecule [72–76].

The weight loss and temperatures associated with phase transformations were determined by thermogravimetric analysis. The TG curve of the boehmite powder obtained in the temperature range between 23 and 600 °C is shown in Fig. 3.8a, while Fig. 3.8b represents the differential TG (DTG) curve.

The TG curve of boehmite powder exhibited three characteristic weight loss stages. The first stage was observed from 23 °C to 160 °C (inset in Fig. 3.8a), with a sharp peak at 160 °C in the DTG curve (Fig. 3.8b). This stage corresponds to the desorption of water molecules adsorbed on the crystallites surface. It can be seen from the inset in Fig. 3.8a that water desorption occurs gradually. The weight loss in this stage in the initial pseudoboehmite was calculated to be 11 wt%. According to the literature [77], the weight loss in pseudoboehmite is higher than the weight loss in boehmite (about 1 wt%) due to the smaller crystallite size and higher surface area available for water adsorption. The observed weight loss can be explained by the boehmite structure. Namely, boehmite crystallites are made of double layers of octahedra sharing edges along the *a* axis and vertexes along the *c* axis, with an aluminum atom near their center and two hydroxyl groups and four oxygen atoms in their vertexes [76, 77]. The octahedra in the double layers interact strongly with each other, so the interaction within a double layer is strong. However, between the double layers it is weak because the interaction is due to hydrogen bonds. Therefore, a boehmite crystallite ends in the interface between two double layers, thus producing surfaces that are full of hydroxyl groups. The crystallite surfaces perpendicular to the double layers have oxygen atoms in low coordination. These oxygen atoms can easily react with hydroxyl groups and hydrogen ions from their environment, allowing these surfaces to become fully covered by hydroxyl groups. Due to the smaller crystallite size in pseudoboehmite, the number of these low coordination oxygen sites increases, thus enabling more water molecules to be adsorbed. Indeed, the first peak in the DTG curve, located at 160 °C (Fig. 3.8b), corresponds to the desorption of water molecules on the boehmite surface.

The second stage of weight loss of the TG curve was observed between 160 and 450 °C (Fig. 3.8a), which corresponds to the transition of boehmite into γ-Al_2O_3. Two peaks in the DTG curve were observed in this temperature interval: a sharp peak at 225 °C and a wide peak at 375 °C (Fig. 3.8b). The peak at 225 °C and a weight loss of 11 wt% between 160 and 225 °C, calculated from the TG curve, indicate the transformation of pseudoboehmite ($Al_2O_3{\cdot}nH_2O$, where $1 < n < 2.5$) into boehmite ($Al_2O_3{\cdot}H_2O$) and corresponds to 0.41 mol H_2O per mol boehmite. Thus, the formula of the starting pseudoboehmite was calculated to be $AlOOH{\cdot}0.41H_2O$ or $Al_2O_3{\cdot}1.82H_2O$. The excess water in pseudoboehmite is not surface layer, but interlayer water. The weight loss of 15 wt% between 225 and 375 °C, calculated from the TG curve (Fig. 3.8a), is associated with further boehmite transformation to γ-Al_2O_3 according to the equation:

$$2AlOOH \rightarrow \gamma\text{-}Al_2O_3 + H_2O \tag{3.5}$$

and corresponds to the theoretical weight loss of 15 wt% . The wide peak at 375 °C in the DTG curve (Fig. 3.8b) corresponds to the transition from boehmite into γ-Al_2O_3, which occurs through partial dehydroxylation as a consequence of broken hydrogen bonds between the double layers in the boehmite crystallites. The total weight loss of 26 wt% in the second stage of the TG curve between 160 and 450 °C confirms the transformation from pseudoboehmite to γ-Al_2O_3.

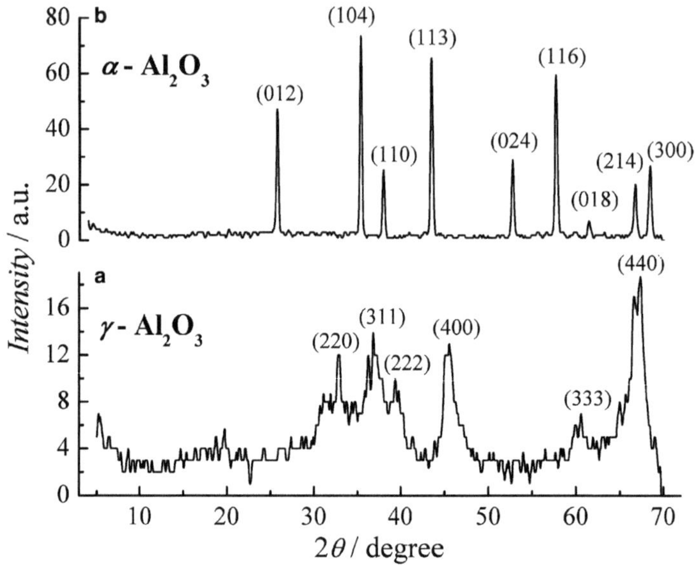

Fig. 3.9 XRD patterns of boehmite powder thermally treated at different temperatures (**a**) 1,000 °C and (**b**) 1,300 °C (reprint from [6] with permission from Elsevier)

The third stage of negligible weight loss at temperatures above 450 °C (Fig. 3.8a) corresponds to the transformation of γ-Al_2O_3 into the most stable α-Al_2O_3 via a sequence of transitional aluminas $\gamma \rightarrow \delta \rightarrow \theta \rightarrow \alpha$-$Al_2O_3$. This transition is a result of the continuous and gradual loss of residual hydroxyl groups and involves a reorganization of the oxygen into a denser, hexagonal closely packed configuration by nucleation and growth [77–82]. The transformation of transitional aluminas into α-Al_2O_3 occurs when their crystallites are completely dehydroxylated. XRD data reveal the presence of γ-Al_2O_3 at 1,000 °C (Fig. 3.9a) and α-Al_2O_3 at 1,300 °C (Fig. 3.9b).

3.4.2 Electrophoretic Deposition of Boehmite Coatings

Boehmite coatings were deposited on titanium from aqueous boehmite sol using the constant voltage method, at different values of constant voltage ranging between 1.0 and 10 V, for various deposition times of 10-30 min at room temperature [6, 12].

The dependence of the deposition current density on the deposition time at constant applied voltage is shown in Fig. 3.10 and corresponds to the deposition of porous films [83, 84]. The current density does not achieve zero value during electrodeposition, suggesting that the coating is still porous and that particle deposition still occurs [85].

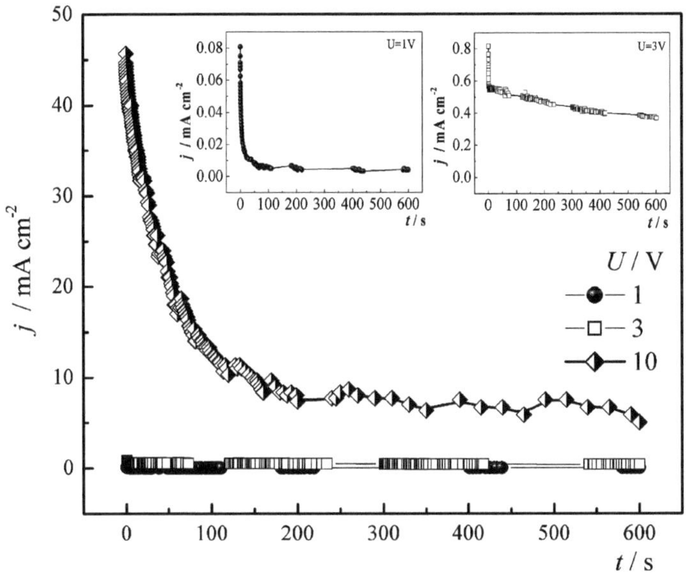

Fig. 3.10 Current density vs. electrodeposition time at different applied voltages, insets: j vs. t curves at 1 V and 3 V (reprint from [6] with permission from Elsevier)

Adsorption and desorption processes in the case of boehmite particles can be represented as follows:

$$> \mathrm{AlOH} + \mathrm{H}^+ \leftrightarrows\ > \mathrm{AlOH_2}^+ \quad \left(\mathrm{pH} < \mathrm{pH_{pzc}}\right) \tag{3.6}$$

$$> \mathrm{AlOH} + \mathrm{OH}^- \leftrightarrows\ > \mathrm{AlOH}^- + \mathrm{H_2O} \quad \left(\mathrm{pH} > \mathrm{pH_{pzc}}\right) \tag{3.7}$$

The electrophoretic deposition of boehmite particles occurs in several steps according to Eqs. (3.1), (3.2), (3.3) and (3.4). During electrodeposition, the hydrogen evolved according to Eq. (3.2) bubbles on the cathode and diffuses throughout the deposited boehmite film. In addition, some hydrogen bubbles may remain trapped between the deposited sol particles, leading, in both cases, to film porosity [77]. Indeed, the dependence of the deposition current density on the deposition time at constant applied voltage (Fig. 3.10) corresponded to the deposition of porous films because current density did not achieve zero value during electrodeposition, suggesting that the coating is still porous and that particle deposition still occurs.

The mass of the boehmite deposit as a function of the applied deposition voltage for different deposition times is shown in Fig. 3.11.

The increase in applied voltage and deposition time increases the boehmite deposit mass. According to Eq. (3.1), the increase in applied voltage increases the rate of particle migration and, consequently, the mass of boehmite deposit. The rate

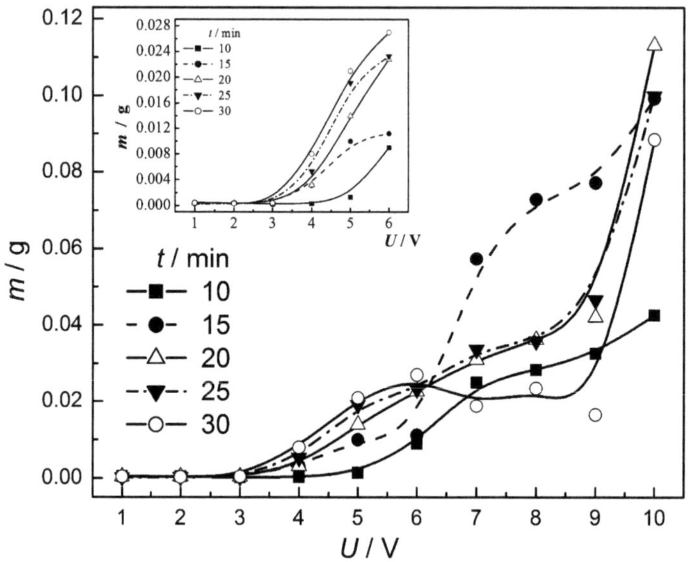

Fig. 3.11 Dependence of the boehmite deposit mass on the applied voltage for different deposition times, *inset*: *m* vs. *U* curves at voltages up to 6.0 V (reprinted from [12] with permission from Serbian Chemical Society)

of hydrogen evolution, as an undesirable process, also increases with increasing applied voltage. At lower applied voltages, up to 3.0 V, the mass of the boehmite deposit is too small and the deposition time, as well as the deposition voltage, does not influence the process (inset in Fig. 3.11). A possible explanation is that the applied electric field is not sufficiently strong and that the rate of colloidal particle migration is too slow. At the same time, the applied voltage is insufficient to cause a destabilization of the boehmite sol near the electrode.

An increase in the applied voltage from 3.0 to 6.0 V, for all deposition times, increases the mass of the boehmite deposit (inset in Fig. 3.11). Also, the rate of hydrogen evolution increases at higher applied voltages greater than 3.0 V, causing greater porosity and cracks in the boehmite deposit during subsequent drying at room temperature. A SEM micrograph (Fig. 3.12a) shows cracks on the boehmite deposit surface, as a consequence of the evolved hydrogen. Figure 3.12b represents a detail from Fig. 3.12a showing that the boehmite deposit has low porosity. Poor adhesion of the boehmite deposit was observed at applied voltages higher than 4.0 V, although the mass of boehmite deposit increased. A decrease in the mass of the boehmite deposit was observed at voltages higher than 6.0 V for a deposition time of 30 min. This behavior can be explained by a larger amount of hydrogen evolved at the cathode at higher voltages and for longer deposition times, and additionally, by poor adhesion of boehmite coating.

The effect of deposition time on the mass of the boehmite deposit at different applied voltages is shown in Fig. 3.13. At lower applied voltages, up to 3.0 V, an

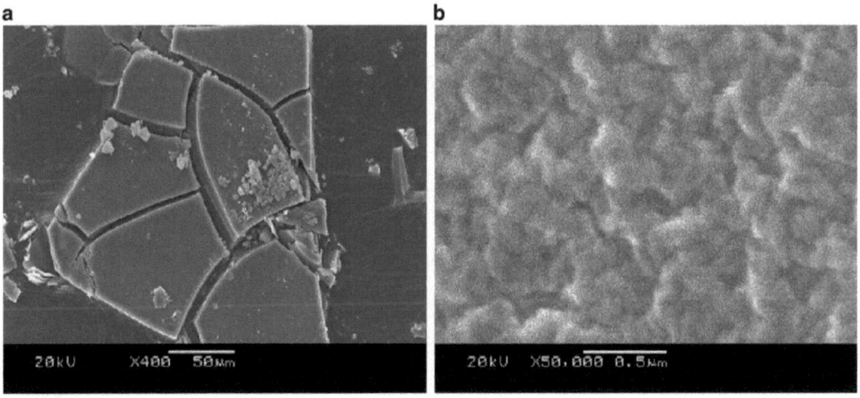

Fig. 3.12 SEM micrographs of the boehmite deposit ($U = 4.0$ V, $t = 30$ min) dried at room temperature in air (**a**) magnification × 400 and (**b**) magnification × 50,000 (reprint from [6] with permission from Elsevier)

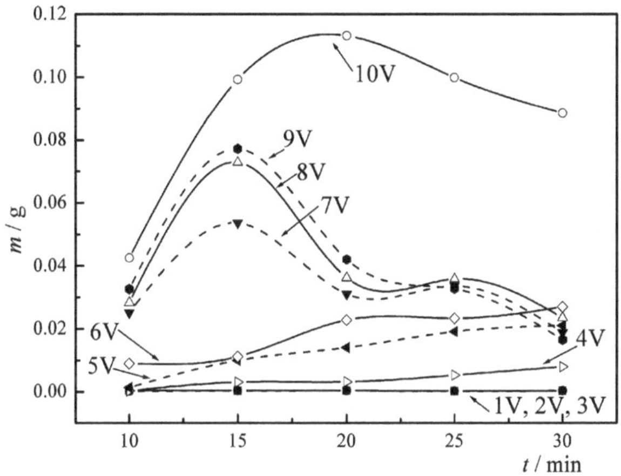

Fig. 3.13 Dependence of the boehmite deposit mass on the deposition time at different applied voltages (reprint from [6] with permission from Elsevier)

increase in the deposition time does not influence the mass of the boehmite deposit. At applied voltages from 3.0 V to 6.0 V, increasing deposition time increases the mass of boehmite deposit due to more particles reaching the cathode. At voltages higher than 6.0 V the mass of the boehmite deposit achieves a maximum and then decreases with increasing deposition time due to the larger amount of evolved hydrogen, which accumulates on the cathode and forms more pores in the deposited film.

Thus, it may be concluded that the optimal deposition voltage for the boehmite coating is 4.0 V, because it is the highest voltage at which a film of low porosity and

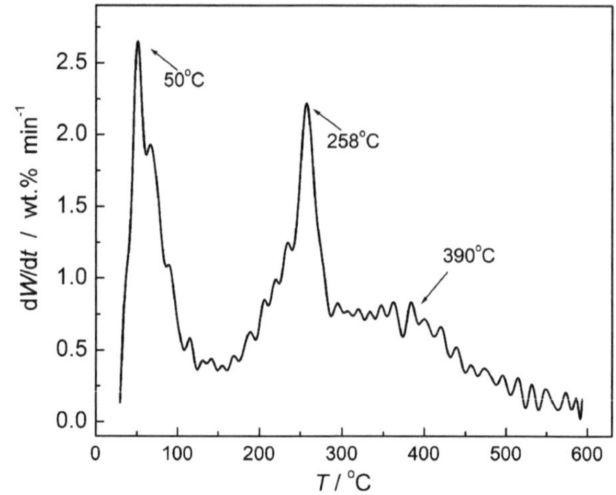

Fig. 3.14 Differential TG (DTG) curve of the boehmite coating ($U = 4.0$ V, $t = 30$ min, dried at room temperature in air) (reprinted from [12] with permission from Serbian Chemical Society)

good adhesion can be formed on titanium. On the other hand, the optimal deposition time is 30 min, since an increase in the deposition time increases the mass of the boehmite deposit. Consequently, boehmite coatings of maximum thickness, low porosity, and good adhesion were formed for longer deposition times and at lower deposition voltages. A coating electrodeposited at 4.0 V and for a deposition time of 30 min was chosen for further thermal treatment at 1,000 and 1,300 °C with a holding period of 1 h at the maximum temperature.

The weight loss and temperatures associated with phase transformations were determined by thermogravimetric analysis. Figure 3.14 represents the differential TG (DTG) curve of the boehmite coating ($U = 4.0$ V, $t = 30$ min, dried at room temperature) obtained in the temperature range between 23 and 600 °C

The DTG curve of boehmite coating exhibited three characteristic weight loss stages. The first stage was observed from 23 °C to 160 °C, with a sharp peak at 50 °C in the DTG curve. This stage corresponds to the desorption of water molecules adsorbed on the crystallites surface. The weight loss in this stage in the initial pseudoboehmite was calculated to be 13 wt%. This is in accordance with data from the literature [77], showing that the weight loss in pseudoboehmite is higher than the weight loss in boehmite (about 1 wt% in boehmite) due to the smaller crystallite size with higher surface area available for water adsorption. The second stage of weight loss of the DTG curve was observed between 160 and 450 °C, which corresponds to the transition of boehmite into γ-Al_2O_3, where two peaks at the DTG curve were observed in this temperature interval: a sharp peak at 258 °C and a wide peak at 390 °C. The peak at 258 °C and a weight loss of 11 wt% between 50 and 258 °C, calculated from the TG curve, indicate the transformation of pseudoboehmite ($Al_2O_3{\cdot}nH_2O$, where $1 < n < 2.5$) into boehmite ($Al_2O_3{\cdot}H_2O$) [75, 76, 78–82] and corresponds to 0.41 mol H_2O per mol boehmite. Consequently, the formula of the starting pseudoboehmite was calculated to be $AlOOH{\cdot}0.41H_2O$ or $Al_2O_3{\cdot}1.82H_2O$. The excess water in pseudoboehmite is not surface layer, but

Fig. 3.15 SEM micrograph of the boehmite coating (U = 4.0 V, t = 30 min) sintered at 1,000 °C (reprinted from [12] with permission from Serbian Chemical Society)

interlayer water. The weight loss of 15 wt% between 258 and 390 °C, calculated from the DTG curve, is associated with further boehmite transformation to γ-Al_2O_3 according to the equation:

$$2AlOOH \rightarrow \gamma\text{-}Al_2O_3 + H_2O \tag{3.8}$$

and is in agreement with the theoretical weight loss of 15 wt%. The wide peak at 390 °C in the DTG curve corresponds to the transition from boehmite into γ-Al_2O_3, occurring through partial dehydroxylation as a consequence of broken hydrogen bonds between the double layers in the boehmite crystallites. The total weight loss of 26 wt% in the second stage of the DTG curve between 50 and 450 °C confirms the transformation from pseudoboehmite to γ-Al_2O_3.

The third stage of negligible weight loss at temperatures above 450 °C corresponds to the transformation of γ-Al_2O_3 into the most stable α-Al_2O_3 via a sequence of transitional aluminas $\gamma \rightarrow \delta \rightarrow \theta \rightarrow \alpha$-$Al_2O_3$. This transition is a result of the continuous and gradual loss of residual hydroxyl groups and involves a reorganization of the oxygen into a denser, hexagonal closely packed configuration by nucleation and growth [76, 77].

The same results of phase transformations and calculated values of weight loss between three characteristic weight loss stages, as well as the calculated formula of the starting pseudoboehmite, were obtained from TGA of boehmite powder, indicating the same structure of boehmite coating and boehmite powder, with no effect of the substrate on the boehmite coating structure.

Figure 3.15 represents the SEM micrograph of the boehmite coating treated at 1,000 °C for 1 h. According to analysis of XRD pattern of boehmite powder treated

Fig. 3.16 SEM micrograph of the boehmite coating (U = 4.0 V, t = 30 min) sintered at 1,300 °C (magnification × 10,000) (reprinted from [12] with permission from Serbian Chemical Society)

at the same way [6], it could be concluded that the boehmite coating is also- γ-alumina, having in mind the results obtained from TGA for both powder and coating.

The SEM micrograph of the boehmite coating treated at 1,300 °C for 1 h (Fig. 3.16) shows the graininess of the coating of α-Al_2O_3, which is in accordance with the XRD data of boehmite powder treated at the same temperature [6], having in mind the results obtained from TGA for both powder and coating. This indicates the significantly lower sintering temperature of the boehmite coating obtained by the sol–gel procedure compared to sintering temperature of 1,600 °C for of the boehmite coatings obtained by classical forming methods [86].

3.5 Monetite Coatings

Calcium phosphate compounds are becoming of increasingly great importance in the field of biomaterials and, in particular, as bone substitutes [87]. Bioactive surface coatings that consist of calcium phosphate play an important role in mediating in osteointegration of orthopedics and dental Ti-based implants [88]. Hence, the good mechanical properties of the metals are combined with the biocompatibility of calcium phosphate coatings. The bioactivity, biocompatibility, and good mechanical properties of calcium phosphate coatings are determined by their morphology, composition, crystallite size, and structure.

Calcium hydroxyapatite, HA or HAP ($Ca_{10}(PO_4)_6(OH)_2$), is both bioactive and osteoconductive, making it a very attractive material for biomedical applications in orthopedics [89]. HA is highly biocompatible material due to its chemical similarity with inorganic part of the human hard tissue. HA is known as a bioactive ceramic because of its capability to improve osteointegration. Biological reaction to a HA ceramic is strongly dependent on its chemical composition, phase purity, and microstructural properties. Besides controlling the stoichiometry of synthetic HA, the control of crystallinity, porosity, particle shape, surface area, and agglomeration characteristics are of the great interest.

Fine-grained nanopowders [90] are required to enhance the sinterability of HA at lower temperatures. The control over the morphology of HA is of great importance for its bioapplication. Different crystal shapes of nanometer HA, such as needle-like, sheet-like, rod-like, spherical, plate-like [91–99], can be controlled directly, by synthesis parameters or by conversion of precursors into HA under specific conditions. Monetite and brushite, as well as octacalcium phosphate, are known as a precursor of HA formation [100–104]. The synthesis of monetite can be provided by hydrothermal synthesis [105, 106], microwave-assisted method [100], precipitation from microemulsion [107], crystallization from solution [108], chemical deposition [109, 110], or by electrodeposition method [111]. The transformation of monetite to apatite in aqueous solution requires the presence of ions in solution and can be converted to HA by alkaline treatment [87, 100, 112, 113] or by immersion to simulated body fluid [114]. Electrophoretic deposition was used to obtain monetite coatings on titanium from ethanol suspension and to convert monetite to hydroxyapatite [7].

3.5.1 Synthesis and Characterization of Monetite Powder

Electrochemical synthesis of nanosized monetite powder was performed galvanostatically from homogeneous solution of $Na_2H_2EDTA{\cdot}2H_2O$, NaH_2PO_4, and $CaCl_2$ at a concentration relationship Ca/EDTA/PO_4^{3-} of 0.25/0.25/0.15 M at current density of 137 mA cm^{-2} and pH value of 5.0 [7]. From XRD pattern (Fig. 3.17) the only detected crystalline phase was monetite, $CaHPO_4$ (JCPDS file 70-0360), while crystallite domain size from (001) reflection was calculated to be 27.1 nm.

The IR spectrum of monetite powder (Fig. 3.18) shows the characteristic bonds for monetite. The wavenumbers and corresponding assignments are represented in Table 3.1.

The values of average monetite particle diameter and particle size distribution (PSD) based on the scattering light intensity are given in Table 3.2.

It is evident that the prepared monetite suspension was a high degree polydisperse one. The average monetite particles diameter is 203 nm. However, based on PSD by intensity, more than 73 % particles have an average diameter of ~ 1.4 nm. Size distribution by intensity for monetite suspension (8 runs) is presented in Fig. 3.19. The large particles are agglomerates of the small ones.

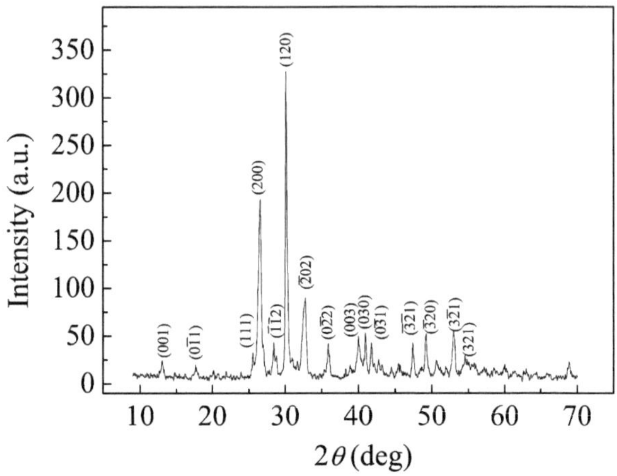

Fig. 3.17 X-ray diffraction pattern of the monetite powder (reprinted from [7] with permission from Elsevier)

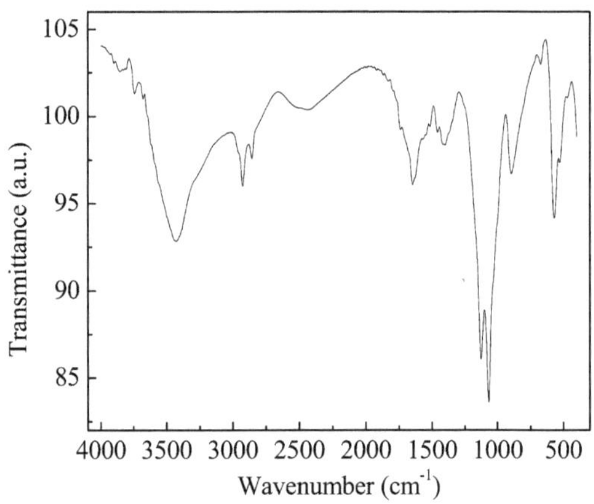

Fig. 3.18 IR spectrum of the monetite powder (reprinted from [7] with permission from Elsevier)

Table 3.1 IR wavenumbers and assignments for monetite ($CaHPO_4$)

IR monetite wavenumbers (cm^{-1})	Assignments
3,431	O–H stretching of residual free water
2,924	PO–H stretching
2,851	
1,644	H–O–H bending and rotation of residual free water
1,406	P–O–H in plane (bending)
1,120	P–O stretching
1,068	
891	P–O(H) stretching
565	O–P–O(H) bending mode
525	

Reprinted from [7] with permission from Elsevier

Table 3.2 The values of average monetite particle diameter and particle size distribution (PSD) by intensity of monetite suspension

Size measurement	Monetite suspension
z-Average diameter (nm)	203
PSD intensity	
Peak 1: *z*, nm (Int., %)	1.42 (73.7)
Peak 2: *z*, nm (Int., %)	3,600 (26.3)

Reprinted from [7] with permission from Elsevier

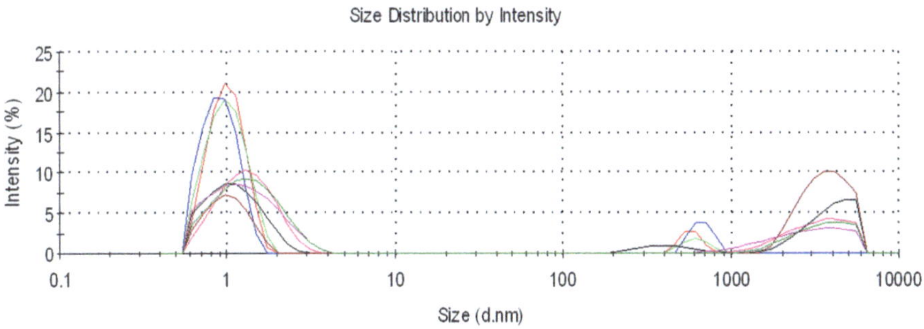

Fig. 3.19 Particle size distribution of the monetite suspension (reprinted from [7] with permission from Elsevier)

DTA curve for monetite powder is shown in Fig. 3.20, where three characteristics endothermic peaks can be observed at temperatures of 332 °C, 414 °C, and 1,260 °C. The first two peaks, at 332 °C and 414 °C, can be attributed to the initial and final temperatures of monetite powder decomposition to calcium pyrophosphate ($Ca_2P_2O_7$) and H_2O. The third peak at 1,260 °C corresponds to the melting of $Ca_2P_2O_7$. The lower melting point compared to value of 1,353 °C reported in [115] is a consequence of larger specific surface area of nanosized particles of synthesized monetite powder. Namely, bulk properties of crystals are basically dependent upon their structure, but at the nanoscale, in addition to the structure, their size is an important factor which influences their properties and they often exhibit properties that are drastically different from those of conventional materials. The most significant characteristic of materials at the nanoscale is their high surface-to-volume ratio affecting their thermodynamic properties. As the size decreases beyond the critical value, due to increase in the surface-to-volume ratio, the melting temperatures deviates from the bulk value and becomes a size-dependent property. Smaller particle sizes create larger interfacial areas and increased number of atoms at the particles interfaces. It was shown that melting temperatures of nanosized aluminum particles [116, 117], ammonium nitrate nanopowder [118], nanosized Sn particles [119], and nanosized Au particles [120] decrease with decreasing particles size.

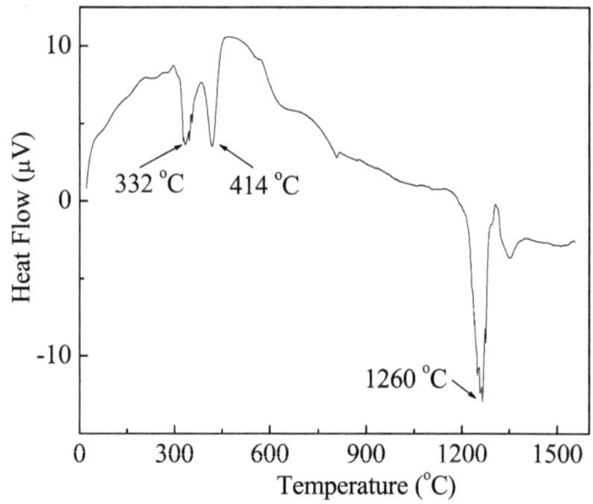

Fig. 3.20 DTA curve of the monetite powder (reprinted from [7] with permission from Elsevier)

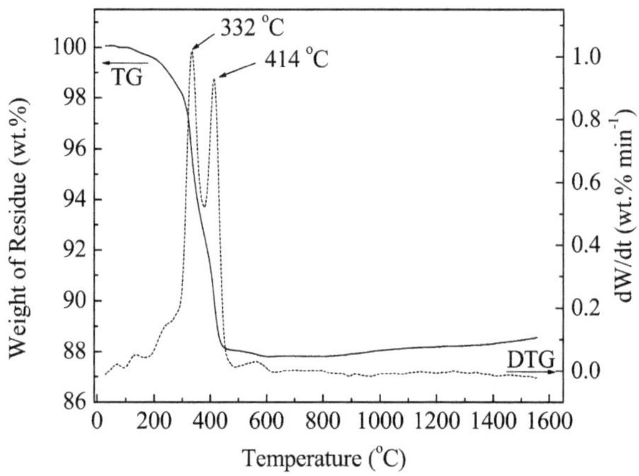

Fig. 3.21 TG/DTG curves of the monetite powder (reprinted from [7] with permission from Elsevier)

The weight loss and temperatures associated with phase transformation were determined by thermogravimetric analysis. The TG curve and differential TG (DTG) curve of monetite powder obtained in the temperature range between 30 °C and 1,600 °C are shown in Fig. 3.21.

The TG curve of monetite powder shows two weight loss stages. The first stage was observed from 30 to about 200 °C. This stage corresponds to desorption of water molecules adsorbed on the crystallite surface. The second stage of weight loss of the TG curve was observed between 200 and 500 °C, with sharp peaks at DTG curve at 332 and 414 °C. The weight loss between 332 and 414 °C is 5.7 wt% and can be attributed to the decomposition of monetite to $Ca_2P_2O_7$ and H_2O. Theoretical weight loss for this reaction is 6.5 %. This discrepancy between theoretical and experimental value could be assigned to the unreacted starting component, $CaCl_2$, in monetite powder. At the temperatures higher than 500 °C no significant changes in weight loss can be observed. The total weight loss for

Fig. 3.22 SEM micrograph of the monetite powder (reprinted from [7] with permission from Elsevier)

investigated temperature interval between 30 and 1,600 °C is about 11.5 wt% for monetite powder, which can be attributed to the desorption of water molecules adsorbed on the crystallite surface and decomposition of monetite to calcium pyrophosphate.

SEM micrograph of monetite powder is shown in Fig. 3.22. It can be seen that powder consists of big agglomerates, which are composed of fine crystallites. The crystallite domain size of 27.1 nm, determined by X-ray diffraction (Fig. 3.17), and 203 nm, determined by PSD measurements (Table 3.2), indicates that agglomerates were formed through the drying process, which was confirmed by SEM analysis (Fig. 3.22). This is in good agreement with nucleation–aggregation–agglomeration–growth mechanism: (a) nucleation and growth to form crystallites in the nanometric size range, (b) aggregation of elemental nanocrystals by physical attractions, and (c) further crystals growth to form stable agglomerates [121].

3.5.2 Electrophoretic Deposition of Monetite Coatings

Monetite coatings were electrophoretically deposited on titanium from ethanol suspension, using constant voltage method. The deposition was performed at constant voltage between 10 and 50 V and for a constant deposition time between 1 and 30 min, with the aim to investigate the influence of the applied voltage and deposition time on the mass and morphology of monetite deposits [7].

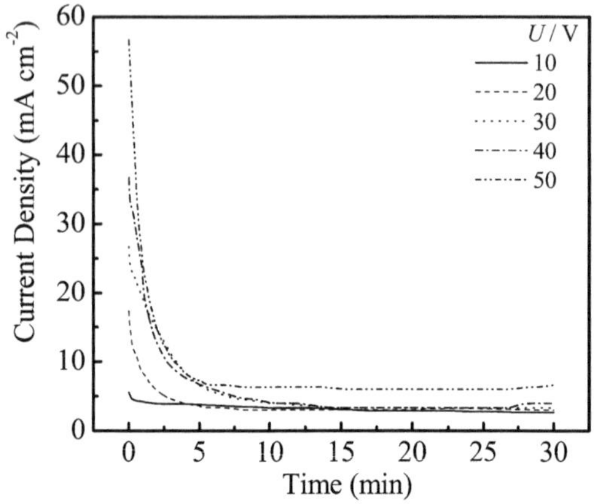

Fig. 3.23 Current density vs. electrodeposition time at different applied voltages (reprinted from [7] with permission from Elsevier)

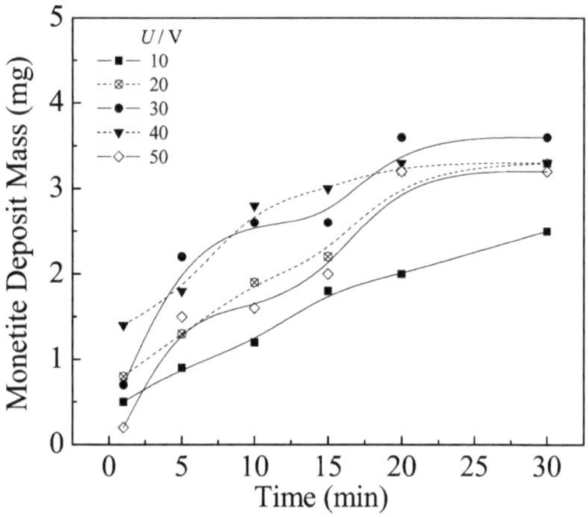

Fig. 3.24 Dependence of the monetite deposit mass on the deposition time at different applied voltages (reprinted from [7] with permission from Elsevier)

The dependences of the deposition current density on the deposition time at constant applied voltage are shown in Fig. 3.23 and correspond to deposition of porous films. The current density does not achieve zero value during electrodeposition, suggesting that the coatings are porous and that particle deposition still occurs.

The effect of deposition time on the mass of the monetite deposit at different applied voltages is shown in Fig. 3.24. At all applied voltages, increasing deposition time up to 20 min increases the mass of the monetite deposit due to more particles reaching the cathode. On the other hand, increase in deposition time over 20 min does not affect significantly the mass of the monetite deposit due to the larger amount of evolved hydrogen, which accumulates on the cathode and forms more pores in the deposited film; thus, the optimal deposition time is 20 min.

The effect of applied voltage on the mass of the monetite deposit for different deposition times is shown in Fig. 3.25.

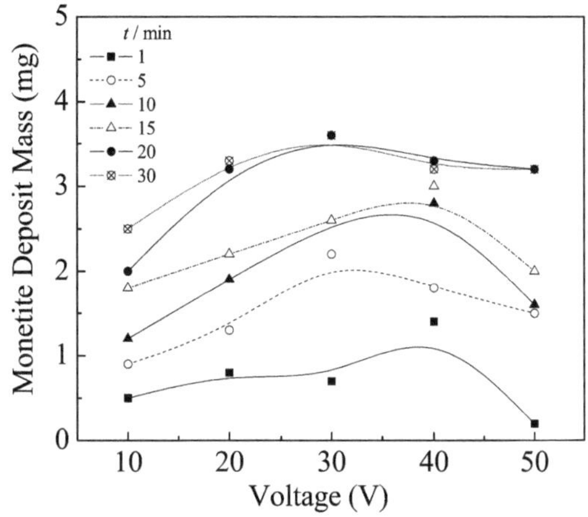

Fig. 3.25 Dependence of the monetite deposit mass on the applied voltage for different deposition times (reprinted from [7] with permission from Elsevier)

Table 3.3 The values of mean pore area, mean pore diameter, and percentage of film surface covered by pores for monetite coatings deposited on titanium at different applied voltages and for deposition time of 20 min

Deposition conditions	Mean pore area (μm^2)	Mean pore diameter (μm)	Percentage of film surface covered by pores (%)
$U = 20$ V $t = 20$ min	0.12	0.24	9.95
$U = 30$ V $t = 20$ min	0.03	0.14	7.32
$U = 50$ V $t = 20$ min	0.13	0.27	10.73

Reprinted from [7] with permission from Elsevier

The increase in applied voltage up to 30 V, for all deposition times, increases the monetite deposit mass. According to Eqs. (3.1) and (3.2), the increase in applied voltage increases the rate of particle migration and, consequently, the mass of monetite deposit. The rate of hydrogen evolution also increases with increasing applied voltage, causing greater porosity of the monetite deposits. Therefore, a decrease in the mass of monetite deposit was observed at voltages higher than 30 V for a deposition time of 20 and 30 min.

Thus, the monetite coating of the highest thickness and low porosity was formed at applied voltage of 30 V and for deposition time of 20 min, which was confirmed by image analysis of SEM micrographs (Table 3.3).

The minimum values of mean pore area, mean pore diameter, and percentage of film surface covered by pores were obtained for coating deposited at 30 V and for deposition time of 20 min, i.e., for the deposit of the highest thickness. This can be explained by the fact that at lower voltages, the smaller is amount of evolved hydrogen accumulated on the cathode, which diffuses throughout the film leaving less vacancies inside the deposited film and causing less porous structure.

Fig. 3.26 SEM micrograph of the monetite coating electrodeposited at 30 V for 20 min (reprinted from [7] with permission from Elsevier)

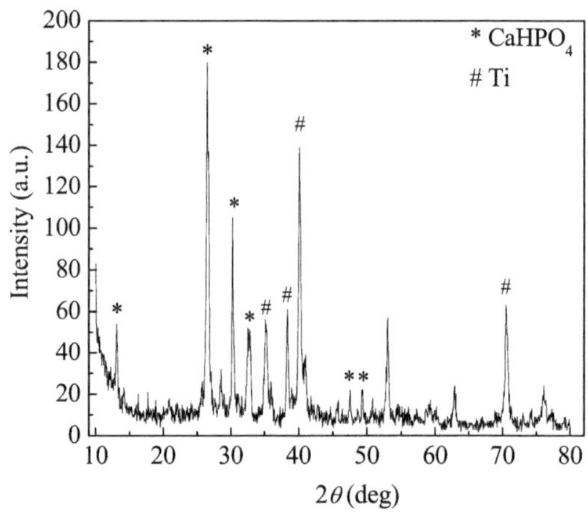

Fig. 3.27 XRD pattern of the monetite coating electrodeposited at 30 V for 20 min (reprinted from [7] with permission from Elsevier)

The evolved hydrogen makes not just the greater number of pores but also the greater pore diameter of the existing pores inside the deposited film, since the release of greater amount of hydrogen may cause unstable, nonuniform, and boiling type evolution of the gas bubbles. These results are in accordance with results obtained from the dependences of monetite deposit mass on deposition time and applied voltage (Figs. 3.24 and 3.25, respectively). Therefore, monetite coating deposited at 30 V for 20 min was chosen for further conversion to hydroxyapatite. SEM micrograph and XRD pattern of this coating are shown in Figs. 3.26 and 3.27, respectively.

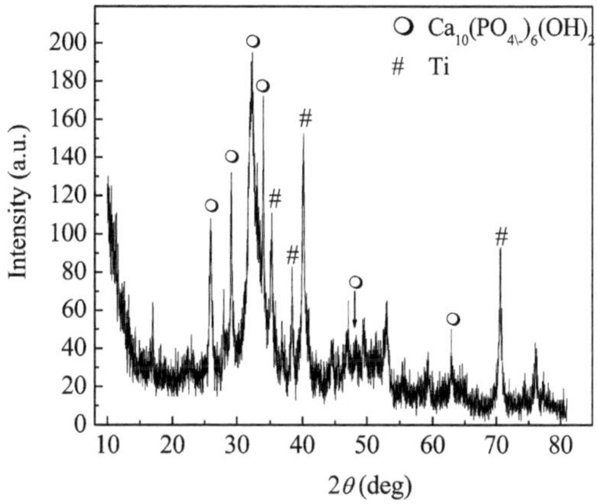

Fig. 3.28 XRD pattern of the hydroxyapatite coating obtained after immersion of monetite coating in 10 wt% NaOH (reprinted from [7] with permission from Elsevier)

Figure 3.28 shows the XRD pattern of product obtained after immersion of monetite coating in 10 wt% NaOH for 2 h. The identified peaks correspond to titanium (originate from the substrate) and hydroxyapatite ($Ca_{10}(PO_4)_6(OH)_2$). The broadening of corresponding peaks can be explained by the fine particle size. Characteristic reflection for monetite phase disappeared, suggesting the complete surface of monetite coatings was converted to hydroxyapatite.

The monetite conversion to hydroxyapatite after alkaline treatment is predicted by the fact that at a pH above 9 the most stable phase is hydroxyapatite, while the monetite ($CaHPO_4$) is stable in acidic conditions [112, 122–125]. According to the literature [112], the proposed conversion mechanism involves dissolution of the precursor monetite crystals followed by precipitation of hydroxyapatite. Namely, at the monetite/solution interface, continuous dissolution of monetite occurs. Through dissolution, the interface enriched with Ca^{2+} and PO_4^{3-} ions and the pH corresponding to hydroxyapatite thermodynamic equilibrium is reached. Consequently, the hydroxyapatite is precipitated at the precursor monetite surface.

Figure 3.29 represents the SEM microphotograph of hydroxyapatite obtained after conversion of monetite coating in alkaline solution. The micrograph shows that the particles surface of monetite is covered by nanosized crystallites of precipitated hydroxyapatite making interconnection between starting particles of monetite. Molar calcium/phosphate ratio (Ca/P = 1.67) obtained by EDS analysis is in accordance with XRD results (Fig. 3.28), indicating that hydroxyapatite obtained after conversion of monetite is stoichiometric.

3.6 Brushite Coatings

Brushite [101, 103, 104, 126–130], as well as monetite and octacalcium phosphate, is known as precursors of HA formation. Brushite (dicalcium phosphate dihydrate (DCPD), $CaHPO_4{\cdot}2H_2O$) is a phase of calcium phosphate that is relatively soluble

Fig. 3.29 SEM micrograph of the hydroxyapatite coating obtained after immersion of monetite coating in 10 wt% NaOH (reprinted from [7] with permission from Elsevier)

in simulated body fluid and can provide rather increased levels of calcium and phosphate ions near the tissue-implant interface. It follows that the application of calcium phosphate coatings could supply increased amount of Ca^{2+} and PO_4^{3-} in early stages of implantation, and thus the transformation to a less soluble, biocompatible, calcium phosphate (as HA) can be achieved. The transformation of brushite to HA can be realized in NaOH solution [103, 126, 127, 131] or simulated body fluid (SBF) [104, 128–130].

Compared to numerous previous papers on electrosynthesis of hydroxyapatite and other calcium phosphate coatings, the new attempt was made to investigate the effect of current density and the deposition time on the mass and morphology of the brushite coatings deposited on titanium, as precursors for HA formation, and to determine the optimal deposition parameters for the preparation of the brushite coatings of the greatest mass and the smallest crystallite domain size. Additionally, the conversion of deposited calcium phosphate coatings to hydroxyapatite coatings in SBF solution (7.996 g dm^{-3} NaCl, 0.350 g dm^{-3} $NaHCO_3$, 0.224 g dm^{-3} KCl, 0.228 g dm^{-3} $K_2HPO_4{\cdot}3H_2O$, 0.305 g dm^{-3} $MgCl_2{\cdot}2H_2O$, 0.278 g dm^{-3} $CaCl_2$, 0.071 g dm^{-3} Na_2SO_4, 6.057 g dm^{-3} $(CH_2OH)_3CNH_2$, 40 dm^3 1 M HCl) was investigated in order to find the relationship between the morphology, porosity, and crystallite domain size of electrochemically deposited brushite coatings and crystallite domain size of converted HA coatings [13].

3.6.1 Electrochemical Deposition of Brushite Coatings

Electrochemical deposition of brushite coatings on titanium was performed galvanostatically from the aqueous solution of 0.042 M $Ca(NO_3)_2$ and 0.025 M $NH_4H_2PO_4$, pH 4.0, at room temperature. The electrodeposition was performed at different values of the current density ranging between 5.0 and 10 mA cm^{-2}, for various deposition times from 1 to 30 min and soaked in SBF, with the aim to evaluate the possibility of conversion of brushite coatings to hydroxyapatite [13].

Depending on experimental conditions, the precipitation of different types of calcium phosphate coatings, e.g., dicalcium phosphate dihydrate (brushite, DCPD), hydroxyapatite (HA), octacalcium phosphate (OCP), and calcium-deficient hydroxyapatite (Ca-def HA) on the cathode is possible. Consequently, the mechanism of electrochemical deposition of calcium phosphate coatings can be described by electrochemical (hydrogen evolution from different ions of the solution and/or water molecules at high pH values as a side reaction), acid–base (consecutive pH-dependent transitions among phosphates), and precipitation reactions (deposition of Ca phosphates) [131–136]. The pH value near the electrode surface appears as crucial factor which determines the structure and composition of the coating. Different phosphates (e.g., $H_2PO_4^-$, HPO_4^{2-}, and PO_4^{3-}) are stable in different solutions and pH ranges [137, 138]. In the case of electrochemical deposition, pH is the function of potential at which the coating is deposited, due to hydrogen evolution reaction [139].

The cathodic polarization curve of Ti substrate in the solution containing 0.042 M $Ca(NO_3)_2$ and 0.025 M $NH_4H_2PO_4$ is shown in Fig. 3.30. Three different stages can be distinguished on the polarization curve. In the first stage (at around −1.15 V vs. SCE), the minor polarization causes the increase in current density for two orders of magnitude. This stage corresponds to the deposition time of about 10 s at nearly constant potential. In the second stage (corresponding to much longer deposition time of approximately 12.5 min), considerable change in potential from −1.2 to −1.9 V (vs. SCE) is followed by the increase in current density by only about half an order of magnitude, with an inflection appearing around −1.5 V (vs. SCE). The third stage is represented by more pronounced increase in current density while potential decreases from −1.9 V to −2.6 V (vs. SCE) (approximately, 18.5 min of deposition).

The initial, first step of the deposition process, corresponding to potential region around −1.15 V (Fig. 3.30, stage I), could be assigned to hydrogen reduction from NH_4^+, originated from starting $NH_4H_2PO_4$:

$$2NH_4^+ + 2e^- \rightarrow H_2 + 2NH_3 \tag{3.9}$$

Reaction (3.9) can cause a local increase in pH in the vicinity of the cathode up to the value of self-buffered ammonia solution (NH_4^+/NH_3), with pH around 9 [140]. As the consequence, initially present $H_2PO_4^-$ (originating from $NH_4H_2PO_4$) are converted into HPO_4^{2-}. In the presence of Ca^{2+} (originating from

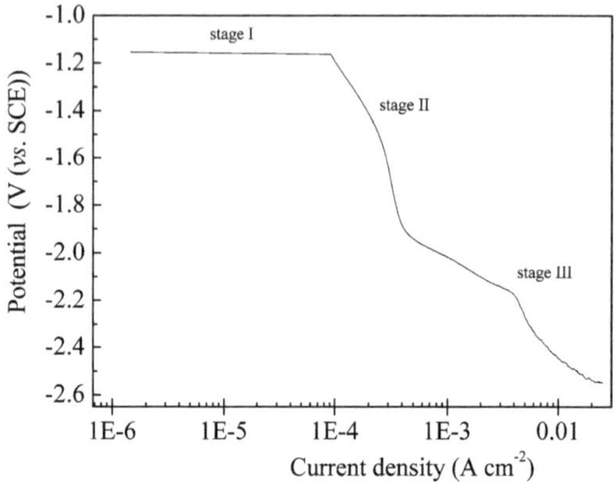

Fig. 3.30 Cathodic polarization curve of Ti substrate in the solution containing 0.042 M Ca $(NO_3)_2$ and 0.025 M $NH_4H_2PO_4$ (reprinted from [13] with permission from Elsevier)

starting $Ca(NO_3)_2$) and at pH ~9, the conditions for the precipitation of brushite are fulfilled, causing the formation of the precipitate at the cathode surface:

$$Ca^{2+} + HPO_4{}^{2-} + 2H_2O \rightarrow CaHPO_4 \cdot 2H_2O \tag{3.10}$$

Since considerable part of the surface is occupied by precipitate, reaction (3.9) should slow down, which is registered as much slower increase in cathodic current density in the potential range from -1.2 to -1.9 V (Fig. 3.30, stage II). It should be noted that in this potential range the hydrogen evolution from $HPO_4{}^{2-}$ can also proceed simultaneously [139]:

$$2HPO_4{}^{2-} + 2e^- \rightarrow 2PO_4{}^{3-} + H_2 \tag{3.11}$$

Since the pH value of self-buffered solution is around 9 in the vicinity of the cathode, the most stable phosphate ion is $HPO_4{}^{2-}$. Since $PO_4{}^{3-}$ is not stable in the mentioned pH range [137, 139], it is converted back to the $HPO_4{}^{2-}$. It can be suggested that the only phosphate ion present near the cathode surface should be $HPO_4{}^{2-}$.

At potentials below -1.9 V (Fig. 3.1, stage III), the hydrogen evolution from water molecules takes place according to Eq. (3.2). However, it can be concluded that, due to self-buffering, the only stable ion is still $HPO_4{}^{2-}$, and brushite is deposited onto the cathode, which was confirmed by XRD analysis.

The effect of the deposition time on the mass of brushite coatings deposited at different applied current densities is represented in Fig. 3.31.

At all applied current densities, increase in deposition time increases the mass of the deposit since more ions reached the cathode, achieving the maximum value between 17 and 20 min depending on applied current density, and then decreases to minimum value for 25 min. This can be explained as follows. The increase in deposition time increases also the amount of hydrogen evolved on the cathode (Eq. (3.2)), meaning that hydrogen bubbles cover in a greater extent the surface of cathode, thus leaving less space for the formation of phosphate deposit. This slows

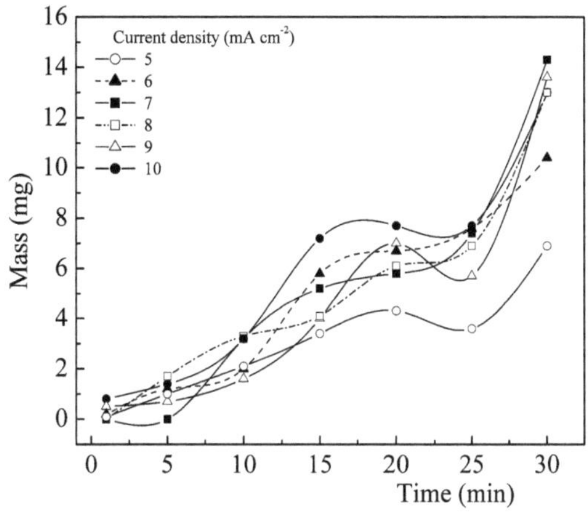

Fig. 3.31 Dependence of the brushite coating mass on the deposition time at different current densities (reprinted from [13] with permission from Elsevier)

down the native coating deposition rate, and consequently the coating mass started to decrease after achieved maximum. This could be related to some depletion of the deposit by evolved hydrogen, which can free the cathode surface (previously occupied by deposit, but also by hydrogen bubbles) for an avalanche of fresh deposit, observed as considerable increase in deposit mass after 25 min. This consideration is supported by the finding that the maxima slightly move toward shorter deposition times at higher current densities (Fig. 3.31). The hydrogen evolution is faster at higher current densities, and hence the release of the fresh parts of the surface takes place somewhat earlier.

To conclude, the increase in coating mass with increasing time is a consequence of two competitive reactions, phosphate deposition and growth and, on the other hand electrochemical reaction of hydrogen evolution. Further increase in deposition time up to 30 min increases the mass of coating for all applied current densities. This is more pronounced at higher current densities because not only the rate of hydrogen evolution is faster but also the generated H_2 bubbles leave more rapidly the surface, thus enabling the further coating deposition. Consequently, the greatest coating mass was observed for the longest deposition time.

The effect of the applied current density on the mass of the brushite coatings for different deposition times is shown in Fig. 3.32. For deposition times up to 25 min, the increase in current density does not affect significantly the mass of coatings. For the longest deposition time (30 min) the increase in current density up to 7 mA cm^{-2} increases the coating mass, while further increase in current density does not affect the mass of deposited coating. So, it can be concluded that the greatest deposited mass was obtained for the longest deposition time of 30 min, while increase in applied current density over 7 mA cm^{-2} does not influence the coating mass.

Figure 3.33 shows the XRD patterns for brushite coatings deposited at different current densities from 5.0 to 10 mA cm^{-2} for the deposition time of 30 min. The identified diffraction maxima in all samples correspond to brushite, $CaHPO_4{\cdot}2H_2O$ (JCPDS file 72-0713), and α-Ti as the substrate (JCPDS file 89-2762). The results

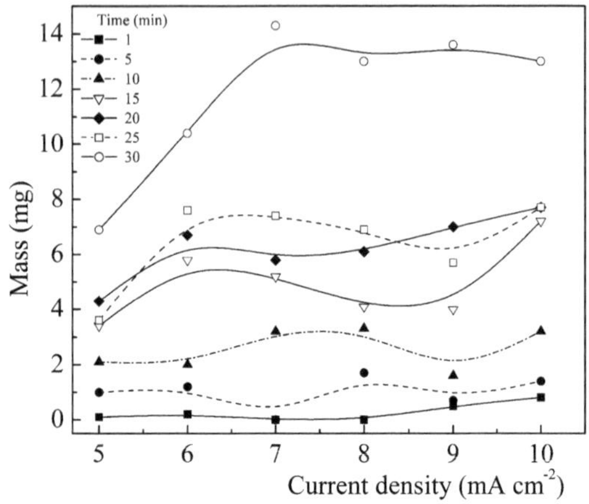

Fig. 3.32 Dependence of the brushite coating mass on the applied current density for different deposition times (reprinted from [13] with permission from Elsevier)

of XRD analysis confirmed the proposed mechanism of brushite coating formation [Eqs. (3.9), (3.10), and (3.11)].

According to literature [137], the ratio between the intensities of the strongest diffraction maximum of the coating and the strongest diffraction maximum of the substrate can point out to porosity or thickness of the coating. Figure 3.34 represents the ratio between the intensity of the strongest diffraction maximum of brushite and the strongest diffraction maximum of Ti substrate for different applied current densities and for constant deposition time of 30 min. The dependence of crystallite domain size calculated for (020) plane on deposition current density is depicted in Fig. 3.35.

If the ratio between observed intensities is higher, the coating is thicker or less porous than the coating with lower value of the ratio. Fig. 3.34 shows that the ratio increases with the increasing current density up to 7 mA cm^{-2}. Since it was shown that increase in current density up to 7 mA cm^{-2} increases the coating mass (Fig. 3.32), it can be proposed that increase in ratio up to 7 mA cm^{-2} is due to the increase in coating mass gained by the increase in current density from 5 to 7 mA cm^{-2}. The crystallite domain size also increases with increasing current density up to 7 mA cm^{-2} (Fig. 3.35).

Although the further increase in current density (7–9 mA cm^{-2}) does not affect the coating mass, as it is shown in Fig. 3.32, this causes considerable decrease in the peak intensity ratio (Fig. 3.34) and crystallite domain size (Fig. 3.35). This suggests that a new effect of the increased current density on the coating morphology should be taken into the consideration: the increase in porosity caused by increased generation of the hydrogen bubbles at higher current density. The increase in current density increases the initial driving force for brushite nucleation (Eq. (3.10)), but it also accelerates the hydrogen evolution (Eq. (3.2)). As a consequence, the nucleation is promoted, but the crystallite growth is simultaneously suppressed by the increase in amount of the H_2 gas bubbles, contrary to the case of the coatings deposited at lower current densities, where the crystallite

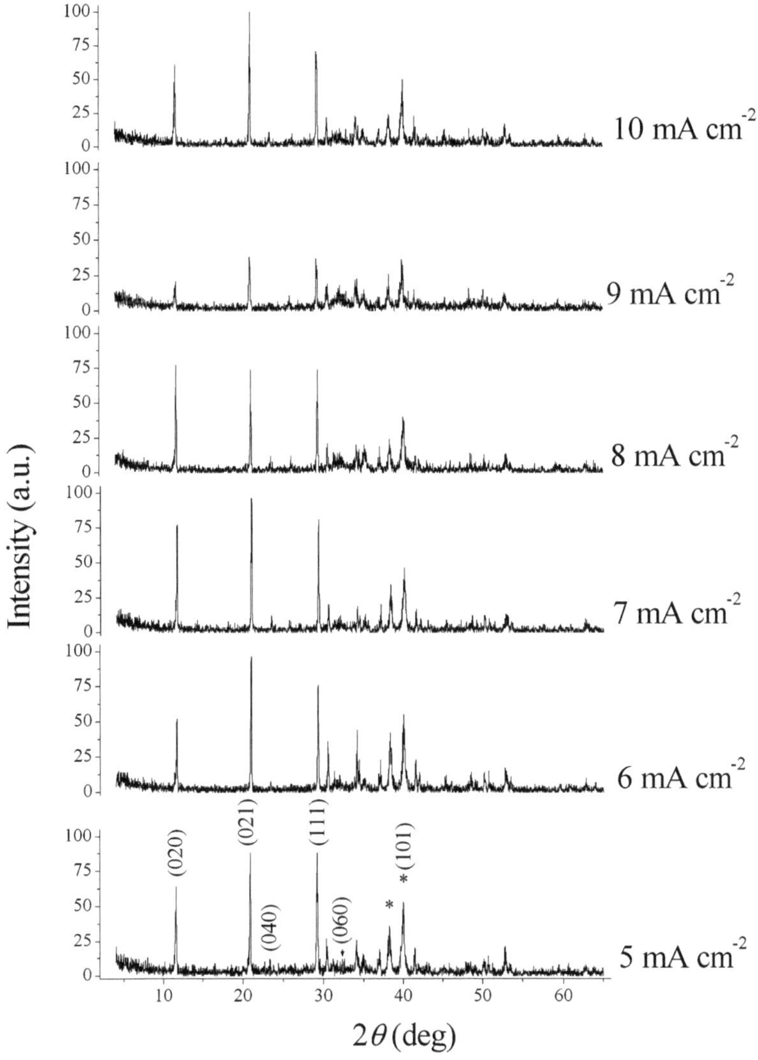

Fig. 3.33 XRD patterns of brushite coatings deposited on titanium at different current densities (deposition time: 30 min); (*asterisk*) titanium diffraction maximum (reprinted from [13] with permission from Elsevier)

growth dominated over hydrogen evolution reaction. This means that in high current region (7–9 mA cm^{-2}) the smaller crystallites were formed (Fig. 3.35), while the coatings become more porous (this is seen in Fig. 3.34 as considerable decrease in peak intensity ratio at 9 mA cm^{-2}). Therefore, the minimum of peak intensity ratio at current density of 9 mA cm^{-2} indicates the most porous coating due to the greatest amount of evolved H_2 gas bubbles, because at that high current density the hydrogen evolution reaction dominates over crystallite growth.

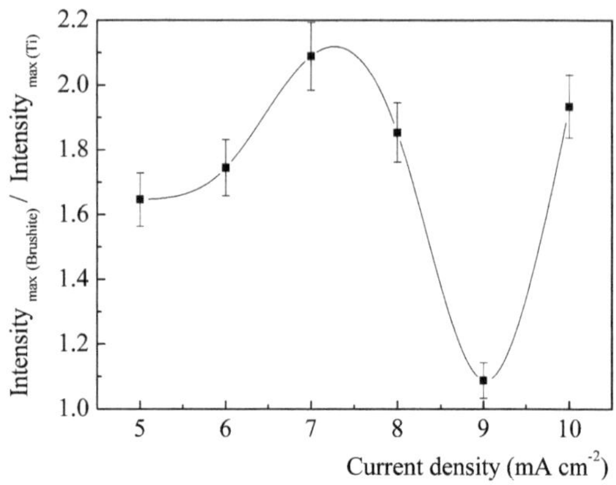

Fig. 3.34 Dependence of the ratio of intensities of the strongest diffraction maxima for brushite and titanium on applied current density (deposition time: 30 min) (reprinted from [13] with permission from Elsevier)

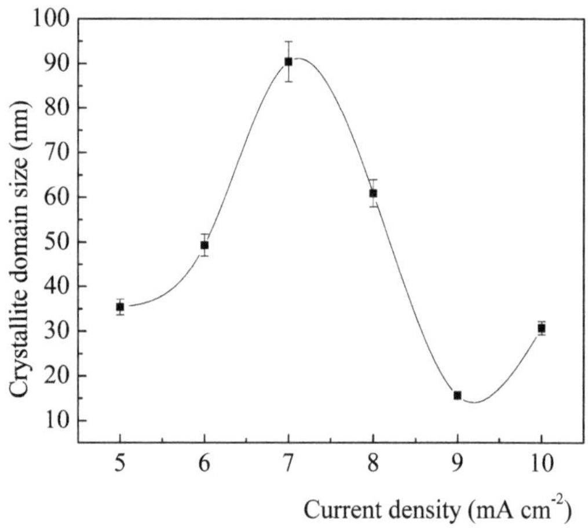

Fig. 3.35 Dependence of crystallite domain size of brushite coatings on applied current density (deposition time: 30 min) (reprinted from [13] with permission from Elsevier)

Finally, at the highest applied current density of 10 mA cm^{-2}, the peak intensity ratio (Fig. 3.34) and crystallite domain size (Fig. 3.35) increase with respect to the data at 9 mA cm^{-2}. Since the coating mass obtained at 9 and 10 mA cm^{-2} is the similar, this means that coating deposited at 10 mA cm^{-2} is less porous. This is a consequence of very fast H_2 gas bubbles evolution at the highest current density of 10 mA cm^{-2}, while generated H_2 bubbles leave more rapidly the electrode surface, enabling further deposition of less porous coating.

SEM micrographs of the coatings deposited at current densities of 7, 9, and 10 mA cm^{-2} are represented in Fig. 3.36a–c, respectively, indicating that brushite coatings have plate-like shape with volcano-like sites, e.g., plate-like crystals are arranged to produce tubular form. The volcano-like structure of the coatings is a consequence of the hydrogen bubbles evolved at the cathode. This statement is supported by the fact that the volcanoes are almost rounded, suggesting that the buildup of brushite plates appears around the gas bubble [129, 130]. Increase in applied current densities increases the number of volcano-like sites on the surface of brushite coatings due to the larger amount of evolved hydrogen (Fig. 3.36a, b).

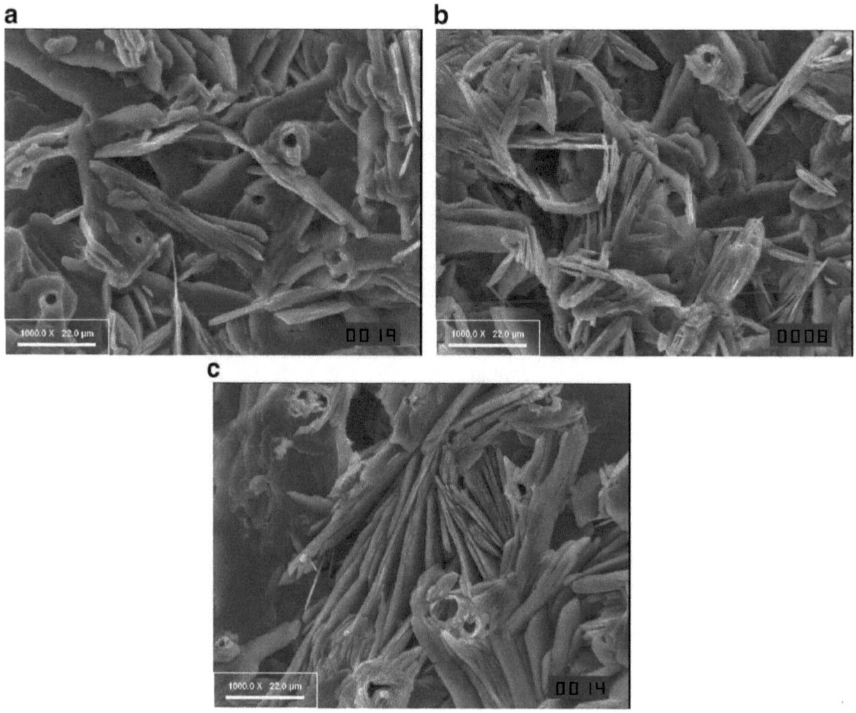

Fig. 3.36 SEM micrographs of brushite coatings electrodeposited at (**a**) 7 mA cm^{-2}, (**b**) 9 mA cm^{-2}, and (**c**) 10 mA cm^{-2} (deposition time: 30 min) (reprinted from [13] with permission from Elsevier)

However, at highest applied current density (Fig. 3.36c) volcano-like structure appears distorted, which could be assigned to the faster leaving of H_2 gas bubbles from the cathode surface. The uniform and rather compact parts of the coating deposited at 10 mA cm^{-2} is represented by ordered dense-packed plates of brushite (Fig. 3.36c), contrary to those deposited at lower current densities (Fig. 3.36a, b) with randomly distributed plates.

Based on all experimental data, it can be concluded that brushite coating with the smallest crystallite domain size of 15.6 nm and the greatest mass was obtained at current density of 9 mA cm^{-2}.

3.6.2 Conversion of Brushite Coatings to Hydroxyapatite Coatings

In order to investigate the effect of applied current density for brushite coating deposition on the morphology of precipitated HA coatings, the coatings deposited at 5 mA cm^{-2} (sample B5) and 10 mA cm^{-2} (sample B10) for 15 min, respectively,

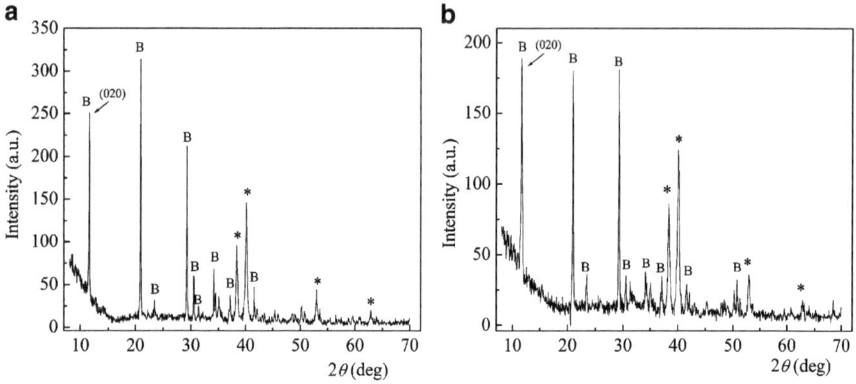

Fig. 3.37 XRD patterns of brushite coatings (before soaking in SBF) (**a**) deposited at 5 mA cm^{-2} and (**b**) deposited at 10 mA cm^{-2} (deposition time: 15 min); *B* brushite, (*asterisk*) titanium diffraction maximum (reprinted from [13] with permission from Elsevier)

were chosen for soaking in SBF solution for 7 and 14 days. Figure 3.37 represents the XRD patterns of brushite coatings B5 (Fig. 3.37a) and B10 (Fig. 3.37b) electrochemically deposited on titanium. The brushite and α-titanium (originate from the substrate) have been detected. The crystallite domain size, calculated for (020) plane, for coatings B5 and B10 amount to 72.5 nm and 71.8 nm, respectively. The ratio between the intensities of the strongest diffraction maxima of the brushite coating and the substrate for coatings B5 and B10 are 2.15 and 1.52, respectively, suggesting that coating deposited at lower current density is less porous than coating deposited at higher current density.

Figure 3.38 shows the XRD patterns of products obtained after soaking of brushite coatings B5 and B10 in SBF for 7 days (Fig. 3.38a, b, respectively), compared with standard XRD pattern for HA (JCPDS, 9-0432, Fig. 3.38c).

Phase composition of converted coatings (Fig. 3.38a, b) corresponds to hydroxyapatite, $Ca_{10}(PO_4)_6(OH)_2$ (JCPDS file 9-0432), and α-Ti (originate from the substrate), which means that complete surface, primarily coated with brushite, is fully converted to hydroxyapatite. The same phase composition (HA) was observed from XRD patterns of converted coatings soaked for 14 days in SBF. The crystallite domain size, calculated for (002) plane, for hydroxyapatite coatings after soaking of brushite coatings in SBF during 7 and 14 days, are represented in Table 3.4.

The results represented in Table 3.4 suggest that increase in soaking period from 7 to 14 days, for both brushite coatings, does not affect significantly the crystallite domain size of converted HA coatings. On the other hand, the applied current density for brushite coating deposition exhibited the great influence on crystallite domain size of converted HA coating. These results indicated that HA coating obtained after conversion of brushite coating deposited at higher current density (10 mA cm^{-2}) has significantly smaller crystallite domain size: 20.2 nm in respect to 45.2 nm, after 7 days, and 22.2 nm in respect to 48.5 nm, after 14 days. It was already pointed out that brushite coating deposited at higher current density is more porous.

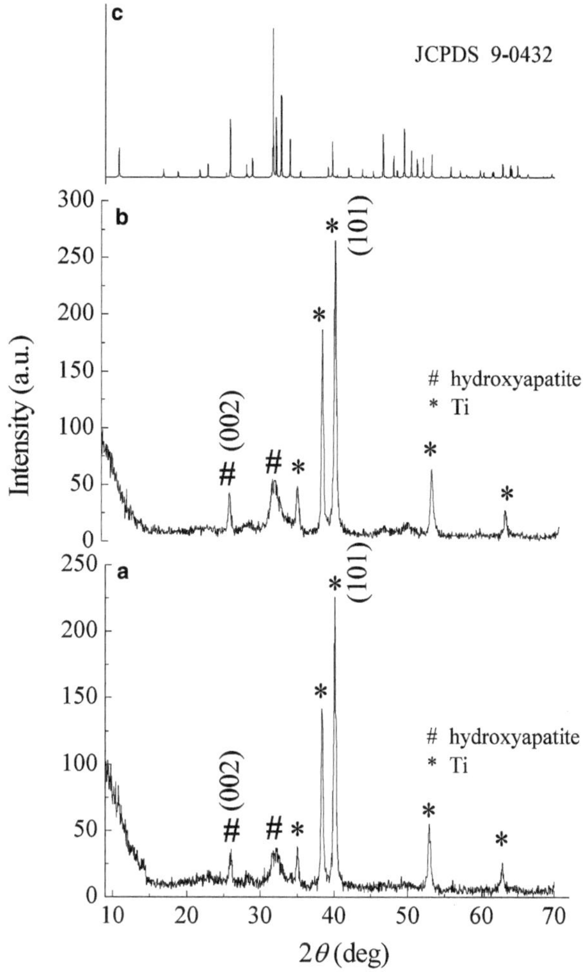

Fig. 3.38 XRD patterns of the hydroxyapatite coatings obtained after soaking of brushite coatings deposited at (**a**) 5 mA cm^{-2} and (**b**) 10 mA cm^{-2}, in SBF for 7 days; (**c**) standard hydroxyapatite (JCPDS 9-0432) (reprinted from [13] with permission from Elsevier)

Table 3.4 The crystallite domain size for HA obtained after soaking of brushite coatings during 7 and 14 days in SBF

Brushite coating/ deposition conditions	HA crystallite domain size (nm) Soaking period (days) 7	14
j = 5 mA cm^{-2} t = 15 min (B5)	45.2	48.5
j = 10 mA cm^{-2} t = 15 min (B10)	20.2	22.2

Reprinted from [13] with permission from Elsevier

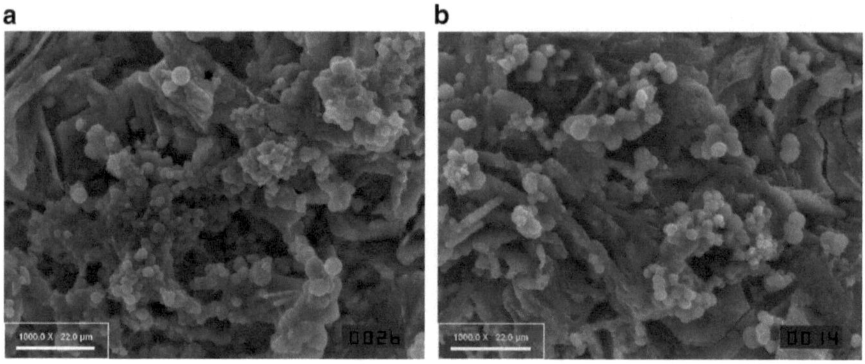

Fig. 3.39 SEM micrographs of the hydroxyapatite coatings obtained after soaking of brushite coatings, deposited at (**a**) 5 mA cm^{-2} and (**b**) 10 mA cm^{-2}, in SBF for 7 days (reprinted from [13] with permission from Elsevier)

So, it can be proposed that the crystallization of HA at more porous brushite coating (sample B10), although the crystallite domain size of brushite coatings before soaking is almost the same (72.5 nm and 71.8 nm, for B5 and B10, respectively), occurs on greater number of crystallization sites, leading to formation of greater number of smaller HA crystallites (crystallite domain size 20.2 nm). On the contrary, the crystallization of HA coating obtained at less porous brushite coating (deposited at lower current density of 5 mA cm^{-2}) occurs on smaller number of crystallization sites, causing the formation of smaller number of bigger HA crystallites (crystallite domain size 45.2 nm).

Figure 3.39 represents the SEM micrographs of hydroxyapatite coatings obtained after conversion of brushite coatings B5 and B10 in SBF for 7 days, indicating that samples are fully covered with agglomerated sphere-like crystallites of HA. The same structures were revealed after 14 days of immersion.

3.7 Hydroxyapatite Coatings

It is well known that hydroxyapatite (HAP, HA) is an important material for biomedical applications, as its chemical composition is similar to that of bone tissue [41, 141–143]. Unlike HAP, titanium and titanium alloys are proven to be potentially very suitable materials for load-bearing bioimplant applications. Titanium and some of its alloys are most often the material of choice for metallic implants due to their attributes of strength, stiffness, toughness, impact resistance, and corrosion resistance [144–149]. Hence, a good combination of the biocompatibility of hydroxyapatite and the excellent mechanical properties of metals is considered a promising approach to fabricate more suitable bone implants. The concept of coating metal implant surfaces with HAP combines the mechanical benefits of metal alloys with the biocompatibility of HAP [42].

However, HAP is very brittle, and for this reason, a great deal of research concentrates on the development of composite HAP coatings. Composite coatings impart other functional properties to the implants, such as chemical stability, bioactivity, biocompatibility, and antimicrobial properties [33]. Electrophoretic deposition cannot be used for the fabrication of dense coatings on metals due to the sintering shrinkage, chemical reactions between the coatings and the substrates, and other problems discussed in the literature [150]. These problems can be eliminated by the fabrication of polymer–ceramic composites. Development of biocomposite HAP/polymer coatings is especially important for applications in medicine, specifically transplantational surgery. The characteristics of composite coating that make it biocompatible are non-toxicity, corrosion stability, and controlled biodegradibility, as well as elastic modulus, therefore suitable for specific biomedical applications. The use of polymers offers the advantage of low temperature processing of composite materials [151]. Significant interest and investigations were focused on the fabrication of composite HAP/chitosan coatings [1, 152–157], HAP/chitosan/carbon nanotube [158], HAP/lignin composites and coatings [2, 3, 159–163], aluminosilicate nanotubes/HAP/hyaluronic acid coatings [164], HAP/alginate coatings [5], Bioglass®/HAP/chitosan and Bioglass®/ HAP/alginate coatings [165], HAP/glucose coatings [166], and Y_2O_3/HAP coatings [167].

3.7.1 *Electrochemical Synthesis of Hydroxyapatite Powder*

The preparation of hydroxyapatite powders with controlled morphology, stoichiometry, crystallinity, and, particularly, size in the nanometer range has the main role in production of biomaterials [168]. Various synthesis methods have been used for HA powder production, e.g., solid-state reaction [169], chemical precipitation method [170], formation of HA crystals on glass [171] or brushite [172], synthesis of nanocrystalline powders in simulated body fluid (SBF) [173], growth of HA crystal using the gel systems [174], sol–gel synthesis [175, 176], and surfactant templating method [177]. The possibilities of application for human tissue depend upon HA crystallite sizes—130 × 30 nm for adult enamel, 20 × 4 nm for dentine, and 25 × 3 nm for bone [178]. The properties such as bioactivity, dissolution range, and resorption are close to those of natural bones.

Electrochemical synthesis of hydroxyapatite (HA) powders was performed galvanostatically at current densities of 137 and 207 mA cm^{-2} from homogeneous aqueous solution of $Na_2H_2EDTA{\cdot}2H_2O$, NaH_2PO_4, and $CaCl_2$ at the concentration relationship Ca/EDTA/PO_4^{3-} of 0.25/0.25/0.15 M, and at different pH values of 3.0, 9.0, and 12.0, during 3 h [91]. The aim was to investigate the effect of applied current density and solution pH value on the phase composition, crystallite size, morphology, and thermal characteristics of obtained hydroxyapatite powders.

The mechanism of HA formation requires the liberation of Ca^{2+} ions from EDTA [123]:

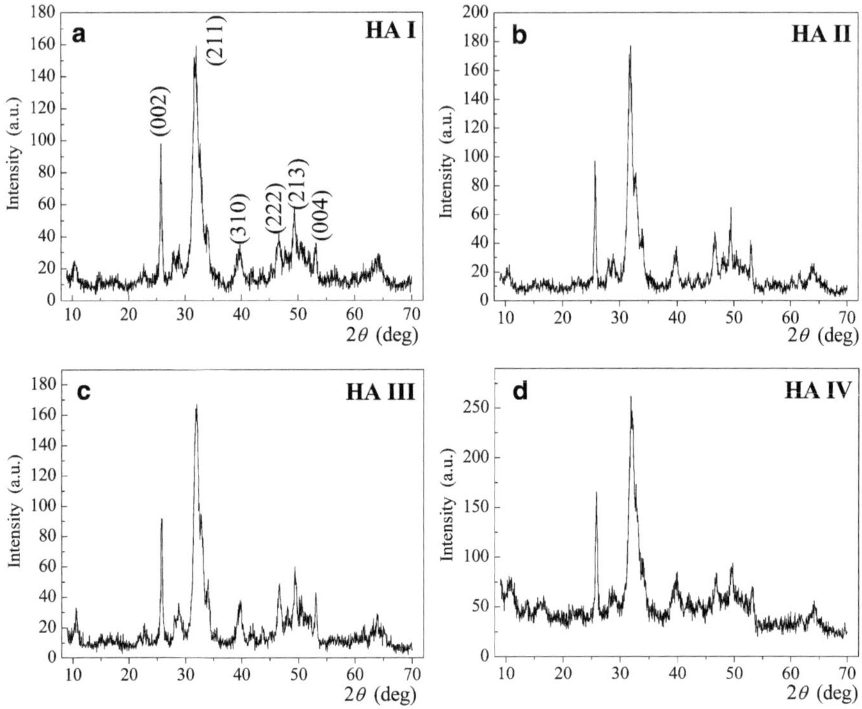

Fig. 3.40 XRD patterns of HA powders synthesized at (**a**) pH 9.0 and $j = 137$ mA cm^{-2} (sample HA I), (**b**) pH 12.0 and $j = 137$ mA cm^{-2} (sample HA II), (**c**) pH 9.0 and $j = 207$ mA cm^{-2} (sample HA III), and (**d**) pH 12.0 and $j = 207$ mA cm^{-2} (sample HA IV) (reprinted from [91] with permission from Elsevier)

$$Ca(EDTA)^{2-} \leftrightarrows Ca^{2+} + EDTA^{4-} \tag{3.12}$$

followed by generation of OH^- ions through water reduction according to Eq. (3.2) and ionic migration toward the electrodes, where precipitation of HA occurs:

$$10Ca^{2+} + 6PO_4{}^{3-} + 2OH^- \leftrightarrows Ca_{10}(PO_4)_6(OH)_2 \tag{3.13}$$

Figure 3.40 represents the XRD patterns of HA powders synthesized at pH 9.0 and $j = 137$ mA cm^{-2} (sample HA I, Fig. 3.40a), pH 12.0 and $j = 137$ mA cm^{-2} (sample HA II, Fig. 3.40b), pH 9.0 and $j = 207$ mA cm^{-2} (sample HA III, Fig. 3.40c), and pH 12.0 and $j = 207$ mA cm^{-2} (sample HA IV, Fig. 3.40d). The only detected crystalline phase in all samples was hydroxyapatite, $Ca_{10}(PO_4)_6(OH)_2$.

The crystallite domain size was calculated using (002) plane and was observed (Table 3.5) that crystallite domain size does not depend significantly on pH value as

Table 3.5 Crystallite domain size of HA powders synthesized at different current densities and different pH values

Sample	pH value	j (mA cm^{-2})	D_{002} (nm)
HA I	9.0	137	21.6
HA II	12.0	137	23.4
HA III	9.0	207	24.3

Reprinted from [91] with permission from Elsevier

Table 3.6 The values of average HA particle diameter and particle size distribution (PSD) by intensity of HA I suspension

Size measurement	HA I suspension
z-Average diameter (nm)	583
PSD intensity	
Peak 1: z, nm (Int., %)	0.938 (95.3)
Peak 2: z, nm (Int., %)	331 (4.7)

Reprinted from [91] with permission from Elsevier

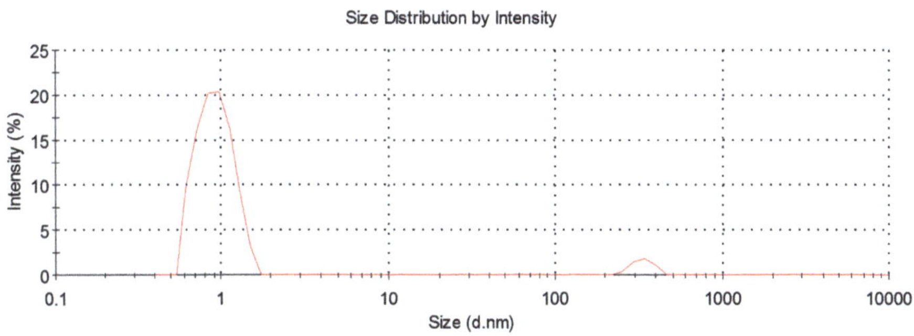

Fig. 3.41 Particle size distribution of HA I suspension (reprinted from [91] with permission from Elsevier)

well as on applied current density. However, the smallest crystallite size was obtained at lower pH value and lower current density (sample HA I).

The measurements of particle size distribution (PSD) by intensity were made using HA I suspension. It is evident that the prepared HA I suspension was a high degree polydisperse one (Table 3.6). The average HA particles diameter is 583 nm. However, based on PSD more than 95 % HA particles have an average diameter of ~ 1 nm. Size distribution by intensity for HA I suspension is presented in Fig. 3.41.

Typical TEM images of permeate obtained after double ultrafiltration of HA I suspension are presented in Fig. 3.42. It can be noticed that the primary particles are of nanometer range, e.g., about 1 nm, according to existence of different contrast on the TEM images. This result is in good agreement with the particle size distribution and XRD analysis, which revealed that the particles were in the nanometer size range.

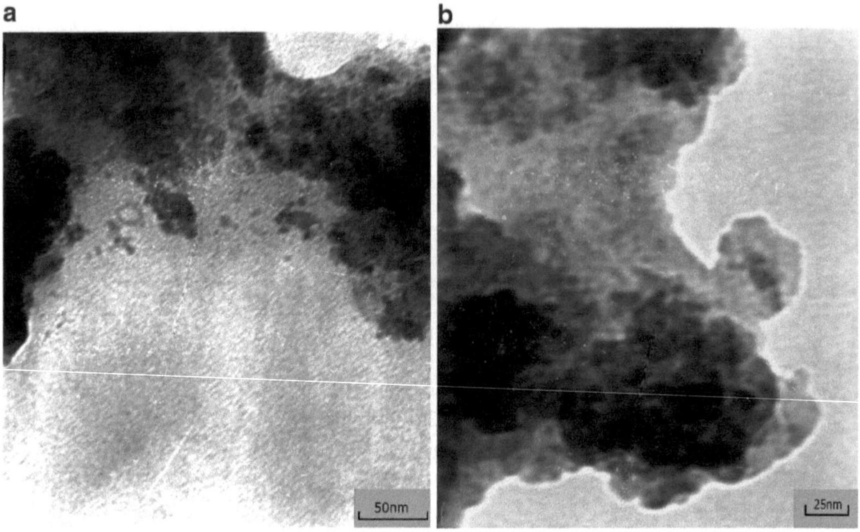

Fig. 3.42 TEM micrographs of permeate obtained after double ultrafiltration of HA I suspension (reprinted from [91] with permission from Elsevier)

SEM micrographs of HA powders are shown in Fig. 3.43. All of the investigated powders consist of big agglomerates which are composed of fine crystallites. The spherule agglomerates can be observed in the case of HA I and HA III (Fig. 3.43a, c), while the HA II and HA IV present plate-like morphology (Fig. 3.43b, d). According to these results it could be proposed that the morphology of agglomerates depends on the pH value of the solution. Namely, HA powders obtained at lower pH values (HA I and HA III, pH 9.0) consist of spherule agglomerates, while HA powders obtained at higher pH (HA II and HA IV, pH 12.0) have the plate-like structure. In addition, it can be suggested that applied current density does not affect the HA powder morphology.

The crystallite domain size of 21.6 nm for HA I, determined by X-ray diffraction (Table 3.5), indicates that agglomerates were formed through the drying process, which is confirmed by SEM analysis (Fig. 3.43a). This is in good agreement with the nucleation–aggregation–agglomeration–growth mechanism [121].

DTA curves for HA powders are represented in Fig. 3.44. The first endothermic peak observed at DTA curves at about 80 °C can be attributed to desorption of water molecules adsorbed on the crystallite surface. The subsequent strong exothermic peak at about 300 °C results from the combustion of organic material [179], e.g., EDTA. Another exothermic peak appears at 1,305 °C, indicating the decomposition of HA to anhydrous calcium phosphates such as tri-calcium phosphate (TCP), $Ca_3(PO_4)_2$. Namely, according to the literature [180, 181] HA decomposes to tri-calcium phosphate at 1,200–1,400 °C. At 1,280 °C well-crystallized HA and β-TCP exists [180], and at temperatures higher than 1,350 °C, β-TCP irreversibly transforms to α-TCP [181].

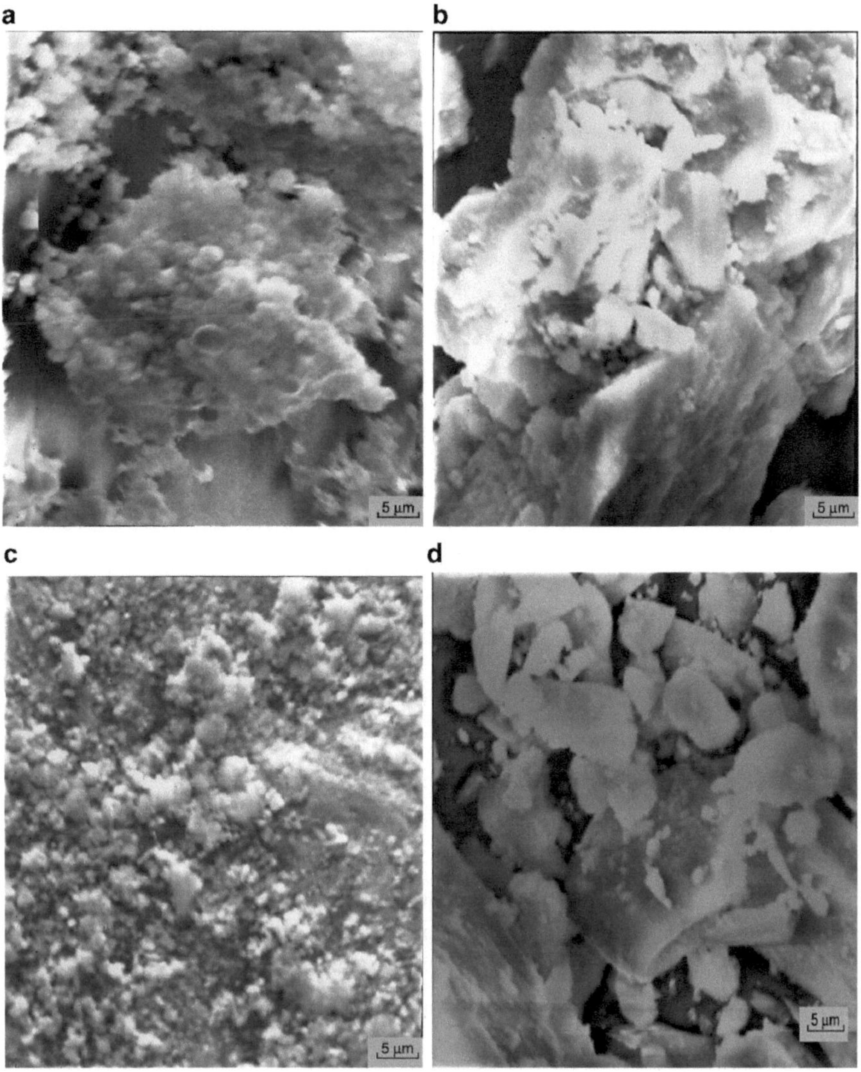

Fig. 3.43 SEM micrographs of hydroxyapatite powders (**a**) HA I, (**b**) HA II, (**c**) HA III, and (**d**) HA IV (reprinted from [91] with permission from Elsevier)

The weight loss and temperatures associated with phase transformation were determined by thermogravimetric analysis. The TG curves of hydroxyapatite powders obtained in the temperature range between 23 and 1,400 °C are shown in Fig. 3.45a, while Fig. 3.45b represents the differential TG (DTG) curves.

The TG curves of HA powders exhibit three weight loss stages. The first stage was observed from 23 to about 80 °C (Fig. 3.45a), with sharp peaks at DTG curves at 71, 63, and 76 °C for samples HA I, HA II, and HA III, respectively (Fig. 3.45b).

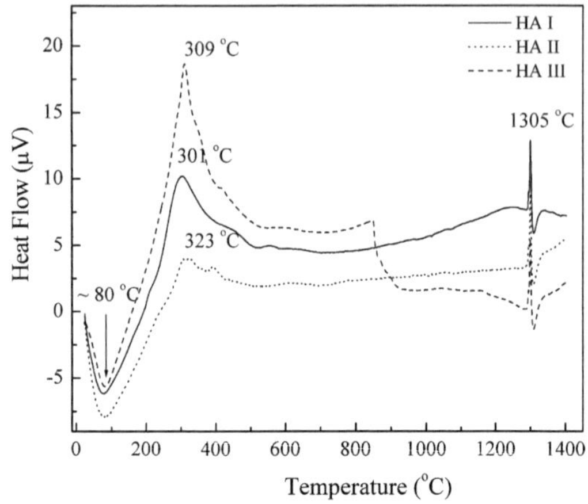

Fig. 3.44 DTA curves for hydroxyapatite powders HA I, HA II, and HA III (reprinted from [91] with permission from Elsevier)

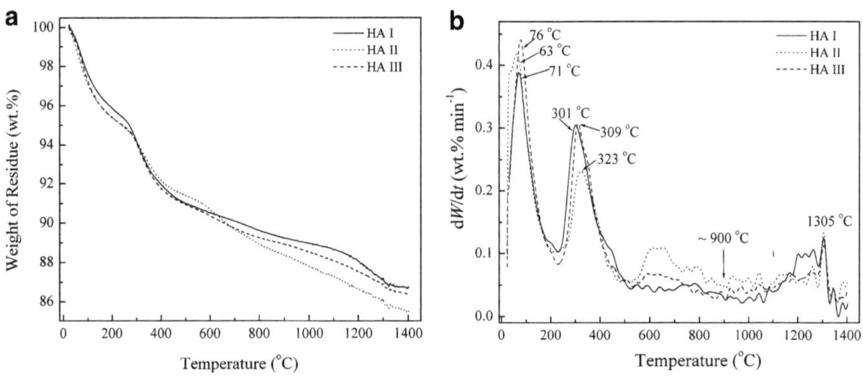

Fig. 3.45 TG curves (**a**) and differential TG (DTG) curves (**b**) of hydroxyapatite powders HA I, HA II, and HA III (reprinted from [91] with permission from Elsevier)

This stage corresponds to desorption of water molecules adsorbed on the crystallite surface.

The second stage of weight loss of the TG curves was observed between 80 and 530 °C (Fig. 3.45a), with sharp peaks at DTG curves at 301, 309, and 323 °C for samples HA I, HA II, and HA III, respectively. This stage corresponds to the combustion of organic material (EDTA) [179, 182].

The third stage of weight loss of the TG curves was observed between 530 and 1,400 °C (Fig. 3.45a), with peaks at DTG curves (Fig. 3.45b) at around 930 and 1,305 °C, for all HA samples. The endothermic peak observed at around 930 °C can be attributed to the dehydroxylation of HA and formation of oxyhydroxyapatite, $Ca_{10}(PO_4)_6(OH)_{2-2x}O_x\square_x$, where $\square$ is vacancy and $x < 1$ [183]. Next exothermic peak, recognized at 1,305 °C, indicates the decomposition of HA to anhydrous calcium phosphate such as tri-calcium phosphate (TCP) [181].

The total weight loss for investigated temperature interval between 23 and 1,400 °C was 13.3, 14.5, and 13.7 wt% for HA I, HA II, and HA III, respectively, which can be attributed to the desorption of water molecules adsorbed on the crystallite surface, dehydroxylation of HA to oxyhydroxyapatite, and decomposition of HA to tri-calcium phosphate.

From DTA/TG analysis it can be concluded that parameters of electrochemical synthesis of HA powders (applied current density and pH of the solution) do not effect significantly the weight loss and temperatures associated with phase transformation during thermal treatment.

3.7.2 Hydroxyapatite and Hydroxyapatite/Lignin Composite Coatings

Based on recent research developments organosolv lignin emerged as a suitable candidate for composite hydroxyapatite/polymer coatings. Lignin (Lig) is a complex natural polymer composed primarily of phenolic moieties connected with a variety of chemical bonds [184]. Lignin possesses antioxidant and antimicrobial properties; therefore, its incorporation is interesting in medical applications because of its thermal stability and biocompatibility in different materials [159, 161–163]. Organosolv lignins are being examined because they show significantly better solubility and thermal properties than sulfite or kraft lignins [160]. Incorporation of lignin is interesting in medical applications because it led to thermal stability, hydrophilicity, biocompatibility, and biodegradability of different materials [159, 161].

As complex natural polymer networks composed primarily of phenolic moieties, lignins have a wide variety of chemical bonds [163, 184]. Among the functional groups present in lignin, the most reactive chemical sites are phenolic hydroxyl groups [185]. Other major chemical functional groups in lignins include methoxyl, carbonyl, and carboxyl groups, depending on the plant origin and the applied pulping processes [186]. Organosolv lignins are being examined because they show significantly better solubility and thermal properties than sulfite or kraft lignins [160]. The organosolv processes are convenient because of the greater simplicity of the chemical recovery system, since only the solvent has to be recovered by rectification of the black liquors [185]. In contrast to Kraft and sulfite lignin, Alcell (organosolv) lignin derives from a process using ethanol as the only pulping chemical and consists of low molecular weight phenol fragments with enhanced hydrophobicity [184]. Hence, this organosolv lignin, in its purified form, possesses a chemical structure different to that of native lignin. The incorporation of lignin is interesting in medical applications because of its thermal stability and biocompatibility in different materials [161, 162, 187].

Table 3.7 The values of average particle diameter and particle size distribution (PSD) by volume of HAP/Lig suspension with 1 wt% Lig

Size measurement	HAP/Lig suspension
z-average diameter, nm	363
PSD volume	
Peak 1: *z*, nm (Int., %)	174 (28.6)
Peak 2: *z*, nm (Int., %)	653 (46.6)

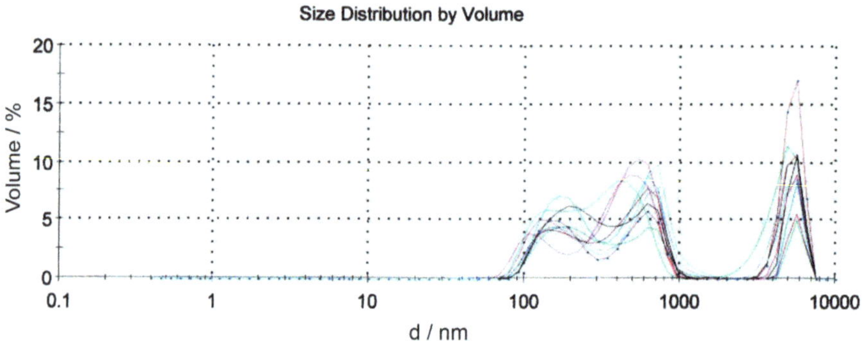

Fig. 3.46 Particle size distribution by volume of the HAP/Lig suspension with 1 wt% Lig

3.7.2.1 Electrophoretic Deposition of Hydroxyapatite and Hydroxyapatite/Lignin Composite Coatings

HAP/Lig biocomposite coatings with 1 wt% Lig were formed on titanium by EPD method, and the effect of applied voltage and deposition time on HAP/Lig deposit mass and morphology was investigated [163]. It was shown that adherent HAP/Lig coatings formed on titanium at lowering sintering temperature, while lignin affect significantly the microstructure, morphology, and thermal behavior of HAP coatings.

Alcell lignin was extracted from a mixture of hardwoods (maple, birch, and poplar) by an organosolv process using aqueous ethanol. The lignin powder was used as received and without further purification. A nanosized HAP powder was obtained using a modified chemical precipitation method, by the reaction of calcium oxide (obtained by calcination of $CaCO_3$ for 5 h at 1,000 °C in air) with phosphoric acid, described earlier [188, 189] according to the following reaction:

$$10Ca(OH)_2 + 6H_3PO_4 \rightarrow Ca_{10}(PO_4)_6(OH)_2 + 18H_2O \qquad (3.14)$$

Using particle size distribution (PSD) by volume measurements, the average particle diameter was calculated to be 363 nm, while 28.6 % particles have an average diameter of 174 nm and 46.6 % particles have an average diameter of 653 nm (Table 3.7 and Fig. 3.46).

The HAP coating and HAP/Lig coating with 1 wt% Lig were deposited on titanium from ethanol suspensions at different values of constant voltage between

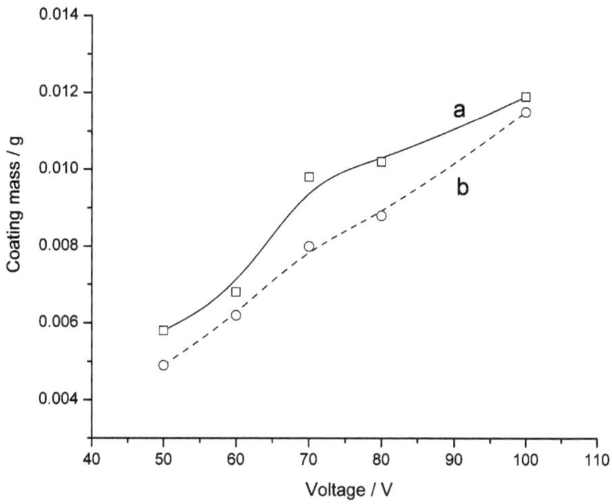

Fig. 3.47 The mass of HAP coating (**a**) and HAP/Lig coating with 1 wt% Lig (**b**) as a function of applied deposition voltage (according to [163])

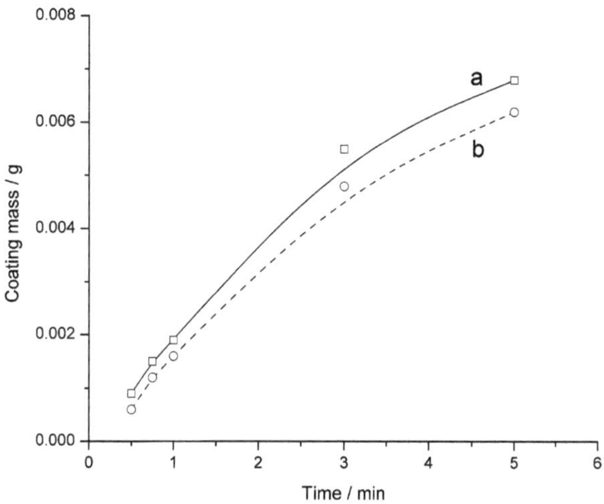

Fig. 3.48 The mass of HAP coating (**a**) and HAP/Lig coating with 1 wt% Lig (**b**) as a function of deposition time (according to [163])

30 and 150 V and for various deposition times from 30 s to 5 min [163]. The deposited HAP and HAP/Lig coatings were air dried at room temperature and then sintered at 900 °C in argon atmosphere. The effect of applied voltage on the mass of HAP and HAP/Lig coatings is represented in Fig. 3.47. The increase in applied voltage up to 100 V increases the HAP and HAP/Lig coating mass due to enhanced rate of particle migration.

The increase in deposition time enhances the mass of HAP and HAP/Lig coatings since more particles are reaching the cathode (Fig. 3.48).

XRD patterns of HAP powder and HAP coating on titanium after sintering at 900 °C are presented in Fig. 3.49a, b, while Fig. 3.50a, b represent XRD patterns of HAP/Lig coatings with 1 wt% Lig before and after sintering at 900 °C.

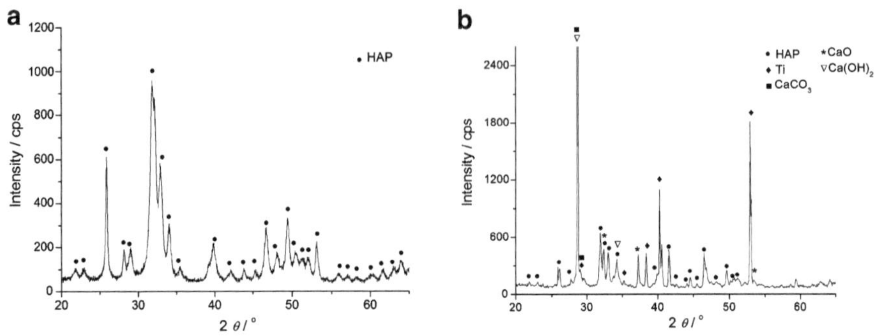

Fig. 3.49 XRD patterns of HAP powder (**a**) and HAP coating on titanium (**b**) after sintering at 900 °C (according to [163])

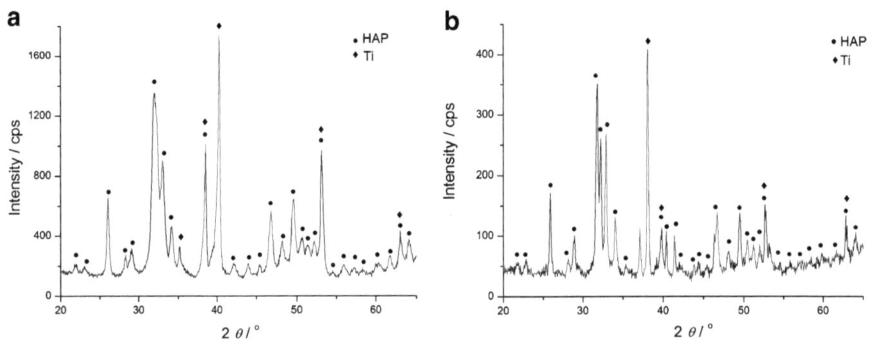

Fig. 3.50 XRD patterns of HAP/Lig coatings with 1 wt% Lig on titanium before (**a**) and after sintering at 900 °C (**b**) (according to [163])

All the peaks in XRD pattern in Figs. 3.49 and 3.50 perfectly matched the JCPDS pattern 09-0432 for hydroxyapatite, meaning that sintered HAP coating and HAP/Lig coating with 1 wt% Lig were obtained at significantly lower sintering temperature of 900 °C comparing to the usually reported to be between 1,000 and 1,300 °C [38]. This could be attributed to the nanosized particles of the initial HAP previously synthesized by the precipitation method [189]. This is very important fact, because a higher degree of crystallinity would make coatings less prone to dissolution in body fluids [190]. Ma et al. [38] reported that the HAP coatings became denser as the temperature increases from 1,100 °C to 1,300 °C. The high sintering temperature will result in the decomposition of HAP coatings with reduced biocompatibility and increased in vitro dissolution. Therefore, sintering of HAP coatings needs to be performed at temperatures lower than 1,050 °C [190]. Since specific peaks indicating the presence of CaO were revealed only in XRD pattern for sintered HAP coating (Fig. 3.49b), the partial HAP decomposition during sintering is confirmed only for pure HAP coating on titanium, while in the case of HAP/Lig coating decomposition was not observed (Fig. 3.50b).

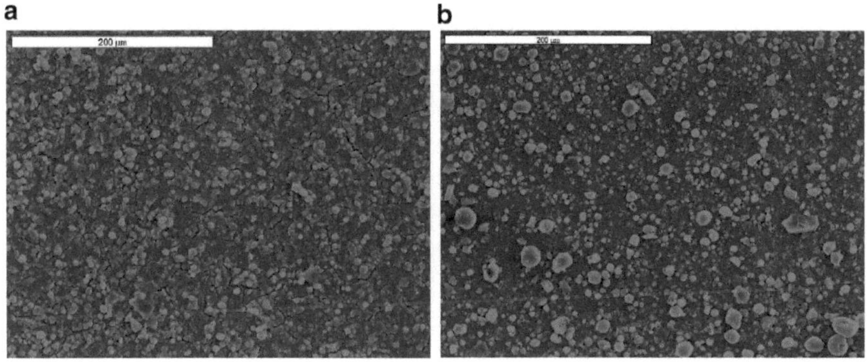

Fig. 3.51 SEM micrographs of sintered HAP coating (**a**) and HAP/Lig coating with 1 wt% Lig (**b**) on titanium

In order to avoid fractured surface, the HAP and HAP/Lig coatings on titanium were electrodeposited at 60 V for deposition times shorter than 1 min [163]. It can be seen from Fig. 3.51a that sintered HAP coating electrodeposited for 45 s still exhibited fractured surface while sintered HAP/Lig coating with 1 wt% Lig revealed homogenous surface without fractures (Fig. 3.51b) due to different hydrogen bonds between (1) phenolic hydroxyl –OH from lignin and –OH from HAP, (2) α-carbonyl from lignin side chain –C=O and –OH from HAP, (3) –OH from lignin and PO_4^{3-} from HAP, and (4) ether bond oxygen from lignin, C–O–C from lignin, and –OH from HAP [163].

Figure 3.52 represents the TGA curves for HAP coating and HAP/Lig coating with 1 wt% Lig on titanium.

TGA curve for HAP coating (curve a) demonstrates three majors weight loss steps: from 40 to 350 °C related to the loss of water adsorbed on the HAP coating, from 350 to 750 °C where no notable weight loss occurs, and from 750 to 1,000 °C may be related to HAP dehydroxylation or decomposition. The total weight loss at 1,000 °C is 4.4 %. However, TGA curve for HAP/Lig coating (curve b) shows continuous weight loss: from 40 to 600 °C the weight loss can be attributed to the loss of absorbed water and thermal decomposition of lignin, from 600 to 750 °C the weight loss can be related to the slow decomposition of HAP and remaining lignin, and from 750 to 1,000 °C it may be attributed to HAP dehydroxylation or decomposition. One can observe that HAP/Lig coating is thermally more stable than HAP coating from 40 to 485 °C. The total weight loss of HAP/Lig coating is 5.5 %. This result also confirms that lignin is well incorporated into the composite coating because the difference in weight loss between HAP and HAP/Lig coatings of 1 % corresponds to the initial addition of lignin during synthesis [163].

Figures 3.53 and 3.54 illustrate the ATR-FTIR spectra in the range of 800–1,200 cm^{-1} and 3,400–3,800 cm^{-1}, respectively, for non-sintered and sintered HAP and HAP/Lig coatings. From the spectrum of non-sintered HAP (Figs. 3.53a and 3.54a), the stretching bands at 3,570 cm^{-1}, originating from OH^- groups, as well as

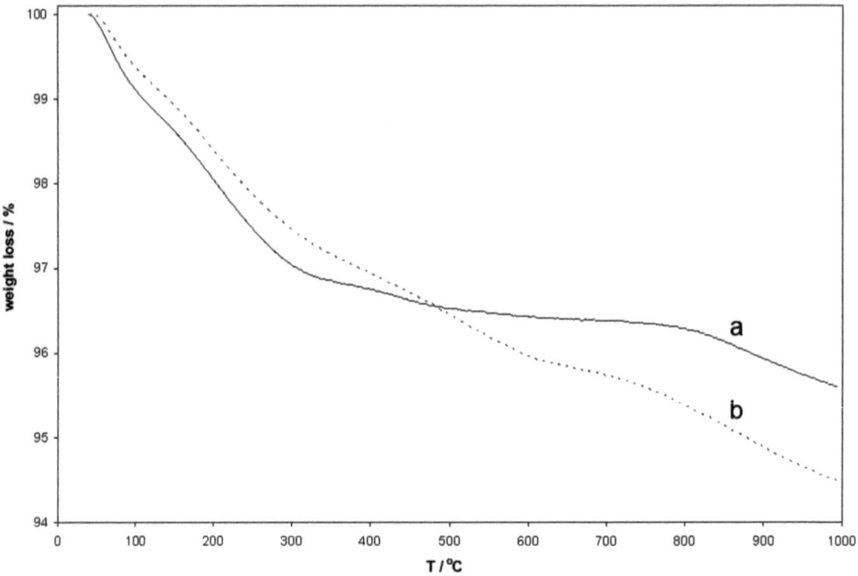

Fig. 3.52 TGA curves of HAP coating (**a**) and HAP/Lig coating with 1 wt% Lig (**b**) on titanium (according to [163])

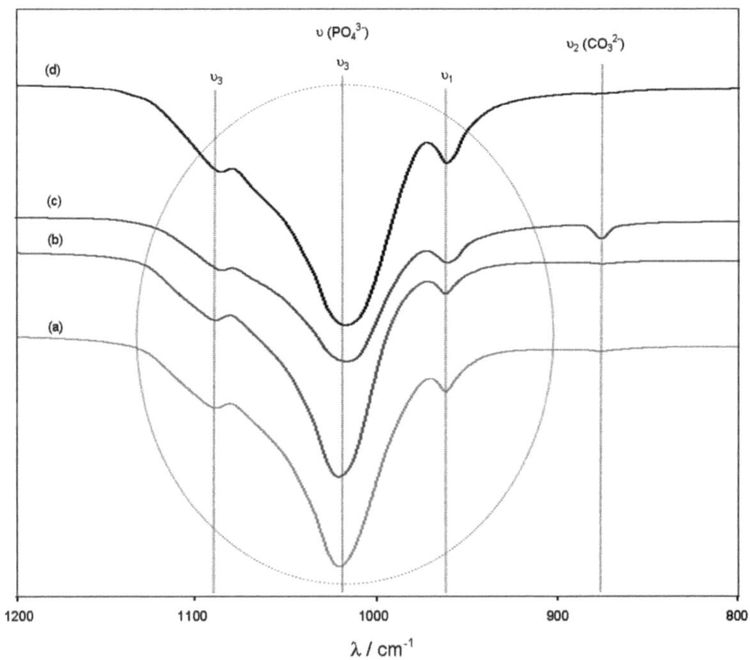

Fig. 3.53 ATR-FTIR spectra of non-sintered HAP coating (**a**), non-sintered HAP/Lig coating with 1 wt% Lig (**b**), sintered HAP coating (**c**), and sintered HAP/Lig coating with 1 wt% Lig (**d**), in the range of 800–1,200 cm^{-1} (according to [163])

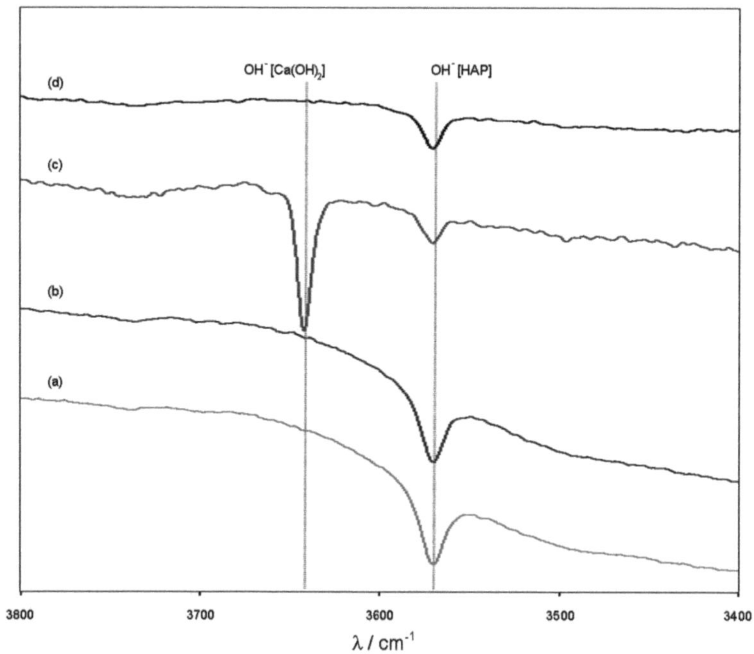

Fig. 3.54 ATR-FTIR spectra of non-sintered HAP coating (**a**), non-sintered HAP/Lig coating with 1 wt% Lig (**b**), sintered HAP coating (**c**), and sintered HAP/Lig coating with 1 wt% Lig (**d**), in the range of 3,400–3,800 cm^{-1} (according to [163])

characteristic bands due to the stretching modes of PO_4^{3-} ions at 960 ($\upsilon 1$), 1,018 ($\upsilon 3$), and 1,088 cm^{-1} ($\upsilon 3$) [191] are clearly visible. A broad peak related to H_2O adsorption was noticed at around 3,375 cm^{-1}. The split bands, at 1,018 and 1,088 cm^{-1}, seem to indicate the formation of a well-crystallized apatite [192] and confirm the XRD pattern of HAP [189]. The spectra of non-sintered HAP/Lig coating (Figs. 3.53b and 3.54b) show quite similar features as the spectra of non-sintered HAP coating, meaning that the implication of lignin in the synthesis of HAP/Lig coating does not change the formation and structure of HAP. However, the band corresponding to the formation of apatite (1,018 cm^{-1}) becomes weaker and sharper than that of pure HAP coating which indicates the formation of HAP/lignin composites [193]. After sintering, changes in the shapes and intensities of the individual bands characteristic for HAP coating can be observed, whereas only changes in the intensities of the individual characteristics for HAP/Lig coating occur (Figs. 3.53c, d and 3.54c, d). The decrease of OH^- stretching band, for both sintered HAP (Fig. 3.54c) and HAP/Lig coatings (Fig. 3.54d), indicates that dehydroxylation occurs during the sintering. In addition, the broad peak corresponding to H_2O adsorption becomes flat, particularly for HAP/Lig coating. However, only in the case of sintered HAP coating $\upsilon 3(PO_4^{3-})$ band at 1,018 cm^{-1} becomes weak, and new bands appear. One can observe in Fig. 3.53c the emergence of carbonate peak at 873 cm^{-1} (CO_3^{2-} in OH^- site), as well as the band at

Table 3.8 Ca/P ratio for the HAP coating and HAP/Lig coating with 1 wt% Lig

Coating		Ca/P
HAP	Non-sintered	1.72
	Sintered	3.00
HAP/Lig (1 wt% Lig)	Non-sintered	1.79
	Sintered	2.10

According to [163]

3,642 cm^{-1} related to OH^- group in $Ca(OH)_2$ (Fig. 3.54c), suggesting the formation of CaO during sintering by the following reaction [194]:

$$Ca_{10}(OH)_2(PO_4)_6 \rightarrow 3Ca_3(PO_4)_2 + CaO + H_2O \tag{3.15}$$

This indicates that pure HAP coating partially decomposes during sintering. It is also reported [195] that one of the predominant phases formed during sintering the Ti/HAP samples is detected as CaO. In addition, Filiaggi et al. [196] reported that during the thermal process, diffusion of calcium (limited) and phosphorous ions into the Ti substrate can take place, resulting in the decomposition of HAP block. However, decomposition during sintering does not occur with HAP/Lig coating with 1 wt% Lig. It seems that lignin protects HAP/Lig coating by binding to the hydroxyls of the HAP lattice, thus preventing decomposition and/or ion diffusion on Ti surface.

The quantitative analysis of the XPS spectra from high-resolution measurements was performed [163]. The results in Table 3.8 show that the non-sintered HAP coating (Ca/P ratio 1.72) and HAP/Lig coating with 1 wt% Lig (Ca/P ratio 1.79) consist of calcium phosphate with a Ca/P ratio close to synthetic apatite conditions (Ca/P ratio 1.67). Even though the ideal Ca/P ratio for stoichiometric HAP is known to be 1.67, stable HAP phases have been found to exist over a range of Ca/P ratios between 1.3 and 1.8 [197]. After sintering, Ca/P ratio increased up to 3.00 and 2.10 for HAP and HAP/Lig coatings respectively, which can be related to the diffusion of phosphorous ions into Ti surface. On the other hand, the increase in carbon content after sintering was noticed, which is more pronounced in the case of pure HAP coating, confirming again the formation of CaO as a result of HAP decomposition [163].

3.7.2.2 The Effect of Lignin on the Structure and Characteristics of Hydroxyapatite/Lignin Composite Coatings

The influence of the lignin concentration in the range of 0.5–10 wt% of lignin on the microstructure, morphology, phase composition, thermal behavior, and cytotoxicity of composite HAP/Lig coatings electrodeposited on titanium was investigated in order to find the optimal lignin concentration for producing HAP/Lig composite coatings [2].

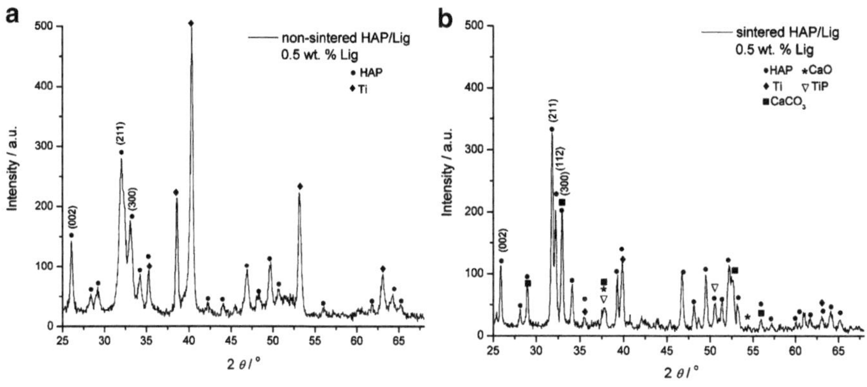

Fig. 3.55 XRD patterns of (**a**) non-sintered and (**b**) sintered HAP/Lig coatings with 0.5 wt% Lig (reprinted from [2] with permission from Elsevier)

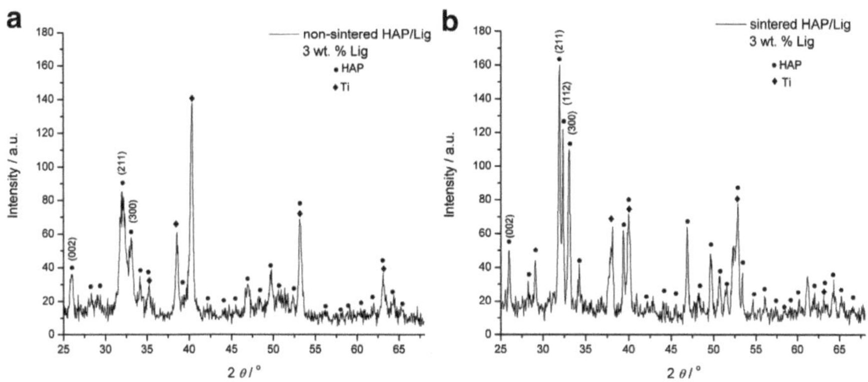

Fig. 3.56 XRD patterns of (**a**) non-sintered and (**b**) sintered HAP/Lig coatings with 3 wt% Lig (reprinted from [2] with permission from Elsevier)

The XRD patterns of non-sintered and sintered HAP/Lig coatings with 0.5, 3, and 10 wt% Lig are represented in Figs. 3.55, 3.56, and 3.57, respectively. All the peaks in XRD patterns of the non-sintered HAP/Lig coatings correspond to the JCPDS pattern No. 09-0432 for hydroxyapatite, confirming the identity of HAP in the studied coatings. After sintering, the diffraction peaks of HAP/Lig coatings become sharper and of higher intensity with a decrease in peak width, which all indicate that the sintered coatings had a better crystallinity. Due to the nanosize of the HAP particles, a sintering temperature of 900 °C was successfully applied for the thermal treatment of the composite HAP/Lig coatings, although the usually applied sintering temperature is in the range between 1,000 and 1,300 °C [38].

The appearance of new peaks in the XRD pattern of the sintered HAP/Lig coating with 0.5 wt% Lig (Fig. 3.55b) indicates that partial decomposition of the HAP occurred during sintering. While the main crystalline phase of sintered coating

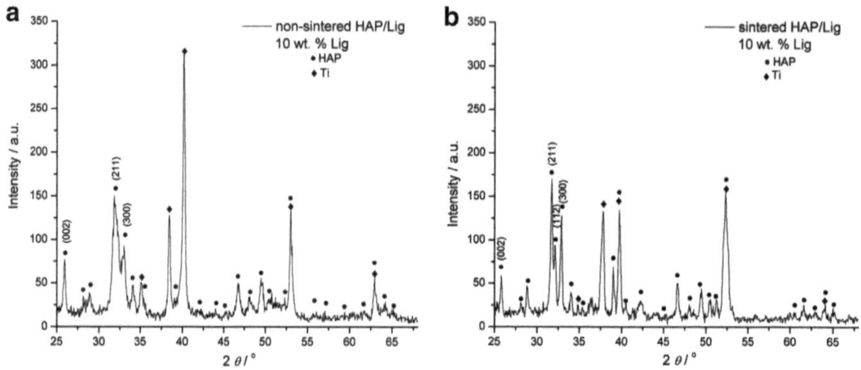

Fig. 3.57 XRD patterns of (**a**) non-sintered and (**b**) sintered HAP/Lig coatings with 10 wt% Lig (reprinted from [2] with permission from Elsevier)

was still HAP, the observed new diffraction peaks indicated the formation of crystalline phases of CaO, $CaCO_3$, and TiP. It can be proposed that reaction of CaO, generated during the sintering of HAP/Lig coating, with traces of atmospheric water and CO_2 yields $Ca(OH)_2$ and $CaCO_3$, respectively [195]. The diffusion of phosphorous ions into the Ti surface, as a result of HAP decomposition, was evidenced by the presence of the not very strong but specific peak of TiP. On the other hand, the specific diffraction peak for β-$Ca_3(PO_4)_2$, (TCP—tricalcium phosphate), at $2\theta = 30.7°$ was absent. The partial HAP decomposition was also observed for a sintered pure HAP coating (Fig. 3.49b).

The XRD patterns of the sintered HAP/Lig coatings containing 3 and 10 wt% Lig did not show any new crystalline phase (Figs. 3.56b and 3.57b, respectively), indicating that the phase composition remained the same as before sintering, as was previously observed for a HAP/Lig coating containing 1 wt% Lig (Fig. 3.50b). Comparing the XRD patterns of all sintered HAP/Lig coatings, it could be concluded that the decomposition of HAP lattice did not occur when the lignin concentration was higher than 0.5 wt%; this means that higher lignin concentrations protect the HAP lattice during sintering.

The mean crystallite domain size of the HAP/Lig coatings with different lignin concentrations was calculated from the diffraction peak that corresponds to (002) reflection at about $2\theta \approx 26°$. The values of mean crystallite domain size were calculated to be 35, 39, and 36 nm for coatings with 0.5, 3, and 10 wt% lignin concentration, respectively, indicating that mean crystallite domain size does not depend on the lignin concentration.

The surface morphologies of the electrodeposited HAP/Lig coatings containing different concentrations of lignin after thermal treatment at 900 °C were analyzed by SEM. The fractured surfaces of sintered HAP/Lig composite coatings containing 0.5, 3, and 10 wt% Lig can be seen in Fig. 3.58a–c, respectively. The cracks on coatings surfaces may be due to mechanical (HAP–lignin interactions) and thermal stresses (sintering). Since the SEM micrograph of the sintered HAP/Lig coating

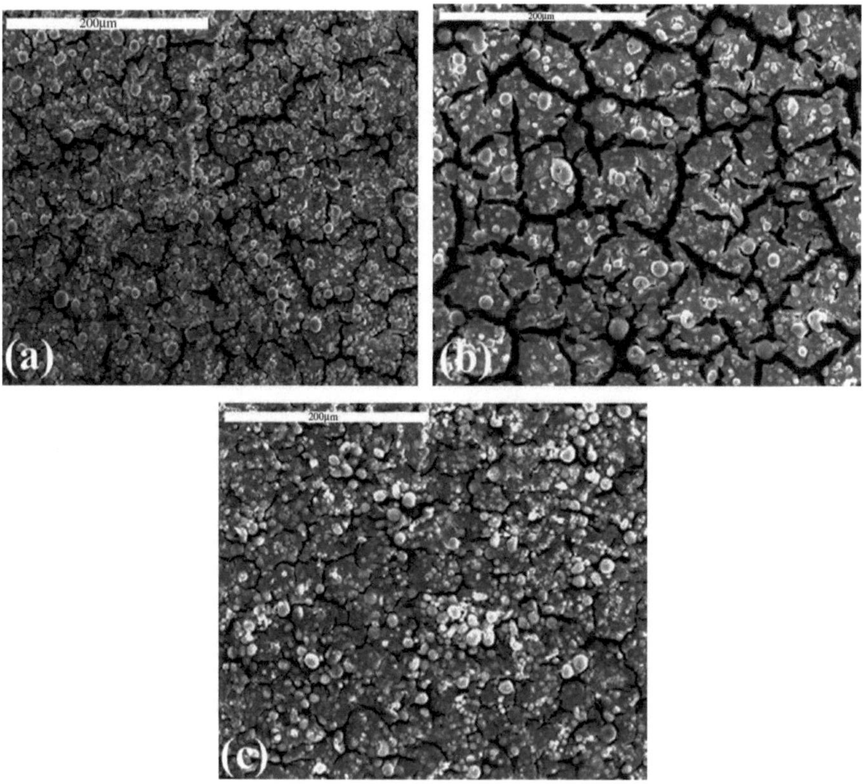

Fig. 3.58 SEM micrographs of sintered HAP/Lig coatings with (**a**) 0.5, (**b**) 3, and (**c**) 10 wt% Lig (reprinted from [2] with permission from Elsevier)

containing 1 wt% Lig revealed a homogenous surface without fractures (Fig. 3.51b), these results indicated that the optimal concentration to obtain coatings with a smooth surface is 1 wt% Lig.

The TG curves of the non-sintered HAP/Lig coatings containing 0.5, 3, and 10 wt% Lig, obtained in the temperature range between 25 and 1,000 °C, are shown in Fig. 3.59a, where three stages can be distinguished. The first stage was observed from 25 to 200 °C with sharp peaks on the DTG curves at 53, 55, and 63 °C for coatings with 0.5, 3, and 10 wt% Lig, respectively (Fig. 3.59b). This stage corresponds to desorption of water molecules adsorbed on the surface of the crystallites of the HAP/Lig coatings [28]. The second stage of weight loss of the TG curves was observed between 200 and 600 °C, with sharp peaks on the DTG curves at 227, 239, and 256 °C, which correspond to the thermal decomposition of lignin [163, 198]. The peaks at DTG curves at around 353, 383, and 392 °C (Fig. 3.59b) can be attributed to dehydroxylation of HAP. The third stage of weight loss of the TG curves was observed between 600 and 1,000 °C without distinguishable peaks on the DTG curves. This phase could still be attributed to dehydroxylation or to the

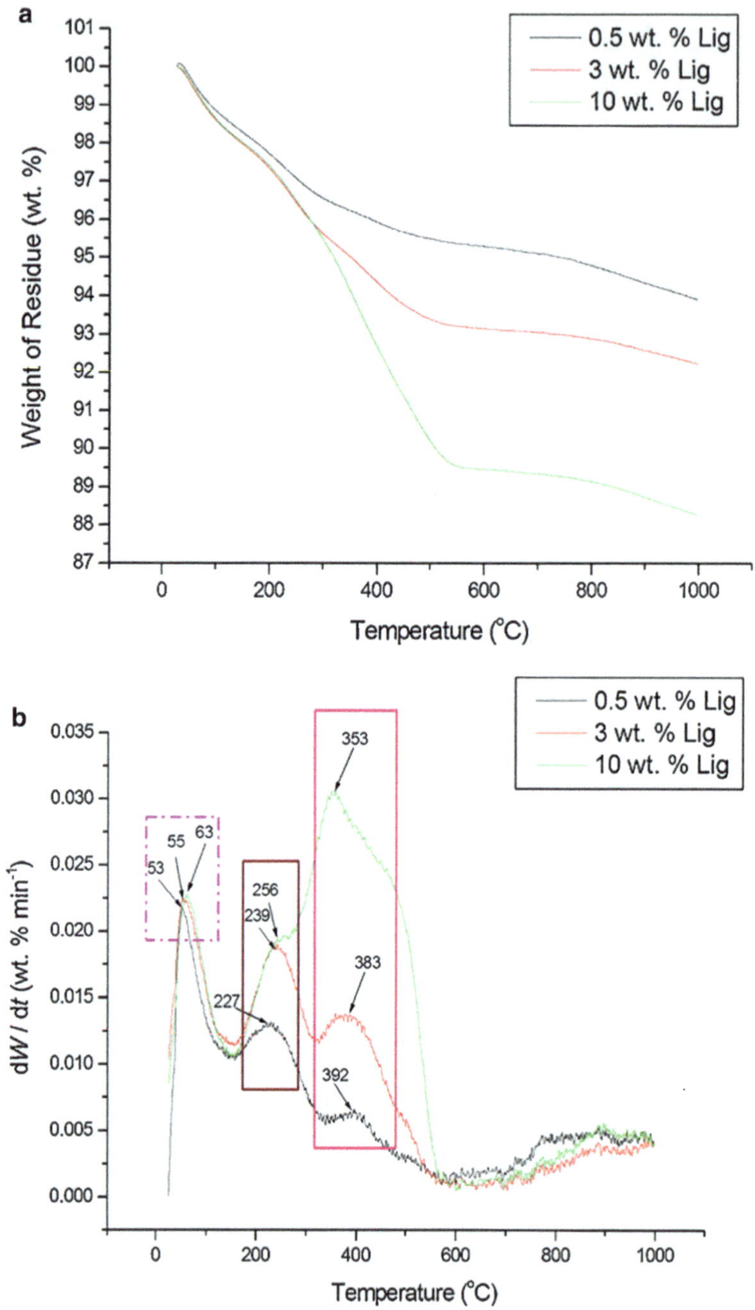

Fig. 3.59 TG curves (**a**) and differential TG (DTG) curves (**b**) of non-sintered HAP/Lig coatings with: 0.5, 3, and 10 wt% Lig (reprinted from [2] with permission from Elsevier)

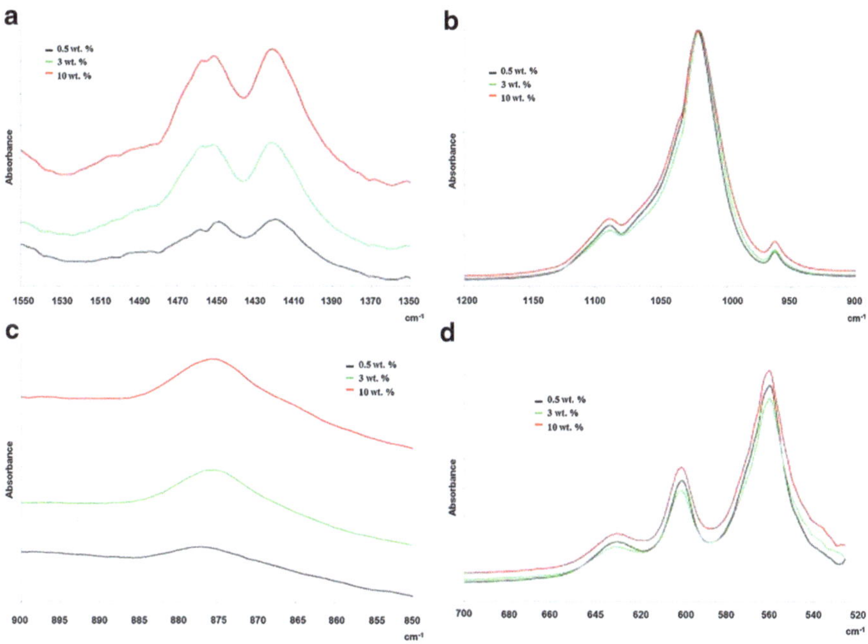

Fig. 3.60 ATR-FT-IR spectra of non-sintered HAP/Lig coatings with 0.5, 3, and 10 wt% Lig, in the wavenumber range 525–1,550 cm^{-1} (reprinted from [2] with permission from Elsevier)

early slow decomposition of HAP and of the remaining lignin. The total weight loss in the temperature range of 25–1,000 °C was 6, 7.5, and 11.8 wt% for the coatings containing 0.5, 3, and 10 wt% Lig, respectively, suggesting that increasing lignin concentration decreased the thermal stability of the HAP/Lig coatings. In other words, increasing HAP concentration increased the thermal stability of the HAP/Lig coatings.

The ATR-FTIR spectra of the non-sintered and sintered HAP/Lig coatings with 0.5, 3, and 10 wt% Lig are presented in Figs. 3.60 and 3.61, respectively.

In the region of 3,000–3,700 cm^{-1}, the broadband in all spectra (not shown) is attributed to water absorption [199]. The spectra of non-sintered HAP/Lig coatings shown in Fig. 3.60 exhibit bands typical for the PO_4^{3-} group. The band at 960 cm^{-1} is assigned to $\nu_1PO_4^{3-}$, and the bands at 1,030 and 1,090 cm^{-1} belong to $\nu_3PO_4^{3-}$ (Fig. 3.60b). In addition, the two bands at 560 and 600 cm^{-1} (Fig. 3.60d) are ascribed to the PO_4^{3-} group ($\nu_4PO_4^{3-}$) [200], and the characteristic band at 630 cm^{-1} corresponds to the vibration of structural OH^- groups [4]. The slight shoulder between 870–885 cm^{-1} (Fig. 3.60c) can be attributed to the C–H deformation vibration of C–H bonds in the aromatic rings, which confirm the presence of lignin [201]. This implies that lignin in HAP/Lig coating does not change the formation and structure of HAP.

In Fig. 3.60a, the peaks detected at 1,420 and 1,450 cm^{-1} are attributed to the methoxy group of lignin [202]. The band at 630 cm^{-1} (Fig. 3.60d), ascribed to

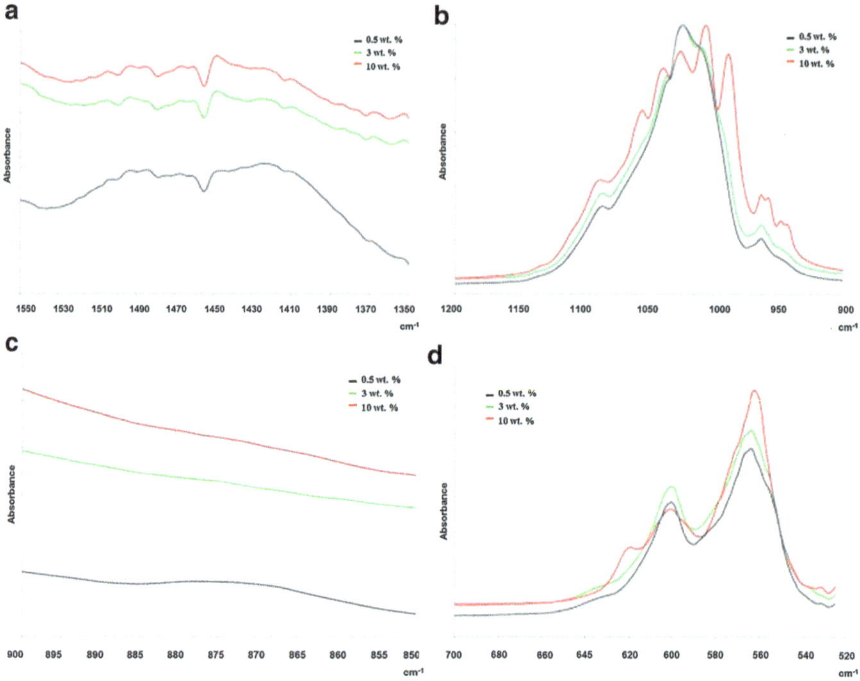

Fig. 3.61 ATR-FT-IR spectra of sintered HAP/Lig coatings with 0.5, 3, and 10 wt% Lig, in the wavenumber range 525–1,550 cm^{-1} (reprinted from [2] with permission from Elsevier)

ν(O–H) vibrations in OH^- groups from HAP, occurs at a lower wavenumber than expected (650 cm^{-1}), confirming that these groups are involved in intermolecular hydrogen bonds between HAP and Lig [203]. According to the literature, an absorption peak at 1,101 cm^{-1} is due to (P–O) stretching vibrations of the phosphorous group [204]. Therefore, it could be proposed that inter-hydrogen bonds (P–O···OH) between OH^- groups from lignin and PO_4^{3-} groups from HAP were established as evidenced by this band appearing at a new wavenumber, 1,090 cm^{-1} (Fig. 3.60b).

After sintering, there were visible differences in the shapes of the spectra in the ranges 1,350–1,550 and 850–900 cm^{-1} between the non-sintered HAP/Lig and the sintered HAP/Lig coatings. Thus, in the case of sintered HAP/Lig coatings (Fig. 3.61a, c), the band intensities at 1,410 and 1,460 cm^{-1} decreased while no band shifts could be observed. Weak carbonate bands in the range 865–880 cm^{-1} were observed only for the sintered HAP/Lig coating with 0.5 wt% Lig (Fig. 3.61c), which is indicative of HAP decomposition during sintering.

Comparing the ATR-FTIR results obtained for 0.5, 3, and 10 wt% Lig, as well as for 1 wt% Lig (Figs. 3.53 and 3.54), it could be concluded that decomposition of HAP during sintering does not occur for coatings with lignin concentration higher than 0.5 wt%. In other words, lignin concentrations higher than 0.5 wt% prevent HAP decomposition and/or diffusion of phosphorous ions into the Ti surface due to

Table 3.9 Quantitative analysis data of the XPS spectra

HAP/Lig (wt% Lig)		Ca	P	O	C
0.5	Non-sintered	19.1	11.3	58.3	8.2
	Sintered	18.4	7.9	56.4	15.9
3	Non-sintered	18.4	12.0	55.5	11.7
	Sintered	18.8	10.8	55.2	12.9
10	Non-sintered	15.8	10.3	51.1	21.3
	Sintered	17.1	9.6	52.2	18.9

Reprinted from [2] with permission from Elsevier

Table 3.10 Ca/P ratio for the HAP/Lig coatings with different Lig concentrations

HAP/Lig (wt% Lig)		Ca/P
0.5	Non-sintered	1.69
	Sintered	2.33
3	Non-sintered	1.53
	Sintered	1.74
10	Non-sintered	1.53
	Sintered	1.78

Reprinted from [2] with permission from Elsevier

the established hydrogen bonds. These results are in accordance with the results of the XRD analysis.

XPS analysis was performed on deposited HAP/Lig coatings, and the qualitative analysis revealed certain differences between the non-sintered and sintered coatings with 0.5, 3, and 10 wt% Lig [2]. The quantitative analysis data of the XPS spectra obtained from high-resolution measurements are presented in Table 3.9.

The carbon content of the non-sintered coatings increased with increasing Lig concentration, which confirmed that lignin was bonded with HAP lattice, as can be seen from the TG curves (Fig. 3.59a). However, the highest increase in the carbon content after sintering was determined for the HAP/Lig coating with 0.5 wt% Lig, which could be explained by carbon from the $CaCO_3$ formed by reaction between CaO with atmospheric CO_2. It could also be noticed that CaO and $CaCO_3$ were detected after sintering in the XRD pattern of the same coatings, confirming the HAP decomposition (Fig. 3.55b). The decrease in carbon content after sintering for the coating with 10 wt% Lig indicates the highest weight loss of lignin, which was also registered by TG analysis (Fig. 3.59a). The Ca/P ratio, presented in Table 3.10, varied in the range of 1.53–1.69 for the non-sintered HAP/Lig coatings, which is similar to Ca/P ratio for stoichiometric HAP (1.67). According to the literature [197], stable HAP phases correspond to a Ca/P ratio within a range of 1.3–1.8. It could be observed that the Ca/P ratios of the sintered HAP/Lig coatings were higher than those of the non-sintered coatings. The highest increase in Ca/P ratio after sintering, recorded for the HAP/Lig coating with 0.5 wt% Lig, refers to HAP decomposition during sintering, which was also confirmed by appearance of a TiP peak in XRD pattern (Fig. 3.55b). It could be concluded that lignin limited the decomposition of the HAP lattice during the sintering of the coatings with 3 and 10 wt% Lig, as well as with 1 wt% Lig, which is indicated by the smaller increase in carbon content (Table 3.9) and smaller Ca/P ratio (Table 3.10), compared to the HAP/Lig coating with 0.5 wt% Lig. This was also confirmed by the XRD and ATR-FT-IR results.

Peripheral blood mononuclear cells (PBMC) stimulated to proliferation with mitogen phytohemagglutinin (PHA) were used as the experimental in vitro cytotoxicity model. MTT results indicated that HAP and both HAP/Lig coatings could induce a mild decrease in survival of healthy immunocompetent PHA-stimulated PBMC but that all the results were similar to that of the control sample ($S = 100\ \%$) [2]. According to literature [205], HAP coating (93.4 % cell survival) and HAP/Lig coating with 1 wt% Lig (90.4 % cell survival) can be classified as non-cytotoxic, while HAP/Lig coating with 10 wt% Lig (83.7 % cell survival) as slightly cytotoxic.

3.7.3 Silver/Hydroxyapatite and Silver/Hydroxyapatite/ Lignin Composite Coatings

Traditional biomedical metallic implants are irreplaceable in repairing the damaged bone tissue, but the greatest concern is their gradual electrochemical degradation [206]. The implants face severe corrosion environment which includes blood and body fluid composed of several constituents (water, sodium, chlorine, proteins, plasma, and amino acids) along with mucin in the case of saliva. These metallic devices are unique in a way that they are exposed to living cells, tissues, and biological fluids, which are not only dynamic, but also a hostile environment for the preservation of the implant. Even though presently available bone implant materials exhibit high corrosion resistance, body fluid environment inevitably results in the release of non-compatible metal ions by the implants into the body. The released ions are found to cause allergic and toxic reactions.

Post-opt infections are the result of bacterial adhesion to the implant surface and subsequently biofilm forms at the implantation site [207]. In order to stop bacterial infection, it is crucial to inhibit bacterial adhesion since biofilm can be very resistant to immune response and antibiotics [208]. The antimicrobial activity of silver and silver ions has been known for a very long time; additionally, silver cation does not develop bacterial resistance and at the same time shows low toxicity to human cells [209, 210]. Therefore, the possibility to prevent the bone implant infections by using antimicrobial properties of Ag has generated great interest in the development of silver doped HAP coatings [156].

3.7.3.1 Electrophoretic Deposition of Silver/Hydroxyapatite and Silver/Hydroxyapatite/Lignin Composite Coatings

The influence of the silver on the microstructure, morphology, phase composition, thermal behavior, and cytotoxicity, as well as on corrosion stability and bioactivity of silver/hydroxyapatite coating (Ag/HAP) and silver/hydroxyapatite/lignin (Ag/HAP/Lig) coating with 1 wt% Lig electrodeposited on titanium, was investigated. A modified chemical precipitation method was used to prepare silver/hydroxyapatite powder by the reaction of calcium oxide (obtained by calcination of $CaCO_3$ for 5 h at 1,000 °C in air), silver nitrate, and phosphoric acid. Lignin powder was used to produce biocomposite silver/hydroxyapatite/lignin coating [3].

The particle size distribution (PSD) measurement of Ag/HAP and Ag/HAP/Lig with 1 wt% Lig suspensions was made by using dynamic light scattering technique. The obtained average particle size is around 200 nm for both suspensions, although agglomerates peaks can be observed also, meaning that silver do not affect the Ag/HAP and Ag/HAP/Lig particle size.

High positive values of ζ-potential of 23 mV and 29 mV indicate positively charged particle surfaces of Ag/HAP and Ag/HAP/Lig, respectively, thus enabling the attraction of particles by negatively charged cathode–titanium plate and therefore successful electrophoretic deposition of coatings.

The contact angle value, as an indicator of surface wettability and its biocompatibility, was determined by sessile drops of SBF solution on coating surfaces. Measured contact angles at Ag/HAP coating and Ag/HAP/Lig coating with 1 wt% Lig on Ti substrate were 35.03° ± 8.07 and 39.19° ± 1.21, respectively, indicating good hydrophilic properties [3]. A low contact angle is usually favorable in order to achieve high surface coverage. Surface wettability (hydrophobicity or hydrophilicity) is one of the most important parameters affecting biological response to implanted biomaterials. Wettability affects protein adsorption, platelet adhesion/activation, blood coagulation, and cell adhesion [211]. HAP is considered to be hydrophilic due to its –OH groups displayed on the surface [212]. Hydrophilic surfaces were shown to promote the highest levels of cell attachment [213].

The measured surface roughness values, R_a, of Ag/HAP and Ag/HAP/Lig coatings with 1 wt% Lig were 2.40 and 0.85 μm, respectively [3]. Costa-Rodrigues et al. [214] reported that the surface roughness values of biomaterial ranging from 0.20 to 2.87 μm enhance the maximum osteoclastic development, thus directly affecting the cell adhesion. They concluded that even though rougher surfaces are better for cell adhesion, at the same time they could become a problem for cellular mobility. Therefore, the surface roughness of Ag/HAP/Lig coating could stimulate cellular response better than the Ag/HAP coating.

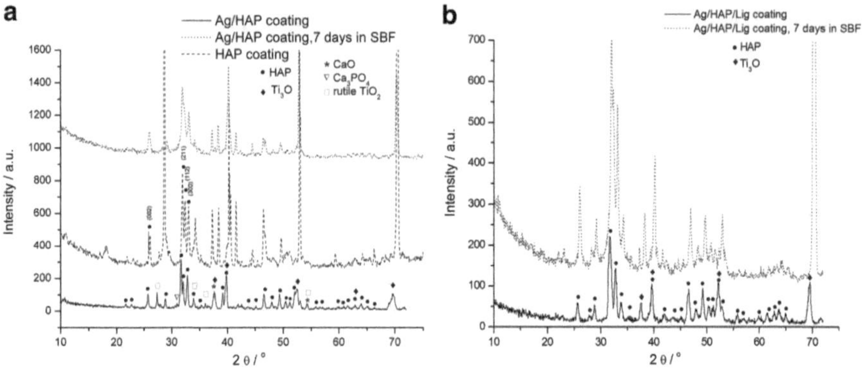

Fig. 3.62 XRD patterns of (**a**) HAP, Ag/HAP, and (**b**) Ag/HAP/Lig coatings, before and after 7 days of immersion in SBF solution at 37 °C (reprinted from [3] with permission from American Chemical Society)

3.7.3.2 Conversion of Silver/Hydroxyapatite and Silver/Hydroxyapatite/Lignin Composite Coatings to Carbonated Apatite

X-ray diffraction analysis was performed to determine the phase composition, and structures of Ag/HAP coating and Ag/HAP/Lig coating with 1 wt% Lig are presented in Fig. 3.62a, b, respectively, before and after 7 days of immersion in SBF solution at 37 °C. As a comparison, the XRD spectrum of pure HAP coating was shown in Fig. 3.62a as well. Labeled XRD peaks match very well to the JCPDS standard XRD card No. 86-1199 for hydroxyapatite, but incorporation of Ag in the hydroxyapatite crystal lattice causes a shift of specific HAP peaks to the left confirming the silver substitution for calcium. The evidence of Ag presence was verified through the shift of characteristic HAP peaks (crystal planes (002), (211), (112), and (300)) toward smaller angles for Ag/HAP coating before immersion in SBF solution compared to pure HAP coating (Fig. 3.62a).

The appearance of new peaks in XRD pattern of Ag/HAP coating indicates the partial HAP decomposition. Although, the diffractogram of Ag/HAP coating corresponds mainly to hydroxyapatite, the observed new diffraction peaks suggested the formation of crystalline phases such as CaO (JCPDS No. 74-1226) and β-tricalcium phosphate (β-TCP, JCPDS No. 32-0176) at $2\theta = 30.7°$ confirming the slight decomposition of HAP during the sintering process. On the other hand, the absence of the specific diffraction peaks for β-TCP and CaO in XRD pattern of Ag/HAP/Lig coating (Fig. 3.62b) confirmed that no partial HAP decomposition occurred. Thus, all peaks matched pure hydroxyapatite. That lead to conclusion that lignin improved thermal stability of Ag/HAP lattice during sintering.

The formation of bone-like apatite layer on the surface of bioactive materials has been reported after soaking of these materials in a biomimetic system such as SBF solution. After 7 days of immersion in the SBF solution, it is important to note the

Fig. 3.63 SEM microphotographs of the Ag/HAP coating (**a**) before and (**c**) after immersion and the Ag/HAP/Lig coating (**b**) before and (**d**) after immersion in SBF solution at 37 °C (reprinted from [3] with permission from American Chemical Society)

shift in characteristic hydroxyapatite peaks for both Ag/HAP and Ag/HAP/Lig coatings toward higher angles (Fig. 3.62a, b). Those findings were attributed to carbonate ions in the lattice and confirmed therefore the growth of carbonate hydroxyapatite onto coating surface. Therefore, the shifting of XRD diffraction peaks is typical for weak crystalline, carbonated HAP, as it is found in bone.

The crystallite domain sizes, calculated at $2\theta \approx 26°$ for Ag/HAP and Ag/HAP/Lig coatings, were 35.2 nm and 20.8 nm, respectively. On the other hand, apatite layer obtained after 7 days of soaking in SBF had almost the same crystallite domain size on Ag/HAP coating (20.2 nm) and Ag/HAP/Lig coating (22.0 nm). Difference between crystallite size before and after immersion is probably due to the incorporation of CO_3^{2-} ions into the apatite lattice by occupying the OH^- sites or the PO_4^{3-} position [215].

The surface morphologies of Ag/HAP and Ag/HAP/Lig coatings before and after 7 days of immersion in SBF at 37 °C are shown in Fig. 3.63a–d, respectively.

The crystallite domain size of Ag/HAP/Lig coatings before and after soaking in SBF was calculated to be almost the same - 20.8 nm and 22.0 nm, respectively, indicating the homogeneous surface (Fig. 3.63b, d). On the contrary, the crystallite domain sizes of Ag/HAP coatings before and after soaking in SBF were calculated to be 35.2 nm and 20.2 nm, respectively, revealing the decrease in crystallite

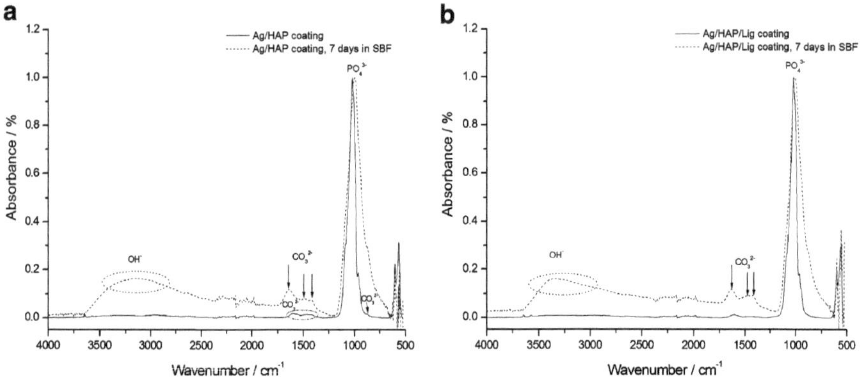

Fig. 3.64 ATR- FTIR spectra of (**a**) Ag/HAP and (**b**) Ag/HAP/Lig coatings, before and after 7 days of immersion in SBF at 37 °C (reprinted from [3] with permission from American Chemical Society)

domain size after soaking and fractured surface (Fig. 3.63a, c). Also, observed high porosity is beneficial for better cell adhesion, as it enables better connection between implant and bone. It has been reported that open interconnected porosity structures facilitate penetration of the surrounding bone tissue and hence lead to better biointegration and mechanical stability [216].

It is obvious that both Ag/HAP and Ag/HAP/Lig coatings formed an apatite film on the surface after just 7 days of immersion in SBF, indicating their potential bioactivity. The apatite composition of newly formed crystals was confirmed by XRD (Fig. 3.62) and ATR-FTIR results (Fig. 3.64), while the formation of apatite can be explained as follows. HAP surface exhibited negative charge that specifically interacted with the positive calcium ions from SBF forming an amorphous Ca-rich apatite. Then, the Ca-rich positively charged surface interacted with the negative phosphate ions in the SBF to form Ca-poor apatite, which gradually crystallizes into bonelike apatite, through which the HAP appeared to stabilize in SBF. Once formed in SBF the apatite grows spontaneously consuming the calcium and phosphate ions, incorporating minor ions such as sodium, magnesium, and carbonate, and thereby developing bone mineral-like compositional and structural feature [217].

The bioactivity of implanted apatite materials can be evaluated by the formation of bone-like apatite on their surface. The presence of carbonate bands in FTIR spectra is clear evidence of its incorporation in the surface of HAP layer. It is known that the biological hydroxyapatite also contain carbonate groups [218]. Thus, the Ag/HAP and Ag/HAP/Lig coatings were investigated by FTIR technique after immersion in SBF.

The FTIR spectra of Ag/HAP coating and Ag/HAP/Lig coating with 1 wt% Lig are presented in Fig. 3.64a, b, respectively. The FTIR spectra of Ag/HAP and Ag/HAP/Lig coatings before immersion in SBF exhibit the typical bands for hydroxyapatite. The broad peak at 3,432 cm^{-1} is related to H_2O adsorption.

The three characteristic bands at 960, 1,016, and 1,089 cm^{-1} belong to vibration of PO_4^{3-} group. Also, the weak characteristic bands at around 3,573 and 627 cm^{-1} correspond to the vibration of structural OH^- groups in the hydroxyapatite.

Finally, in FTIR spectrum of Ag/HAP before immersion in SBF solution (Fig. 3.64a), a very low intensity wide band centered around 1,400–1,585 cm^{-1} can be noticed, which may be attributed to carbonates (CO_3^{2-}) [218]. The appearance of a small vibrational band at 875 cm^{-1} also indicates a trace amount of carbonates as a confirmation of Ag/HAP partial decomposition. Although the carbonates were not detected by XRD results, these peaks in FTIR spectrum can be attributed to the reaction between CaO, confirmed in XRD pattern (Fig. 3.62a), and CO_2 from atmosphere [195]. On the other hand, the absence of low intensity wide band (1,400–1,585 cm^{-1}) in FTIR spectrum of Ag/HAP/Lig before immersion in SBF solution (Fig. 3.64b) confirmed that there was no decomposition of hydroxyapatite, as detected also by XRD (Fig. 3.62b).

During 7 days of immersion in SBF at 37 °C, carbonated apatite was formed on the surface of coatings, as revealed by ATR-FTIR analysis. In biological apatite some PO_4^{3-} ions are substituted by CO_3^{2-} ions. As a sensitive technique IR can detect even a trace amount of carbonate [219]. FTIR spectra of the Ag/HAP and Ag/HAP/Lig coatings after 7 days immersion in SBF solution (Fig. 3.64a, b, respectively) revealed the broad absorbance at 3,382 cm^{-1} attributed to the –OH stretching mode, and the intensity of the –OH band was observed to be higher for coatings after immersion indicating newly formed bone-like apatite on the coating surface [220]. Also, another confirmation was triplet peaks at 1,640, 1,476, and 1,420 cm^{-1} attributed to CO_3^{2-}, which belongs to the vibrational bands of CO_3^{2-} groups of B-type carbonated apatite owing to the formation of bone-like apatite after soaking in SBF solution [221]. The spectrum of bone-like apatite showed a high concentration of hydroxyl and phosphate groups compared to peaks appearing in spectra obtained for coatings before immersion in SBF (Fig. 64a, b), which makes the surface of coatings exhibit negative surface potentials required for apatite nucleation. Therefore, FTIR results indicate to the bioactivity of both coatings and that their surfaces represent favorable substrates for apatite nucleation, which was also confirmed by SEM.

Surface analysis of Ag/HAP and Ag/HAP/Lig coatings before soaking in SBF performed by XPS are illustrated in Figs. 3.65 and 3.66, respectively. The measured binding energy values were calibrated by the C1s (hydrocarbon C–C, C–H) of 285 eV. Deconvoluted spectra corresponding to different elements (Ca, P, and O) are discussed. The relative concentrations of most prominent elements and XPS transitions (C1s, O1s, Ca2p, P2p) for Ag/HAP and Ag/HAP/Lig coatings are presented in Table 3.11.

As shown in Figs. 3.65 and 3.66, the deconvolution XPS spectra of C 1s, O 1s, Ca 2p, and P 2p core level spectra of Ag/HAP and Ag/HAP/Lig coatings. From both XPS spectra the Ca 2p spectrum of the coatings reveals a doublet with Ca 2p3/2 (BE of 347.5 eV) and Ca 2p1/2 (BE of 351.1 eV), and the P 2p spectrum of the coating reveals a single P 2p3/2 peak at BE position at 133.6 eV, for Ag/HAP coating (Fig. 3.65), and 133.8 eV, for Ag/HAP/Lig coating (Fig. 3.66) indicating

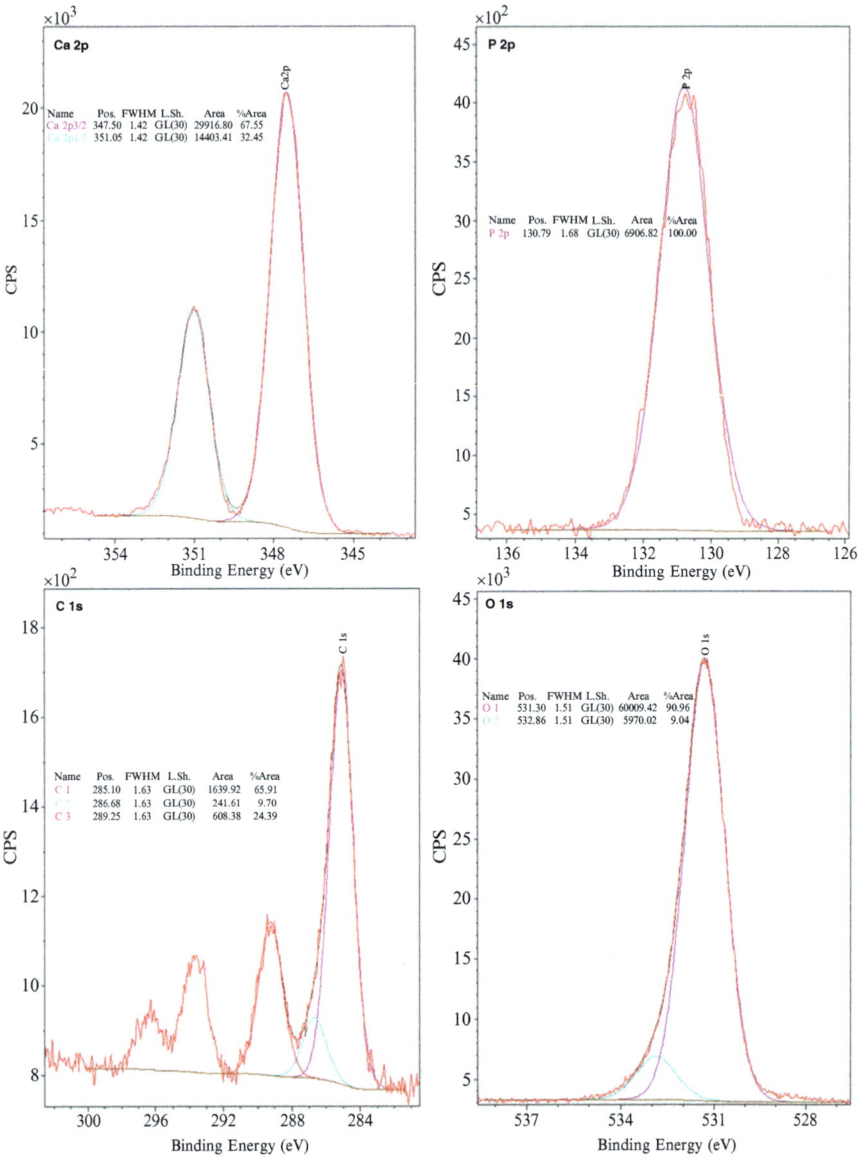

Fig. 3.65 Deconvoluted XPS photoelectron lines of component elements (Ca, P, C, and O) for Ag/HAP coating (reprinted from [3] with permission from American Chemical Society)

the presence of hydroxyapatite [222, 223], but no Ag peak was detected in XPS spectra because the amount of doped silver in crystal lattice of HAP is very small. Although Ca2p spectrum of Ag/HAP coating shows doublet of Ca 2p3/2 and Ca 2p1/2 with the same binding energies as for Ag/HAP/Lig coatings, according to literature [223] Ca 2p3/2 and Ca 2p1/2 (347.5 and 351.1 eV, respectively) can also

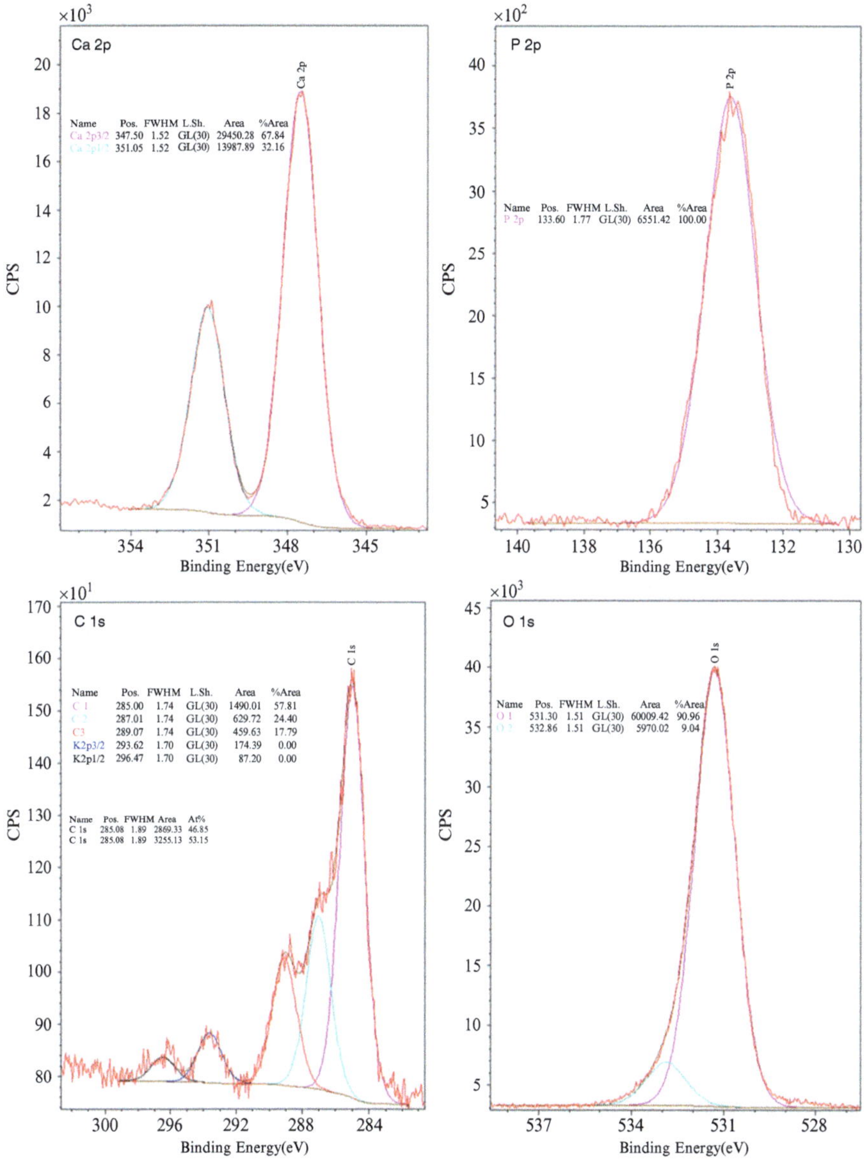

Fig. 3.66 Deconvoluted XPS photoelectron lines of component elements (Ca, P, C, and O) for Ag/HAP/Lig coating (reprinted from [3] with permission from American Chemical Society)

correspond to Ca2p in $Ca_3(PO_4)_2$ and $Ca_{10}(PO_4)_6(OH)_2$. These transitions for Ca 2p make differentiation between chemical environments impossible to identify with outmost certainty by XPS analysis alone. The Ca might exist in various phases, such as CaO and $Ca_3(PO_4)_2$, that occur during hydroxyapatite sintering. These results

Table 3.11 Atomic percentages and Ca/P ratio at sample surface of Ag/HAP and Ag/HAP/Lig coatings determined by XPS

	Element				
Sample	C 1s (%)	O 1s (%)	Ca 2p3/2 (%)	P 2p (%)	Ca/P
Ag/HAP	6.9	62.0	19.1	11.2	1.70
Ag/HAP/Lig	5.8	62.3	18.7	11.5	1.62

can confirm HAP decomposition after sintering Ag/HAP coating, which was also indicated by XRD analysis and ATR-FTIR measurement. The main O1s peak component at BE = 531.3 eV for both coatings may be attributed to PO_4^{3-} groups, as well as second O 1 s peak at 532.8 eV is attributed to the contribution from –OH group of HAP (Figs. 3.65 and 3.66) [223]. As a result of XPS peak fitting, the C 1s spectrum, presented in Figs. 3.65 and 3.66, consists of three components at positions: 285.0, 286.6–287, and 288.0–289.3 eV, which are attributable to aromatic hydrocarbons, alcoxy and $RCOO^-$ groups, respectively [224]. According to literature, the C 1s peak with BE = 286–286.3 eV and 287.7–288.4 eV indicates the presence of carbonyl groups [225], and the third deconvoluted peak occurring at 289.3 eV, which corresponded to C–O bonds, suggests the presence of the polymer lignin, linked to the hydroxyapatite.

The chemical composition of the outermost coating surface is important because it will be in direct contact with the bone tissue and dissolve first at the initial stage of implantation. It was shown that the optimum Ca/P ratio is 1.67–1.76 [226]. The increase of Ca/P ratio confirms the HAP decomposition in the Ag/HAP coating after sintering, while the Ca/P ratio for Ag/HAP/Lig coating remains constant and close to the stoichiometric value of 1.67 (Table 3.11).

Electrochemical impedance spectroscopy (EIS) was used in order to study the corrosion stability of investigated coatings in the physiological environment such as SBF solution [3]. The Nyquist plots for the impedance of Ag/HAP coating and Ag/HAP/Lig coating with 1 wt% Lig deposited on titanium, as well as of thermally treated bare titanium as reference, after prolonged exposure time in SBF solution at 37 °C are presented in Fig. 3.67. The high-frequency range of impedance response is attributed to the coating, while the low-frequency range describes the characteristics of the passive oxide layer on titanium.

The fitting of experimental data obtained from Nyquist plots was accomplished by using the equivalent electrical circuits shown in Fig. 3.68. The equivalent circuit in Fig. 3.68a was used for the fitting of impedance data obtained for the Ag/HAP and Ag/HAP/Lig coatings and consists of electrolyte resistance, R_s, coating pore resistance, i.e., resistance of the electrolyte inside the pores of Ag/HAP and Ag/HAP/Lig coatings, R_p, constant phase elements, CPE_c and CPE_{pf}, which include all the frequency-dependent electrochemical phenomena: coating capacitance, C_c, and passive titanium oxide film capacitance, C_{pf}, for Ag/HAP and Ag/HAP/Lig coatings and bare titanium, respectively, and diffusion processes. CPE is used in

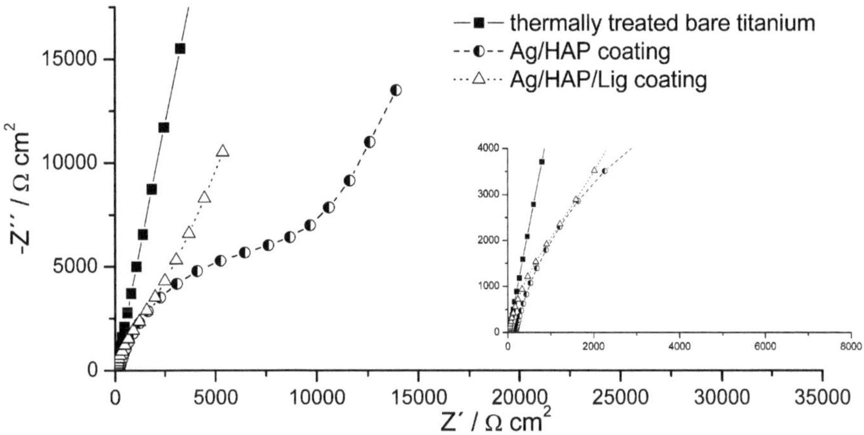

Fig. 3.67 Nyquist plots of Ag/HAP and Ag/HAP/Lig coatings on titanium and thermally treated bare titanium after 10 days of exposure to SBF solution at 37 °C (*Inset*: high-frequency range) (reprinted from [3] with permission from American Chemical Society)

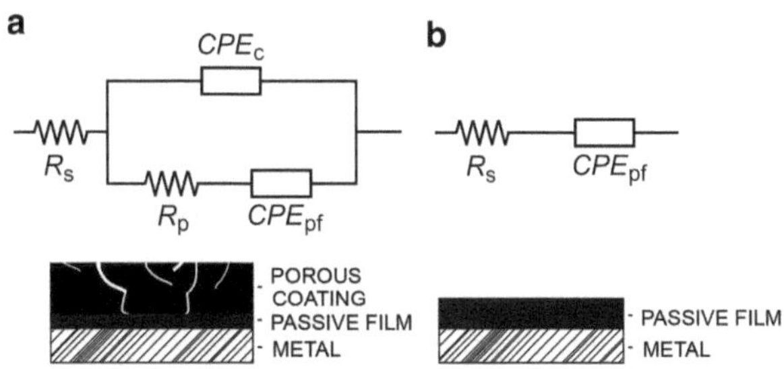

Fig. 3.68 Equivalent electrical circuits and physical models for (**a**) Ag/HAP and Ag/HAP/Lig coatings on titanium and (**b**) thermally treated bare titanium during prolonged exposure time to SBF solution at 37 °C (reprinted from [3] with permission from American Chemical Society)

this model to compensate non-homogeneity in the system and is defined by two parameters, Y_0 and n. The impedance of CPE is represented by the following equation [227]:

$$Z_{\mathrm{CPE}} = {Y_0}^{-1} \cdot (j\omega)^{-n} \tag{3.16}$$

where $j = (-1)^{1/2}$, $\omega = 2\pi f$ is frequency in rad s^{-1}, and f is the frequency in Hz. If n values range from 0.8 to 1, the impedance of CPE can be considered to be one of the pure capacitor:

$$Z_{\mathrm{CPE}} = (j\omega C)^{-n} \tag{3.17}$$

and in this case Y_0 gives a pure capacitance (C). For impedance analysis of bare titanium, the equivalent circuit in Fig. 3.69b was used, consisting of electrolyte resistance, R_s, and constant phase element for passive oxide film, $\mathrm{CPE_{pf}}$.

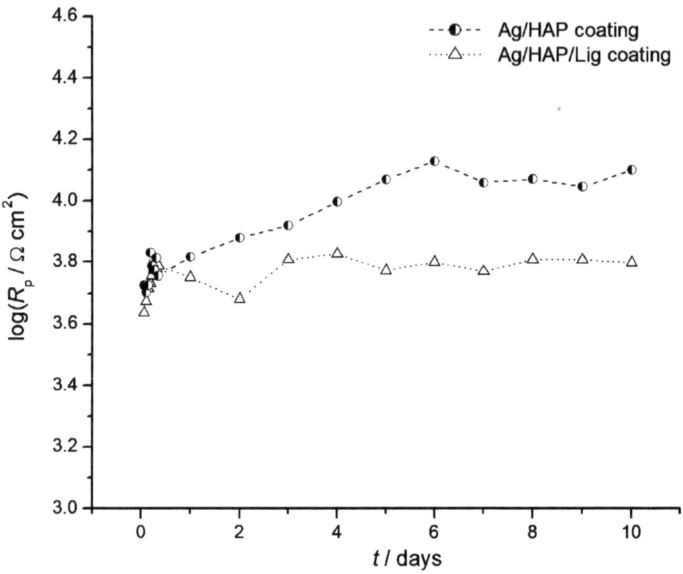

Fig. 3.69 Time dependence of coating pore resistance, R_p, for Ag/HAP and Ag/HAP/Lig coatings during exposure to SBF solution at 37 °C (reprinted from [3] with permission from American Chemical Society)

The pore resistance, R_p, of Ag/HAP and Ag/HAP coatings is plotted as a function of time of exposure to SBF solution in Fig. 3.69. It can be noticed that the pore resistance, R_p, slightly increased for both coatings during the exposure of 10 days, which has a direct relation with the amount of the apatite layer formed on the coating surfaces. The continuous increase in R_p reflects the process of the apatite nucleation after one week, which could also be clearly seen from SEM, XRD, and ATR-FTIR analysis.

Peripheral blood mononuclear cells (PBMC) stimulated to proliferation with mitogen phytohemagglutinin (PHA) were used as the experimental in vitro cytotoxicity model. MTT results indicated that both Ag/HAP and Ag/HAP/Lig coatings (90.4 % cell survival and 84.9 % cell survival, respectively) can be classified as non-cytotoxic [3].

3.7.4 Hydroxyapatite Coatings Modified by Ion Beam Irradiation

Surface modification by ion beam irradiation method has been identified as a good technique to improve bioactivity of hydroxyapatite [228, 229] and its mechanical properties [230, 231], and to reduce particle size [228], without detrimentally

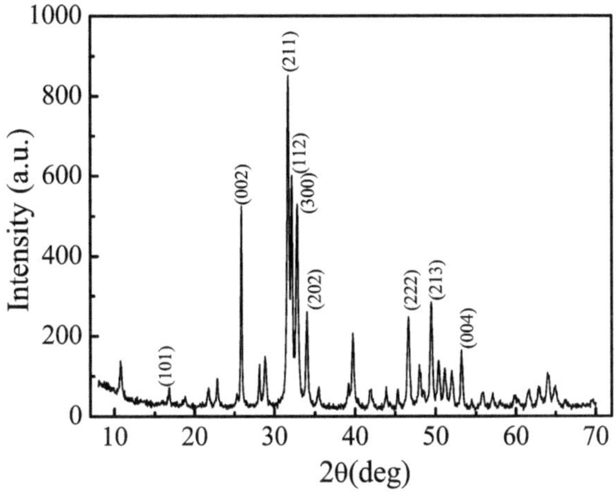

Fig. 3.70 XRD pattern of HA powder (reprinted from [232] with permission from National Institute of Research and Development for Optoelectronics of Romania)

affecting bulk properties. Namely, the surface modification induced by irradiation can give the bioactive materials of improved ion-exchange properties depending upon the particular material and the type of ions used. This can be controlled by (a) varying the depth to which the ions penetrate by varying the energy of ion bombardment or the angle of the material surface related to the ion beam; (b) varying the dose (or ions per unit area) of the ion beam; and (c) varying the type and combination of ions used.

Hydroxyapatite (HA) thin films electrodeposited on titanium were implanted with N^{4+} and Ar^{6+} ions with the aim to qualitatively explain the mechanism of HA surface modification after irradiation with N^{4+} and Ar^{6+} ions [232].

3.7.4.1 Hydroxyapatite Powder and Non-implanted Hydroxyapatite Films

HA powder was synthesized from $CaCl_2$, $Na_2H_2EDTA{\cdot}2H_2O$, NaH_2PO_4, and urea. Figure 3.70 represents the XRD pattern of the HA powder (Card JCPDS No. 09-0432), while the crystallite domain size from the (002) reflection was calculated to be 43.4 nm.

Figure 3.71 shows the SEM micrograph of HA powder. It can be observed that HA particles are spherical, and each particle consists of great number of agglomerated nanosized subparticles. Namely, the crystallite domain size for HA powder from the (002) reflection was calculated to be 43.4 nm while SEM micrograph represents the spherical HA particles, having the size in the range of 1–3 μm.

Hydroxyapatite coatings were deposited on titanium from ethanol HA suspension using the constant voltage method. Figure 3.72 represents the XRD pattern of non-implanted HA film electrodeposited on titanium. Peaks for HA and titanium

Fig. 3.71 SEM micrograph of HA powder (reprinted from [232] with permission from National Institute of Research and Development for Optoelectronics of Romania)

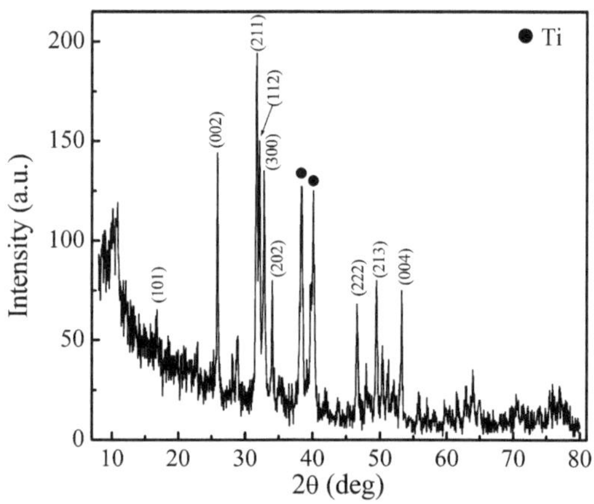

Fig. 3.72 XRD pattern of non-implanted HA film electrodeposited on titanium (reprinted from [232] with permission from National Institute of Research and Development for Optoelectronics of Romania)

(originated from the substrate) are identified. Comparison of Fig. 3.72 and Fig. 3.70 suggests that crystal structure of HA powder was not changed during electrophoretic deposition.

The SEM micrograph (Fig. 3.73) represents the homogenous, crack-free morphology of non-implanted HA film.

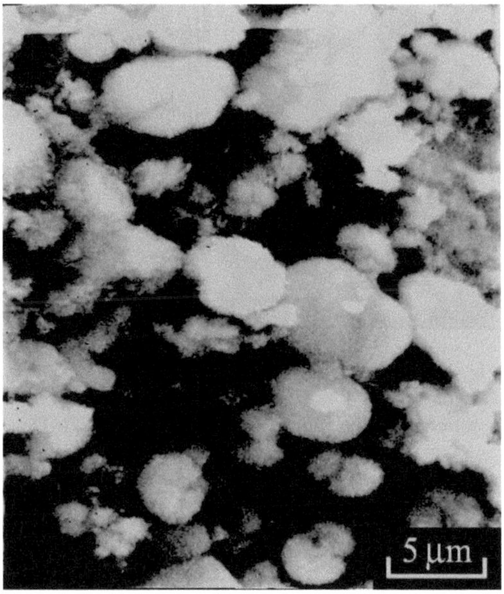

Fig. 3.73 SEM micrograph of non-implanted HA film (reprinted from [232] with permission from National Institute of Research and Development for Optoelectronics of Romania)

Table 3.12 Summary of the calculated values of ion ranges, R_p, statistical spread-standard deviation, ΔR_p, and electron energy loss $(dE/dx)_e$ and nuclear energy loss $(dE/dx)_n$ for nitrogen and argon ions

Ion	N^{4+}	Ar^{6+}
Energy (keV)	60	90
R_p (nm)	134	78
ΔR_p (nm)	47	20
$(dE/dx)_e$ (eV/nm)	2.6×10^2	3.2×10^2
$(dE/dx)_n$ (eV/nm)	1.1×10^2	6.7×10^2

Reprinted from [232] with permission from National Institute of Research and Development for Optoelectronics of Romania

3.7.4.2 Hydroxyapatite Films Implanted With $N^{4+}$and Ar^{6+} Ions

The HA films implanted with N^{4+} and Ar^{6+} at different fluences of 1×10^{15}, 1×10^{16}, and 1×10^{17} ions cm^{-2}, with energies of 60 keV and 90 keV for nitrogen and argon ions, respectively. SRIM calculation was used for determination of ion penetration into HA films and the values of estimated ion ranges, R_p, statistical spread-standard deviation, ΔR_p, and electron energy loss $(dE/dx)_e$ and nuclear energy loss $(dE/dx)_n$ for nitrogen and argon ions are given in Table 3.12.

The results clearly show that argon ions will penetrate into HA film less than nitrogen ions. Whereas the nitrogen range distribution is determined by electronic stopping, that of argon ions is determined by nuclear loss. It can be said that argon slows down and undergoes more frequent nuclear collisions, indicating that the

Fig. 3.74 SEM micrographs of HA films implanted with nitrogen ions at fluences of (**a**) 1×10^{15} ions cm^{-2}, (**b**) 1×10^{16} ions cm^{-2}, and (**c**) 1×10^{17} ions cm^{-2} (reprinted from [232] with permission from National Institute of Research and Development for Optoelectronics of Romania)

heavier ions (Ar) induce much larger energy deposition density than lighter (N) ions [233]. The distribution of the displaced target atoms for argon ion implantation is considerably larger than that for nitrogen implantation into HA films. Consequently, the surface erosion of HA films by electronic energy loss (N ions irradiation) seems to be absent and only erosion by elastic collision between the projectiles and near surface atoms contribute (collisional sputtering). Explicitly, the sputtering coefficient is more pronounced for argon ion irradiation (2.3 at/ion comparing to 0.43 at/ion for nitrogen ion), as calculated by the SRIM code.

In addition, ion beams can alter the topography of surfaces on which they impinge [234, 235]. SEM was used for the analysis of surface topography of HA films before and after implantation with nitrogen and argon ions. Figure 3.74a–c shows the SEM micrographs of HA surface after irradiation with 60 keV N^{4+} ions at fluences of 1×10^{15} ions cm^{-2}, 1×10^{16} ions cm^{-2}, and 1×10^{17} ions cm^2, respectively. No changes of the surface topography after nitrogen implantation as compared to non-implanted sample (Fig. 3.73) were observed. It is evident that the development of the surface topography during nitrogen implantation is negligible. On the other hand, the results show that implantation with argon ions alters the topography of the HA surface. The SEM micrographs shown in Fig. 3.75a–c were taken from the samples implanted with 90 keV Ar^{6+} ions at fluences of 1×10^{15} ions cm^{-2}, 1×10^{16} ions cm^{-2}, and 1×10^{17} ions cm^{-2}, respectively. The surface topography was not changed significantly by argon ions implantation at lower fluences (Fig. 3.75a, b). A slightly higher surface roughness can be noticed compared with the non-implanted sample. On the other hand, the surface topography was changed significantly by argon ions implantation at higher argon ion fluence (1×10^{17} ions cm^{-2}), as shown in Fig. 3.75c. A more pronounced surface topography was developed and formation of cone-like structure occurred. The observed cone-like shapes are denoted by arrows in Fig. 3.75c. This is due to the sputtering effects of argon ions during interaction with HA films. Thus, the main route of surface modification of HA films is a consequence of Ar ions induced damage and surface erosion.

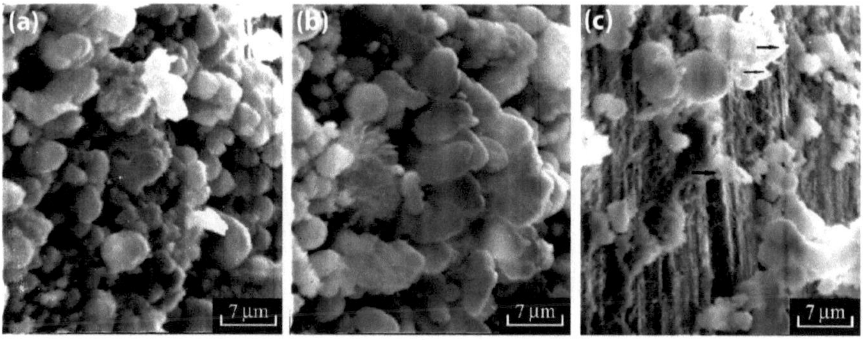

Fig. 3.75 SEM micrographs of HA films implanted with argon ions at fluences of (**a**) 1×10^{15} ions cm^{-2}, (**b**) 1×10^{16} ions cm^{-2}, and (**c**) 1×10^{17} ions cm^{-2} (reprinted from [232] with permission from National Institute of Research and Development for Optoelectronics of Romania)

Figures 3.76 and 3.77 represent the XRD patterns of HA films electrodeposited on titanium after implantation by different fluences of N^{4+} and Ar^{6+} ions, respectively. The X-ray diffraction pattern corresponding to HA phase was recorded for both non-implanted (Fig. 3.72) and implanted films, meaning that initial crystalline structure of HA films was preserved after ion implantation.

The effect of ion fluences on the relative intensities of XRD reflection at $2\theta = 26°$ of implanted HA films can be noticed in Fig. 3.78a for implantation with nitrogen ions and in Fig. 3.78b for implantation with argon ions. Figure 3.79 shows the dependence of crystallite domain size for (002) plane on fluences of implanted N^{4+} and Ar^{6+} ions. The increase in applied fluences decreases the HA crystallite domain size for both implanted ions.

It can be observed from Fig. 3.78a that relative intensities of the reflection at $2\theta = 26°$ decrease with increasing nitrogen ion fluence. This behavior indicates that the decrease in crystallinity of HA film, as well as the decrease in crystallite domain size (Fig. 3.79), are the consequence of electron energy loss mechanism. This is in accordance with the results from literature [228, 236] which confirmed that energy transfer promoted the decrease in crystallinity. Figure 3.78b shows that relative intensities of the reflection at $2\theta = 26°$ decrease after Ar^{6+} ions irradiation to a 1×10^{15} ions cm^{-2}. At higher fluences, the relative intensities slightly increase when argon ion fluences increase to 1×10^{16} and 1×10^{17} ions cm^{-2}. These structural changes, i.e., decrease in crystallinity of HA film followed by further increase, as well as the decrease in crystallite domain size (Fig. 3.79), can be due to the fact that nuclear energy loss dominates for argon ions irradiation. At lower fluences the decrease in crystallinity is more pronounced, while with increasing fluences the heat exchange during irradiation may induce recrystallization in HA structure.

Figures 3.80 and 3.81 represent the dependences of *a* and *c* HA unit cell parameters, as well as unit cell volume, on fluence of implanted nitrogen and

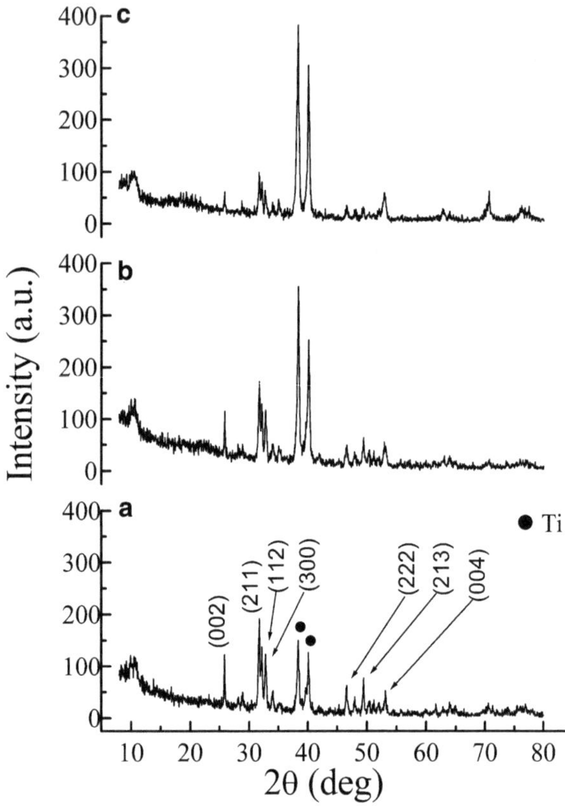

Fig. 3.76 XRD patterns of HA films implanted with nitrogen ions at fluences of (**a**) 1×10^{15} ions cm^{-2}, (**b**) 1×10^{16} ions cm^{-2}, and (**c**) 1×10^{17} ions cm^{-2} (reprinted from [232] with permission from National Institute of Research and Development for Optoelectronics of Romania)

argon, respectively. The values of unit cell parameters and unit cell volume for non-implanted HA film (which correspond to fluence zero) are given with the aim to compare the results.

It can be seen from Fig. 3.80 that unit cell parameters *a* and *c*, as well as unit cell volume of HA, increase with increasing nitrogen ion fluence. Similarly, the unit cell parameters *a* and *c*, as well as unit cell volume of HA, increase after Ar^{6+} ions irradiation to a 1×10^{15} ions cm^{-2} (Fig. 3.81). At higher fluences, the unit cell parameters and unit cell volume decrease when argon ion fluences increase to 1×10^{16} and 1×10^{17} ions cm^{-2}. These structural changes are consequence of different mechanisms of energy transfer of nitrogen and argon ions to HA film. For nitrogen ion irradiation electron energy loss is predominant, while for argon ion irradiation nuclear energy loss prevails.

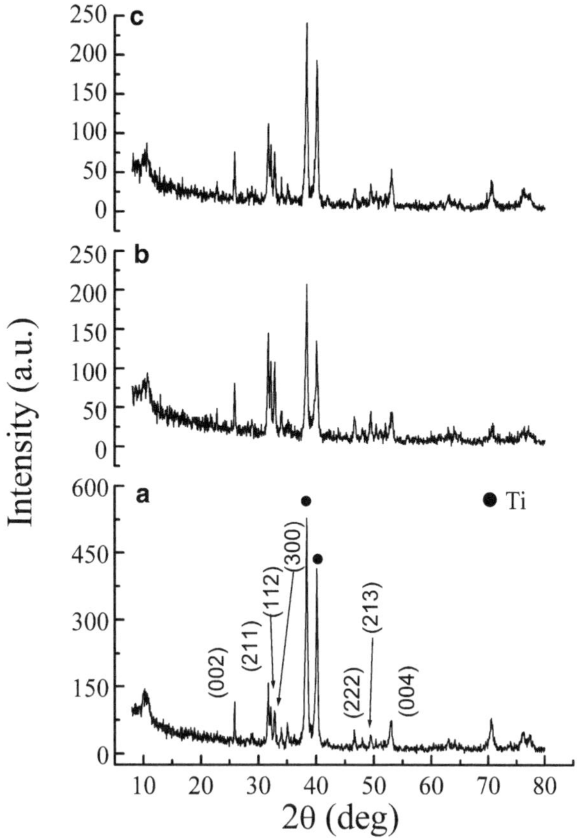

Fig. 3.77 XRD patterns of HA films implanted with argon ions at fluences of (**a**) 1×10^{15} ions cm^{-2}, (**b**) 1×10^{16} ions cm^{-2}, and (**c**) 1×10^{17} ions cm^{-2} (reprinted from [232] with permission from National Institute of Research and Development for Optoelectronics of Romania)

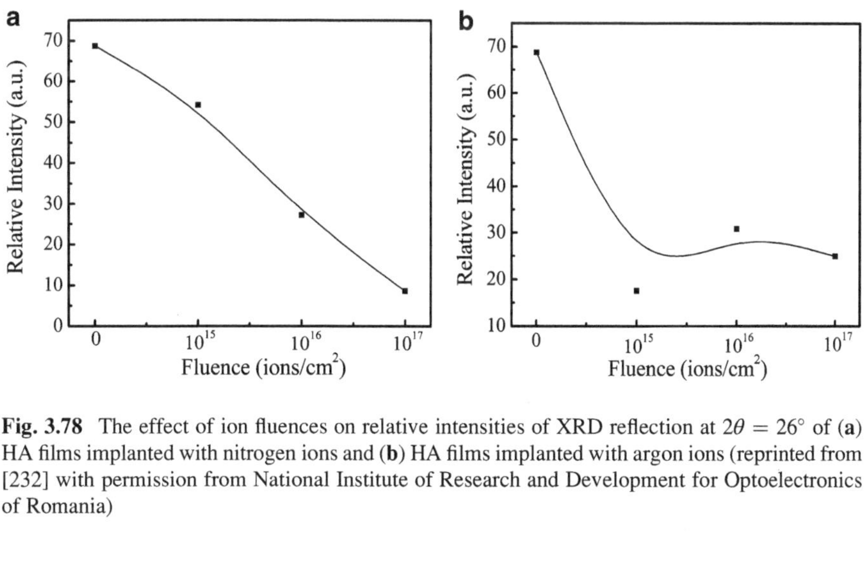

Fig. 3.78 The effect of ion fluences on relative intensities of XRD reflection at $2\theta = 26°$ of (**a**) HA films implanted with nitrogen ions and (**b**) HA films implanted with argon ions (reprinted from [232] with permission from National Institute of Research and Development for Optoelectronics of Romania)

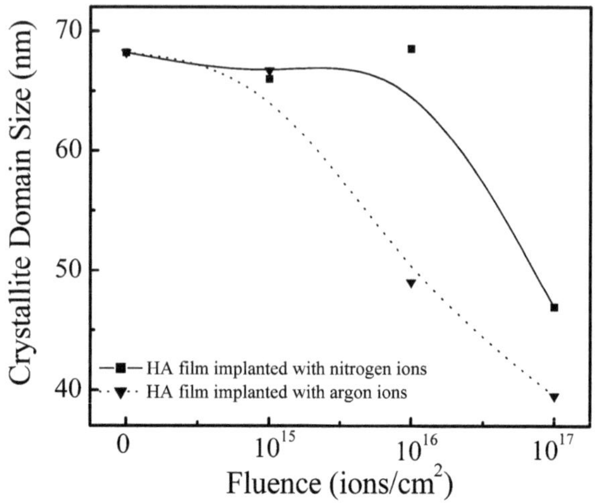

Fig. 3.79 Dependence of crystallite domain size for (002) plane on fluences of implanted nitrogen and argon ions (reprinted from [232] with permission from National Institute of Research and Development for Optoelectronics of Romania)

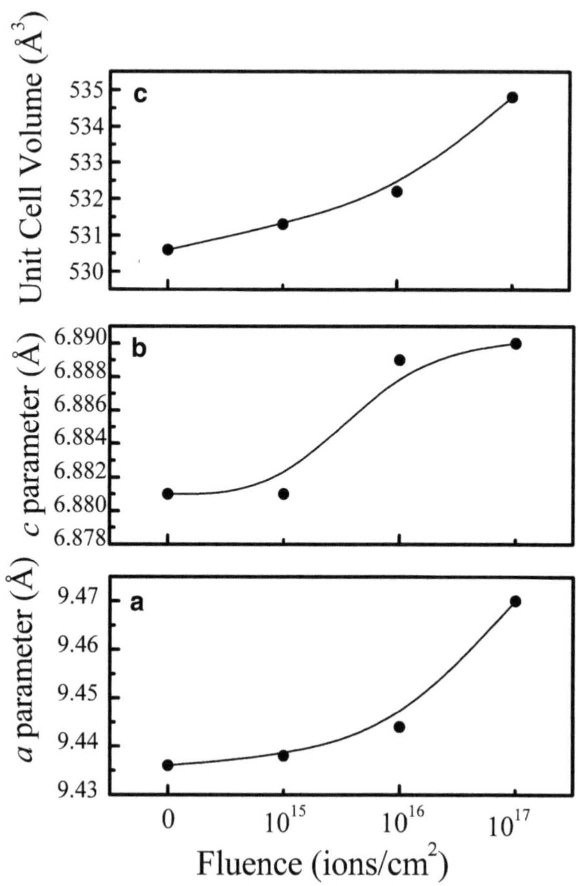

Fig. 3.80 Dependence of (**a**) *a* unit cell parameter, (**b**) *c* unit cell parameter, and (**c**) unit cell volume of HA on implantation fluence of nitrogen ions (reprinted from [232] with permission from National Institute of Research and Development for Optoelectronics of Romania)

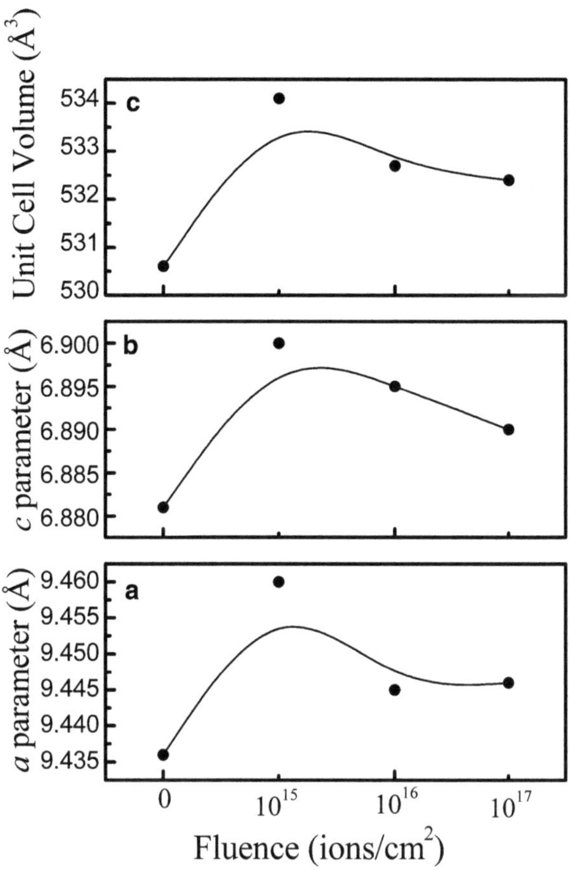

Fig. 3.81 Dependence of (**a**) *a* unit cell parameter, (**b**) *c* unit cell parameter, and (**c**) unit cell volume of HA on implantation fluence of argon ions (reprinted from [232] with permission from National Institute of Research and Development for Optoelectronics of Romania)

3.8 Conclusions

Electrodeposition is powerful method to design the structure, morphology, thickness, porosity, and thermal stability of ceramic coatings on metal surfaces. The mechanism of electrodeposition of alumina, boehmite, monetite, brushite, hydroxyapatite, hydroxyapatite/lignin, silver/hydroxyapatite, and silver/hydroxyapatite/lignin coatings on metal surfaces from aqueous suspensions was explained. The use of sol–gel routes enabled significantly lower sintering temperature due to sinter-active ceramic particles on a nanometer scale. The determination of the optimal deposition parameters, i.e., applied voltage, deposition time, bath temperature, and particles concentration in suspension, in order to control the thickness, morphology, and structure of deposited films is also discussed. It was shown that coatings of the greatest thickness and the lowest porosity were formed at lower applied voltages and for longer deposition time, i.e., when two processes, particle migration and coagulation and hydrogen evolution, are in equilibrium, which results in the minimum pore number, mean pore area, and percentage of film surface covered by pores. Thus, the greatest thickness and the lowest porosity of alumina coatings were obtained at the deposition voltage of 30 V and deposition time of 10 min; boehmite coatings at 4 V and 30 min and treated at 1,000 °C and 1,300 °C, forming γ-Al_2O_3 and α-Al_2O_3 alumina coatings, respectively, at significantly lower sintering temperature; monetite coating at 30 V and for a deposition time of 20 min then was transformed to hydroxyapatite by soaking in an alkaline solution; and hydroxyapatite, hydroxyapatite/lignin, silver/hydroxyapatite, and silver/hydroxyapatite/lignin coatings at 60 V for 45 s and sintered at significantly lower sintering temperature of 900 °C, as a result of nanosized hydroxyapatite particle application.

In the case of composite hydroxyapatite/lignin, the lignin concentrations higher than 0.5 wt% protected the hydroxyapatite lattice during sintering, so composite hydroxyapatite/lignin and silver/hydroxyapatite/lignin coatings with 1 wt% of lignin are promising nontoxic bioactive material for bone tissue engineering, because microcrystalline structure of newly formed plate-shaped carbonate-hydroxyapatite was detected after only 7 days in SBF solution.

The effect of nitrogen and argon irradiation on hydroxyapatite coatings on titanium and the mechanism of surface modification by ion implantation were discussed. The increase in applied fluences decreases the hydroxyapatite crystallite domain size for both implanted ions. In the case of implantation with nitrogen ion, the increase in applied fluences decreases the crystallinity of hydroxyapatite coatings. In the case of implantation with argon ion, the increase in applied fluences causes initial decrease in crystallinity of hydroxyapatite coatings followed by further slight increase, i.e., recrystallization in hydroxyapatite structure. For nitrogen ion irradiation electron energy loss is predominant, while for argon ion irradiation nuclear energy loss prevails.

Using electrochemical synthesis under galvanostatic conditions, the nanosized monetite powder (crystallite domain size 27.1 nm) and hydroxyapatite powders (crystallite domain size 21.6–24.3 nm), as well as brushite coatings were produced, and the influence of applied current density and pH value on the phase composition, crystallite size, morphology, and thermal characteristics of obtained powders were proved. In the case of brushite coatings, it was shown that the brushite morphology, porosity, and crystallite domain size can be easily controlled by applied current density because the applied current density not only increases the rate of crystallite growth but also increases the rate of electrochemical hydrogen evolution reaction. Thus, the brushite coating with the smallest crystallite domain size of 15.6 nm and the greatest porosity was obtained at higher current density of 9 mA cm^{-2}, since the crystallite growth is suppressed by hydrogen evolution reaction. On the contrary, the less porous brushite coatings with bigger crystallite domain size were obtained at lower current densities (5–7 mA cm^{-2}) because crystallite growth dominates over hydrogen evolution reaction. The applied current density for brushite coating deposition control also the crystallite domain size of hydroxyapatite coatings after soaking in SBF solution. The hydroxyapatite coating with significantly smaller crystallite domain size (20.2 nm) was obtained at more porous brushite coating deposited at higher current density of 9 mA cm^{-2}, due to greater number of pores which serve as crystallization sites for hydroxyapatite formation. On the other hand, the hydroxyapatite coating of larger crystallite domain size (45.2 nm) was obtained at less porous brushite coating deposited at lower current density of 5 mA cm^{-2}, as a consequence of smaller number of pores as crystallization sites for hydroxyapatite formation.

Acknowledgments The author wishes to thank all coworkers who contributed with their work to obtain the experimental results presented in this chapter. Their names can be seen in our joint papers listed in the references.

This research was financed by the Ministry of Education, Science and Technological Development, Republic of Serbia, contract No. III 45019.

References

1. Boccaccini AR, Keim S, Ma R, Li Y, Zhitomirsky I (2010) J R Soc Interface 7:S581
2. Eraković S, Veljović Dj, Diouf PN, Stevanović T, Mitrić M, Janaćković Dj, Matić IZ, Juranić ZD, Mišković-Stanković V (2012) Prog Org Coat 75:275
3. Eraković S, Janković A, Veljović Dj, Palcevskis E, Mitrić M, Stevanović T, Janaćković Dj, Mišković-Stanković V (2013) J Phys Chem B 117:1633
4. Kaya C, Singh I, Boccaccini AR (2008) Adv Eng Mater 10:131
5. Cheong M, Zhitomirsky I (2008) Colloids Surf A 328:73
6. Djošić MS, Mišković-Stanković VB, Janaćković ĐT, Kačarević-Popović ZM, Petrović RD (2006) Colloids Surf A 274:185
7. Djošić MS, Mišković-Stanković VB, Kačarević-Popović ZM, Jokić BM, Bibić N, Mitrić M, Milonjić SK, Jančić-Heinemann R, Stojanović J (2009) Colloids Surf A 341:110
8. Song YW, Shan DY, Han EH (2008) Mater Lett 62:3276
9. Corni I, Ryan MP, Boccaccini AR (2008) J Eur Ceram Soc 28:1353
10. Stojanovic D, Jokic B, Veljovic Dj, Petrovic R, Uskokovic PS, Janackovic Dj (2007) J Eur Ceram Soc 27:1595
11. Simović K, Mišković-Stanković V, Kićević D, Jovanić P (2002) Colloid Surf A 209:47
12. Djošić MS, Mišković-Stanković VB, Srdić VV (2007) J Serb Chem Soc 72:275
13. Djošić MS, Panić V, Stojanović J, Mitrić M, Mišković-Stanković V (2012) Colloid Surf A 400:36
14. Sarkar P, Haung X, Nicholson PS (1992) J Am Ceram Soc 75:2907
15. Zhitomirsky I, Gal-Or L, Kohn A, Hennicke HW (1995) J Mater Sci 30:5307
16. Abolmaali SB, Talbot JB (1992) J Electrochem Soc 140:443
17. Park JM, Lee SI, Kim KW, Yoon DJ (2001) J Colloid Interface Sci 237:80
18. de Beer E, Duval J, Meulenkamp EA (2000) J Colloid Interface Sci 222:117
19. Ferrari B, Morreno R (1997) J Eur Ceram Soc 17:549
20. Choudhary JY, Ray HS, Rai KN (1982) Trans J Br Ceram Soc 81:189
21. Fischer R, Fischer E, de Portu G, Roncari E (1995) J Mater Sci Lett 14:25
22. Boccaccini AR, Kaya C (2002) Ceram Int 28:893
23. Van der Biest OO, Vandeperre LJ (1999) Annu Rev Mater Sci 29:327
24. Sarkar P, Nicholson PS (1996) J Am Ceram Soc 79:1987
25. Dalzell WJ, Clark DE (1986) Ceram Eng Sci Proc 7:1014
26. Lazić M, Mišković-Stanković VB, Jovanić P, Simović K (2002) Interceram 51:328
27. Simović K, Mišković-Stanković VB, Kićević D, Jovanić P (2003) Interceram 52:162

28. Boccaccini AR, Cho J, Subhani T, Kaya C, Kaya F (2010) J Eur Ceram Soc 30:1115
29. Singh I, Kaya C, Shaffer MSP, Thomas BC, Boccaccini AR (2006) J Mater Sci 41:8144
30. Zhitomirsky I (1998) J Mater Sci Lett 17:2101
31. Guglielmi M, Licculli A, Mazzarelli S (1994) Ceram Acta 6:19
32. Nishimori H, Tatsumisago M, Minami T (1996) J Mater Sci 31:6529
33. Pang X, Zhitomirsky I (2007) Mater Charact 58:339
34. Kaya C, Kaya F, Boccaccini AR (2002) J Mater Sci 37:4145.
35. Zhang Z, Huang Y, Jiang Z (1994) J Am Ceram Soc 77:1946
36. Zhitomirsky I (2000) Mater Lett 42:262
37. Wang ZC, Chen F, Huang LM, Lin CJ (2005) J Mater Sci 40:4955
38. Ma J, Liang CH, Kong LB, Wang C (2003) J Mater Sci: Mater Med 14:797
39. Wang C, Ma J, Cheng W, Zhang R (2002) Mater Lett 57:99
40. Kuo MC, Yen SK (2002) Mater Sci Eng C 20:153
41. Kar A, Raja KS, Misra M (2006) Surf Coat Technol 201:3723
42. Stoch A, Brozek A, Kmita G, Stoch J, Jastrzebski W, Rakowska A (2001) J Mol Struct 596:191
43. Zhitomirsky I (2011) Surf Eng 27:403
44. Bauer S, Schmuki P, von der Mark K, Park J (2013) Prog Mater Sci 58:261
45. Wang Y, Deen I, Zhitomirsky I (2011) J Colloid Interface Sci 362:367
46. Yoshioka T, Chávez-Valdez A, Roether JA, Schubert DW, Boccaccini AR (2013) J Colloid Interface Sci 392:167
47. Erol-Taygun M, Zheng K, Boccaccini AR (2013) Int J Appl Glass Sci 2:136
48. Cao W, Hench LL (1996) Ceram Int 22:493
49. Pishbin F, Mouriño V, Gilchrist JB, McComb DW, Kreppel S, Salih V, Ryan MP, Boccaccini AR (2013) Acta Biomater 9:7469
50. Ananth KP, Suganya S, Mangalaraj D, Ferreira JMF, Balamurugan A (2013) Mater Sci Eng C 33:4160
51. Chen Q, Cordero-Arias L, Roether JA, Cabanas-Polo S, Virtanen S, Boccaccini AR (2013) Surf Coat Technol 233:49
52. Lin W-C, Chen C-H, Tang H-Y, Hsiao Y-C, Pan JR, Hu C-C, Huang C (2013) Appl Catal B 140–141:32
53. Ipeksac T, Kaya F, Kaya C (2013) Mater Lett 100:11
54. Benehkohal NP, Gomez MA, Gauvin R, Demopoulos GP (2013) Electrochim Acta 87:169
55. Hossain MM, Mamun AHA, Hahn JR (2012) J Phys Chem C 116:23153
56. Hadraba H, Drdlik D, Chlup Z, Maca K, Dlouhy I, Cihlar J (2013) J Eur Ceram Soc 33:2305
57. Cihlar J, Drdlik D, Cihlarova Z, Hadraba H (2013) J Eur Ceram Soc 33:1885
58. Horri BA, Selomulya C, Wang H (2012) Int J Hydrogen Energy 37:19045
59. Cherng JS, Ho MY, Yeh TH, Chen WH (2012) Ceram Int 38S:S477
60. Abdoli H, Alizadeh P (2012) Mater Lett 80:53
61. Hadar R, Golodnitsky D, Mazor H, Ripenbein T, Ardel G, Barkay Z, Gladkich A, Peled E (2013) J Phys Chem B 117:1577
62. Wang W, Li C, Li J, Fan J, Zhou X (2013) J Rare Earth 31:289
63. Maleki-Ghaleha H, Rekabeslami M, Shakeri MS, Siadati MH, Javidi M, Talebian SH, Aghajani H (2013) Appl Surf Sci Appl Surf Sci 280:666
64. Ren C, He Y, Wang D (2012) Appl Surf Sci 258:5739
65. Borrell A, Dolores Salvador M, Rocha VG, Fernández A, Molina T, Moreno R (2013) J Alloy Compd 577:195
66. Uchikoshi T, Matsunaga C, Suzuki TS, Sakka Y, Matsuda M (2013) J Ceram Soc Jpn 121:370
67. Sadeghi AA, Ebadzadeh T, Raissi B, Ghashghaie S (2013) Ceram Int 39:7433
68. Sun Y, Wang Y, Zhitomirsky I (2013) Colloid Surf A 418:138
69. Ma LA, Guo TL (2013) Ceram Int 39:6923
70. Liau LC-K, Chen Y-P (2013) Colloid Surf A 429:121

71. Fori B, Taberna P-L, Arurault L, Bonino J-P, Gazeau C, Bares P (2012) Colloid Surf A 415:187
72. Okada K, Nagashima T, Kameshima Y, Yasumori A, Tsukada T (2002) J Colloid Interface Sci 253:308
73. Petrović R, Milonjić S, Jokanović V, Kostić-Gvozdenović Lj, Petrović-Prelević I, Janaćković Đ (2003) Powder Techn 133:185
74. Souza Santos H, Kiohara PK, Souza Santos P (1996) Mater Res Bull 31:799
75. Tsukada T, Segawa H, Yasumori A, Okada K (1999) J Mater Chem 9:549
76. Bokhimi X, Toledo-Antonio JA, Guzmán-Castillo ML, Mar-Mar B, Hernández-Beltrán F, Navarrete J (2001) J Solid State Chem 161:319
77. Guzmán-Castillo ML, Bokhimi X, Toledo-Antonio A, Salmones-Blásquez J, Hernández-Beltrán F (2001) J Phys Chem B 105:2099
78. Temuujin J, Jadambaa T, Mackenzie KJD, Angerer P, Porte F, Riley F (2000) Bull Mater Sci 23:301
79. Buining PA, Pathmamanoharan C, Jansen JBH, Lekkerkerker HNW (1991) J Am Ceram Soc 74:1303
80. Morgado E, Lam Jr. YL, Nazar LF (1997) J Colloid Interface Sci 188:257
81. Keysar S, Shter GE, Hazan Y, Cohen Y (1997) Chem Mater 9:2464
82. Rousseaux JM, Weisbecker P, Muhr H, Plasari E (2002) Ind Eng Chem Res 41:6059
83. Mišković VB, Maksimović MD (1985) Surf Technol 26:353
84. Pierce PE, Kovac Z, Higginbotham C (1978) Ind Eng Chem Prod Res Dev 17:317
85. Aries L (1994) J Appl Electrochem 24:554
86. Kingery WD, Bowen HK, Uhlmann DR (1975) Introduction to Ceramics, John Wiley & Sons, New York
87. Bohner M (2000) Int J Care Injured 31:D37
88. LeGeros RZ (2002) Clin Orthop Rel Res 395:81
89. Pham MT, Maitz MF, Grambole D, Herrmann F, Reuther H, Richter E (2001) Mater Sci Lett 20:295
90. Cengiz B, Gocke Y, Yildiz N, Aktas Z, Calimli A (2008) Colloid Surface A 322:29
91. Djošić MS, Mišković-Stanković VB, Milonjić S, Kačarević-Popović ZM, Bibić N, Stojanović J (2008) Mater Chem Phys 111:137
92. Tsuchida T, Kuboa J, Yoshioka T, Sakuma S, Takeguchi T, Ueda W (2008) J Catalysis 259:183
93. Luo P, Nieh T.G (1996) Biomaterials 17:1959
94. Santos C, Martins MA, Franke RP, Almeida MM, Costa MEV (2009) Ceram Int 35:1587
95. Xiao X, He D, Liu F, Liu R (2008) Mater Chem Phys 112:838
96. Huang LY, Xu KW, Lu J (2000) J Mater Sci Mater Med 11:667
97. Liu J, Ye X, Wang H, Zhu M, Wang B, Yan H (2003) Ceram Int 29:629
98. Chen JD, Wang YJ, Wei K, Zhang SH, Shi XT (2007) Biomaterials 28:2275
99. Viswanath B, Ravishankar N (2008) Biomaterials 29:4855
100. Ma MG, Zhu YJ, Chang J (2006) J Phys Chem B 110:14226
101. Štulajterova R, Medvecky L (2008) Colloid Surf A 316:104
102. Lebugle A, Sallek B, Tai Tai A (1999) J Mater Chem 9:2511
103. Shih WJ, Chen YH, Wang SH, Li WL, Hon MH, Wang MC (2005) J Cryst Growth 285:633
104. Juhasz JA, Best SM, Auffret AD, Bonfield W (2008) J Mater Sci Mater Med 19:1823
105. Jinawatah S, Pongkao D, Sunchanek W, Youshimura M (2001) Int J Inorg Mater 3:997
106. Hsu S, Chang E, Liub HS (1998) Ceram Int 24:249
107. Kong XD, Sun XD, Lu JB, Cui FZ (2005) Curr Appl Phys 5:519
108. Sivakumar GR, Girija EK, Kalkura SN, Subramanian C (1998) Cryst Res Technol 33:197
109. Rohanizadeh R, LeGeros RZ, Harsono M, Bendavid A (2005) J Biomed Mater Res 72A:428
110. Rohanizadeh R, LeGeros RZ (2008) Key Eng Mater 361–363:617
111. Bonfield W, Best SM, Nobuyuki A (1998) Method for coating a calcium phosphate compound onto a metallic material. WIPO WO9813539

112. Prado Da Silva MH, Lima JHC, Soares GA, Elias CN, De Andrade MC, Best SM, Gibson IR (2001) Surf Coat Tech 137:270
113. Takagi O, Azuma K, Iwamura T (1991) Process for producing hydroxyapatites. US Patent 5 073 357
114. Tas CA, Bhaduri SB (2005) Preparation of brushite ($CaHPO_4{\cdot}2H_2O$) powders and their *in vitro* conversion to nanoapatites. In: Sundar V, Rusin RP, Rutiser CA (eds) Bioceramics: Materials and Applications V, Ceramic Transactions, vol. 164, Wiley-Blackwell, p119
115. HSC Chemistry (1993) Outokumpu Research Oy, Finland
116. Puri P, Yang V (2007) J Phys Chem C 111:11776
117. Alavi S, Thompson DL (2006) J Phys Chem A 110:1518
118. Pivkina A, Ulyanova P, Frolov Y, Zavyalov S, Schoonman J (2004) Propellants, Explosives, Pyrotechnics 29:39
119. Lai SL, Guo JY, Petrova V, Ramanath G, Allen LH (1996) Phys Rev Lett 77:99
120. Buffat P, Borel JP (1976) Phys Rev A 13:2287
121. Mostafa NY (2005) Mater Chem Phys 94:333
122. Rodríguez-Lorenzo LM, Vallet-Regí M (2000) Chem Mater 12:2460
123. Fujishiro Y, Fujimoto A, Sato T, Okuwaki A (1995) J Colloid Interface Sci 173:119
124. Kandori K, Horigami N, Yasukawa A, Ishikawa T (1997) J Am Ceram Soc 80:1157
125. Arce H, Montero ML, Sáenz A, Castaño VM (2004) Polyhedron 23:1897
126. Abdel-Aal EA (2010) Int J Nano and Biomaterials 3:187
127. Abdel-Aal EA (2011) Int J Nanoparticles 4:77
128. Tas AC, Bhaduri SB (2004) J Am Ceram Soc 87:2195
129. Kumar M, Xie J, Chittur K, Riley C (1999) Biomaterials 20:1389
130. Kumar M, Dasarathy H, Riley C (1999) J Biomed Mater Res 45:302
131. Abdel-Aal EA, Dietrich D, Steinhaeuser S, Wielage B (2008) Surf Coat Tech 202:5895
132. Montero-Ocampo C, Villegas D, Veleva L (2008) Surf Eng 24:18
133. Zhang JM, Lin CJ, Feng ZD, Tian ZW (1998) J Electroanal Chem 452:235
134. Dumelié N, Benhayoune H, Rousse-Bertrand C, Bouthors S, Perchet A, Wortham L, Douglade J, Laurent-Maquin D, Balossier G (2005) Thin Solid Films 492:131
135. Kuo MC, Yen SK (2002) Mater Sci Eng C 20:153
136. Yen SK, Lin CM (2002) Mater Chem Phys 77:70
137. Eliaz N, Sridhar TM (2008) Cryst Growth Des 8:3965
138. Okawa S, Homma K, Kanatani M, Watanabe K (2009) Dental Mater J 28:513
139. Erne BH, Maroun F, Ozanam F, Chazalviel JN (1999) Electrochem Solid-State Lett 2:231
140. Jeremiasse AW, Hamelers HVM, Kleijn JM, Buisman CJN (2009) Environ Sci Technol 43:6882
141. Grandfield K, Zhitomirsky I (2008) Mater Charact 59:61
142. Albayrak O, El-Atwani O, Altintas S (2008) Surf Coat Tech 202:2482
143. Gomes JF, Granadeiro CC, Silva MA, Hoyos M, Silva R, Vieira T (2008) Int J Chem React Eng 6:A103
144. Baker MA, Assis SL, Higa OZ, Costa I (2009) Acta Biomater 5:63
145. Siddharthan A, Sampath Kumar TS, Seshadri SK (2010) Surf Coat Tech 204:1755
146. Eliaz N, Sridhar TM, Kamachi Mudali U, Raj B (2005) Surf Eng 21:238
147. Ciobanu G, Carja G, Ciobanu O, Sandu I, Sandu A (2009) Micron 40:143
148. Wei D, Zhou Y, Wang Y, Meng Q, Jia D (2008) Surf Coat Tech 202:5012
149. Krishna BV, Bose S, Bandyopadhyay A (2007) Acta Biomater 3:997
150. Xiao X, Liu R, Tang X (2009) J Mater Sci Mater Med 20:691
151. Pang X, Zhitomirsky I (2005) Mater Chem Phys 94:245
152. Wang J, De Boer J, De Groot K (2004) J Dent Res 83:296
153. Martin H, Schulz KH, Bumgardner JD, Schneider JA (2008) Thin Solid Films 516:6277
154. Xianmiao C, Yubao L, Yi Z, Li Z , Jidong L, Huanan W (2009) Mater Sci Eng C 29:29
155. Gebhardt F, Seuss S, Turhan MC, Hornberger H, Virtanen S, Boccaccini AR (2012) Mater Lett 66:302

156. Simchi A, Tamjid E, Pishbin F, Boccaccini AR (2011) Nanomed-Nanotechnol 7:22
157. Mahmoodi S, Sorkhi L, Farrokhi-Rad M, Shahrabi T (2013) Surf Coat Tech 216:106
158. Batmanghelich F, Ghorbani M (2013) Ceram Int 39:5393
159. Raschip IE, Vasile C, Ciolacu D, Cazacu G (2007) High Perform Polym 19:603
160. Park Y, Doherty WOS, Halley PJ (2008) Ind Crop Prod 27:163
161. Mansur HS, Mansur AAP, Bicallho SMCM (2005) Key Eng Mater 284–286:745
162. Martinez M, Pacheco A, Vargas M (2009) Rev. MVZ Córdoba 14:1624
163. Erakovic S, Veljovic Dj, Diouf PN, Stevanovic T, Mitric M, Milonjic S, Miskovic-Stankovic VB (2009) Int J Chem React Eng 7:A62
164. Deen I, Zhitomirsky I (2014) J Alloy Compd 586:S531
165. Zhitomirsky D, Roether JA, Boccaccini AR, Zhitomirsky I (2009) J Mater Process Tech 209:1853
166. Li P, Huang Z, Xiao X (2009) J Chin Ceram Soc 37:1864
167. Parente P, Sanchez-Herencia AJ, Mesa-Galan MJ, Ferrari B (2013) J Phys Chem B 117:1600
168. Montero ML, Sáenz A, Rodríguez JG, Arenas J, Castaño VM (2006) J Mater Sci 41:2141
169. Arita IH, Castano VM, Wilkinson DS (1995) J Mater Sci Mater Med 6:19
170. Zhitomirsky I, Gal-Or L (1997) J Mater Sci Mater Med 8:213
171. Wang Q, Huang W, Wang D, Darvell BW, Day DE, Rahaman MN (2006) J Mater Sci Mater Med 17:641
172. Ashok M, Kalkura SN, Sundaram NM, Arivuoli D (2007) J Mater Sci Mater Med 18:895
173. Wang F, Li MS, Lu YP, Qi YX, Liu YX (2006) Mater Chem Phys 95:145
174. Kamiya K, Yoko T, Tanaka K, Fujiyama Y J (1989) Mater Sci 24:827
175. Sáenz A, Montero ML, Castaño VM (2002) Surf Rev Lett 9:1799
176. Eshtiagh-Hosseini H, Housaindokht MR, Chahkandi M (2007) Mater Chem Phys 106:310
177. Liu C, Ji X, Cheng G (2007) Appl Surf Sci 253:6840
178. Rodriguez-Lorenzo LM, Vallet-Regi M (2000) Chem Mater 12:2460
179. Bogdanoviciene I, Beganskiene A, Tonsuaadu K, Glaser J, Jurgen Mayer H, Kareiva A (2006) Mater Res Bull 41:1754
180. Tkalcec E, Sauer M, Nonninger R, Schmidt H (2001) J Mater Sci 36:5253
181. Sridhar TM, Mudali UK, Subbaiyan M (2003) Corros Sci 45:2337
182. Tampieri A, Celloti G, Szontag F, Landi E (1997) J Mater Sci Mater Med 8:29
183. Wang PE, Chaki TK (1993) J Mater Sci Mater Med 4:150
184. Baurhoo B, Ruiz-Feria CA, Zhao X (2008) Anim Feed Sci Tech 144:175
185. Domínguez JC, Oliet M, Alonso MV, Gilarranz MA, Rodríguez F (2008) Ind Crop Prod 27:150
186. Gosselink RJA, Abächerli A, Semke H, Malherbe R, Käuper P, Nadif A, van Dam JEG (2004) Ind Crop Prod 19:271
187. Kolli Sahithi MS, Moorthi A, Srinivasan N, Ramasamy K, Selvamurugan N (2010) Int J Biol Macromol 47:1
188. Palcevskis E, Dindune A, Kuznecova L, Lipe A, Kanepe Z (2005) Latvian J Chem 2:128
189. Veljovic Dj, Jokic B, Petrovic R, Palcevskis E, Dindune A, Mihailescu I N, Janackovic Dj (2009) Ceram Int 35:1407
190. Kwok CT, Wong PK, Cheng FT, Man HC (2009) Appl Surf Sci 255:6736
191. Bhattarai SR, Aryal S, Bahadur KCR, Bhatarai N, Hwang PH, Yi HK, Kim HY (2008) Mater Sci Eng C 28:64
192. Yang L, Wei WZ, Gao XH, Xia JJ, Tao H (2005) Talanta 68:40
193. Aryal S, Bahadur KCR, Dharmaraj N, Kim KW, Kim HY (2006) Scripta Mater 54:131
194. Zhou JM, Zhang XD, Chen JY, Zeng SX, Degroot K (1993) J Mater Sci Mater Med 4:83
195. Ye HZ, Liu XY, Hong HP (2009) J Mater Sci Mater Med 20:843
196. Filiaggi MJ, Pilliar RM, Coombs NA (1993) J Biomed Mater Res 27:191
197. Victoria EC, Gnanam FD (2002) Trends Biomater Artif Organs 16:12
198. Alriols MG, Tejado A, Blanco M, Mondragón I, Labidi J (2009) Chem Eng J 148:106
199. Yan L, Leng Y, Weng L-T (2003) Biomaterials 24:2585

200. Zhao J, Lu X, Duan K, Guo LY, Zhou SB, Weng J (2009) Colloid Surf B 74:159
201. Scholze B, Meier D (2001) J Anal Appl Pyrol 60:41
202. Rodrigues PC, Cantao MP, Janissek P, Scarpa PCN, Mathias AL, Ramos LP, Gomes MAB (2002) Eur Polym J 38:2213
203. El-Hendawy A-NA (2006) J Anal Appl Pyrol 75:159
204. Mendes LC, Rodrigues RC, Silva EP (2010) J Therm Anal Calorim 101:899
205. Sjogren G, Sletten G, Dahl JE (2000) J Prosthet Dent 84:229
206. Manivasagam, Dhinasekaran G, Rajamanickam D (2010) Recent Pat Corros Sci A 2:40
207. Mourino V, Cattalini JP, Boccaccini AR (2012) J R Soc Interface 9:401
208. Rameshbabu N, Sampath Kumat TS, Prabhakar TG, Sastry VS, Murty KVGK, Prasad Rao K (2007) J Biomed Mater Res A 80:581
209. Lee I-S, Whang C-N, Oh K-S, Park J-C, Lee K-Y, Lee G-H, Chung S-M, Sun X-D (2006) Nucl Instrum Meth B 242:45
210. Pang X, Zhitomirsky I (2008) Surf Coat Tech 202:3815
211. Kim J-U, Jeong Y-H, Choe H-C (2011) Thin Solid Films 520:793
212. Wang Y, Zhang X, Yan J, Xiao Y, Lang M (2011) Appl Surf Sci 257:6233
213. Vasilescu C, Drob P, Vasilescu E, Demetrescu I, Ionita D, Prodana M, Drob SI (2011) Corros Sci 53:992
214. Costa-Rodrigues J, Fernandes A, Lopes MA, Fernandes MH (2012) Acta Biomater 8:1137
215. Gu YW, Khor KA, Cheang P (2004) Biomaterials 25:4127
216. Garcia C, Cere S, Duran A (2006) J Non-Cryst Solids 352:3488
217. Kim H-M, Himeno T, Kokubo T, Nakamura T (2005) Biomaterials 26:4366
218. Pecheva EV, Pramatarova LD, Maitz MF, Pham MT, Kondyuirin AV (2004) Appl Surf Sci 235:176
219. Stoch A, Jastrzebski W, Brozek A, Trybalska B, Cichocinska M, Szarawara E (1999) J Mol Struct 511–512:287
220. Sun R, Li M, Lu Y, Wang A (2006) Mater Charact 56:250
221. Wang L-N, Luo J-L (2011) Mater Sci Eng C 31:748
222. Roguska A, Pisarek M, Andrzejczuk M, Dolata M, Lewandowska M, Janik-Czachor M (2011) Mater Sci Eng C 31:906
223. Yao ZQ, Ivanisenko Yu, Diemant T, Caron A, Chuvilin A, Jiang JZ, Valiev RZ, Qi M, Fecht H-J (2010) Acta Biomater 6:2816
224. Watling KM, Parr JF, Rintoul L, Brown CL, Sullivan LA (2011) Spectrochim Acta A 80:106
225. Viornery C, Chevolot Y, Leonard D, Aronsson B-O, Pechy P, Mathieu HJ, Descouts P, Gratzel M (2002) Langmuir 18:2582
226. Bai X, Sandukas S, Appleford MR, Ong JL, Rabiei A (2009) Acta Biomater 5:3563
227. Hsu CH, Mansfeld F (2001) Corrosion 57:747
228. Prakash Parthiban S, Suganthi RV, Girija EK, Elayaraja K, Kulriya PK, Katharria YS, Singh F, Sulania I, Tripathi A, Asokan K, Kanjilal D, Yadav S, Singh TP, Yokogawa Y, Narayana Kalkura S (2008) Nucl Instr and Meth B 266:911
229. Girija EK, Parthiban SP, Suganthi RV, Elayaraja K, Joshy MIA, Vani R, Kularia P, Asokan K, Kanjilal D, Yokogawa Y, Narayana Kalkura S (2008) J Ceram Soc Jpn 116:320
230. Pelletier H, Nelea V, Mille P, Muller D (2004) Nucl Instr and Meth B 216:269
231. Pelletier H, Nelea V, Mille P, Muller D (2004) Nucl Instr and Meth B 216:275
232. Djošić MS, Bibić N, Mitrić MN, Šiljegović M, Stojanović JN, Jokić B, Janaćković DjT, Mišković-Stanković VB (2009) J Optoelectron Adv M 11:1848
233. Williams JS (1986) Rep Prog Phys 49:491
234. Ohring M (1991) The Materials Science of Thin Films. Academic Press, London
235. Bibić N, Wilson IH, Milosavljević M, Perusko D (1992) J Mater Sci 27:4945
236. Miro S, Grebille D, Chateigner D, Pelloquin D, Stoquert J-P, Grob J-J, Constantini J-M, Studer F (2005) Nucl Instr and Meth B 227:306

Chapter 4
Electrochemical Synthesis of Metal Oxides for Energy Applications

Lok-kun Tsui and Giovanni Zangari

4.1 Introduction

Metal oxides are an important class of materials, which benefit from the large electronegativity of oxygen to induce strong electron correlations among nearby atoms, thus enabling fine-tuning of the material parameters of the system [1]. The ability to create materials with a wide range of electronic properties has resulted in metal oxides finding applications in a diverse set of fields including semiconductor devices for computing and data storage [2–6], magnetic materials [7–9], and catalysis [10, 11]. To address the grand challenge of providing carbon-free, renewable energy to the world in the twenty-first century, there is an urgent need to develop efficient, inexpensive devices which could be used for the conversion and storage of energy.

Metal oxides are promising candidates for energy conversion and storage applications because of their wide range of properties, low cost and availability [12]. For example, TiO_2 and ZnO are wide band gap semiconductors that have been widely investigated for direct energy conversion from sunlight, primarily for use in dye sensitized solar cells [13–15]. Narrower bandgap oxides that have been investigated for photoelectrochemical or photovoltaic devices include Cu_2O [16–18], hematite (α-Fe_2O_3) [19, 20], and $BiVO_4$ [21–23]. With regard to energy storage, oxides of transition metals such as Mn, Co, Ni, and V are used as cathodes for lithium ion batteries and electrochemical supercapacitors [24–26]. Preparative techniques available to prepare metal oxides include hydrothermal synthesis [27], thermal oxidation [28], chemical vapor deposition [29], sputtering [30], and spray-pyrolysis [31]. Electrodeposition of metal oxides however presents several advantages in

L.-k. Tsui • G. Zangari (✉)
Department of Materials Science and Engineering, University of Virginia,
Charlottesville, VA 22904, USA
e-mail: gz3e@virginia.edu

S.S. Djokić (ed.), *Electrodeposition and Surface Finishing: Fundamentals and Applications*, Modern Aspects of Electrochemistry 57,
DOI 10.1007/978-1-4939-0289-7_4, © Springer Science+Business Media New York 2014

comparison with vapor phase, vacuum, or other high temperature routes; it can be achieved at lower cost, since it does not require high capital investment while allowing the synthesis of materials over a large area, and can be conducted at low temperature and low energy intensity [32]. These features make the electrodeposition route attractive for fabricating metal oxides in a cost-effective and scalable manner, essential features when planning technology deployment at a global scale. Precise control of the morphology of these oxides is possible through the adjustment of deposition conditions and the chemistry of the deposition solution [33–35]. Electrodeposition also permits ready fabrication of mixed oxides, potentially accessing an even wider range of properties. These investigations in many cases have used a combinatorial approach, where an array of small cells on a single substrate was used to rapidly fabricate and test devices of varying compositions [12].

This chapter discusses the fundamentals of the synthesis of metal oxides. The conditions necessary to achieve the electrochemical oxides are described, as well as the conditions to control the morphology and properties of the deposited films. Selected examples of metal oxide systems widely used in energy applications are highlighted and discussed in detail. These materials include ZnO, Cu_2O, α-Fe_2O_3 and related iron oxides, and MnO_2.

4.2 Fundamental Concepts

4.2.1 Stability of Metal Oxides: Pourbaix Diagrams

The use of a Pourbaix diagram can provide a simple first step in determining at what potential and pH the formation of the oxide of interest becomes thermodynamically possible. Pourbaix diagrams compile the thermodynamically stable phases in aqueous solutions of the elements as a function of pH and potential. The stability regions listed include dissolved species, metallic phases, and oxide phases, and they are divided by a set of lines that represent the equilibria between two phases. Consider the case of the Zn Pourbaix diagram (Fig. 4.1). The equilibria of interest to the formation of the ZnO phase are given in the following equations (4.1), (4.2), (4.3), (4.4), (4.5), (4.6), and (4.7) [36]. All potentials in the following are referred to the standard hydrogen electrode (SHE).

$$Zn^{2+} + H_2O = ZnO + 2H^+; \quad pH = \frac{10.96 - \log(Zn^{2+})}{2} \tag{4.1}$$

$$Zn = Zn^{2+} + 2e^-; \quad E_0 = -0.763 + 0.0295\log(Zn^{2+}) \tag{4.2}$$

$$Zn + H_2O = ZnO + 2H^+ + 2e^-; \quad E_0 = -0.439 - 0.0591\text{pH} \tag{4.3}$$

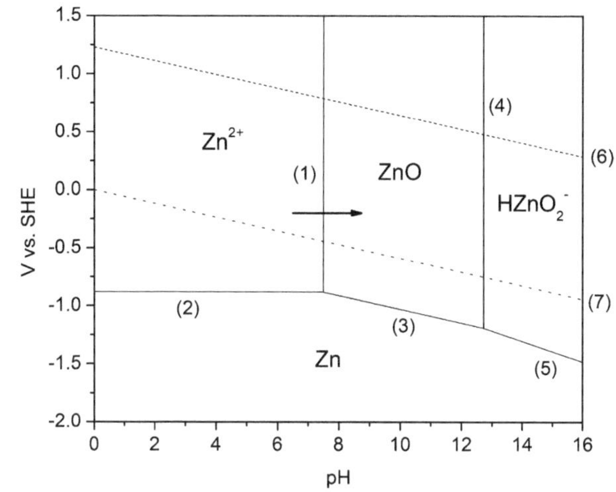

Fig. 4.1 Pourbaix diagram showing the stable Zn species, constructed assuming dissolved species concentrations of 10^{-4} M. Lines labeled 1–7 correspond to the equilibria described above in their respective equations (4.1), (4.2), (4.3), (4.4), (4.5), (4.6), and (4.7). The *arrow* represents the path where a local increase of pH near the working electrode causes formation of ZnO

$$ZnO + H_2O = HZnO_2^- + H^+; \quad pH = \log\left(HZnO_2^-\right) + 16.68 \tag{4.4}$$

$$\begin{aligned} Zn + 2H_2O &= HZnO_2^- + 2H^+ + 2e^-; \\ E_0 &= 0.054 - 0.0886pH + 0.0295\log\left(HZnO_2^-\right) \end{aligned} \tag{4.5}$$

$$O_2/H_2O; \quad E_0 = 1.23 - 0.059pH \tag{4.6}$$

$$H_2/H_2O; \quad E_0 = -0.059pH \tag{4.7}$$

The Pourbaix diagram constructed from these equilibria is shown in Fig. 4.1. Identification of the conditions at which the stable phase changes from an oxide to a metal (line 3) allows one to determine the potential pH window where oxide deposition may occur. ZnO is shown to be stable about a pH range from approximately 8–12 and a potential range down to -1.0 V_{SHE}. Electrodeposition of some metal oxides may require deposition at elevated temperatures in order to accelerate process kinetics, and under these conditions, corrections need to be made to the Pourbaix diagrams from knowledge of thermodynamic parameters [37]. Additionally, in some cases complexing agents and other compounds are added to the electrolyte in order to facilitate formation of the material of interest; in this case also the Pourbaix diagram should be recalculated to include all the relevant equilibria.

Mixed oxides are formed when the electrochemical reactions induced by the applied potential push the system into a region where the two species involved are simultaneously favored to form solid precipitates [38]. A suitable potential pH

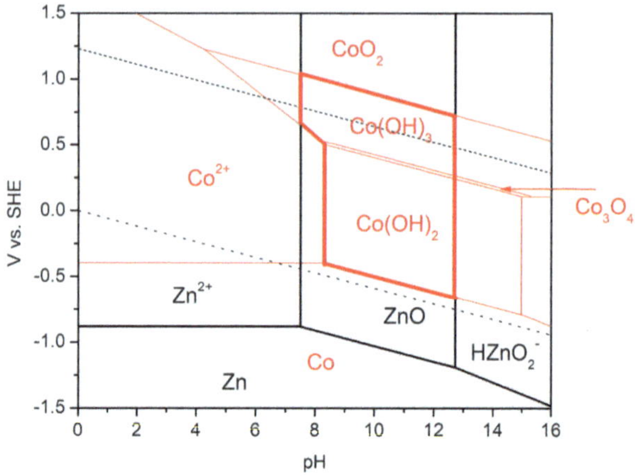

Fig. 4.2 Overlay of the Co and Zn Pourbaix diagrams, showing the areas of stability of oxides and hydroxides of both elements. The *bold contoured area* represents the region where Zn oxides may coexist with Co hydroxides

range can be identified by overlaying the Pourbaix diagrams of the two species. This is exemplified in Fig. 4.2 by overlaying the region of stability for Co oxide/hydroxide with that of Zn oxide, following the same procedure used to derive Fig. 4.1 [36]. The region highlighted with bold lines in Fig. 4.2 represents the conditions where both species tend to form oxides (hydroxides) and where a mixed oxide may therefore be deposited. If the deposition of a mixed oxide of Co and Zn is desired, the window for deposition narrows from approximately $-1.0\ V_{SHE}$ to $-0.5\ V_{SHE}$ due to the more positive potential of the equilibrium line between Co and Co$(OH)_2$ relative to the Zn and ZnO line. The equilibria of interest can be shifted by using suitable complexing agents and additives, allowing the formation of mixed oxides that would not form from simple solutions due to the lack of any intersecting region of the respective metal oxide/metal equilibria.

4.2.2 Synthesis Routes for the Formation of Metal Oxides

Electrochemical synthesis of oxides can proceed by either anodic or cathodic paths. Anodic processes may involve either the direct oxidation of the metal substrate (anodization) or the anodic electrodeposition of the oxide from metal ions in solution. The latter involves oxidizing the metal ions in solution into an oxidation state where they readily react with water or hydroxide anions to yield an insoluble deposit. This route is used for example to prepare Mn, Co, Ni, and V oxides [25, 39]. The relevant reaction sequence for MnO_2 involves the oxidation of Mn from the 2+ to 4+ oxidation state and reaction with water to form MnO_2 [40]:

$$Mn^{2+} + 2H_2O \rightarrow MnO_2 + 4H^+ + 2e^- \qquad (4.8)$$

With regard to the possible cathodic synthesis routes, deposition of the oxide may occur either via (1) the partial reduction of the metal ion to a valence state where a subsequent reaction with water forms an oxide, or (2) the electrogeneration of a base near the working electrode. In the first case, the generic reaction proceeds as given in reaction (4.9) where metal M with an initial valence +n is reduced to the oxidation state +2*x* [38]:

$$M^{n+} + (n - 2x)e^{-} + xH_2O \rightarrow MO_x + 2xH^{+} \tag{4.9}$$

An example of this process is the electrochemical formation of Cu_2O, through the reaction described in Eq. (4.10):

$$2Cu^{2+} + 2e^{-} + H_2O \rightarrow Cu_2O + 2H^{+} \tag{4.10}$$

Deposition of Cu_2O requires an initial reduction of the Cu ions in solution from Cu^{2+} to Cu^{+}, which is then followed by its precipitation as Cu_2O upon reaction with water [41].

The other route for cathodic synthesis exploits a reaction which produces hydroxide ions via for example the reduction of water (4.11), reduction of dissolved oxygen (4.12), or reduction of nitrate (4.13) [36, 42].

$$2H_2O + 2e^{-} \rightarrow H_2 + 2OH^{-}; \quad E = -0.059\text{pH} \tag{4.11}$$

$$O_2 + 2H_2O + 4e^{-} \rightarrow 4OH^{-}; \quad E = 1.23 - 0.059\text{pH} \tag{4.12}$$

$$NO_3^{-} + H_2O + 2e^{-} \rightarrow NO_2^{-} + 2OH^{-}; \quad E = 0.835 - 0.059\text{pH} \tag{4.13}$$

The production of hydroxide near the substrate locally increases the pH and causes the metal ions in solution to react with the hydroxide ions, precipitating out as the desired metal oxide or metal hydroxide material [32]. This reaction in general proceeds according to the schematic reaction [Eq. (4.14)] for oxides or hydroxides [32, 42]. This final precipitation process occurs without any change in the oxidation state of the metal species. The case of ZnO is illustrated in Fig. 4.1. In the bulk solution under acidic conditions, the Zn^{2+} ions are stable, but a local increase in pH drives the system into the portion of the Pourbaix diagram where ZnO is stable [Eq. (4.15)].

$$M^{n+} + nOH^{-} \rightarrow M(OH)_n(\text{Hydroxide}) \tag{4.14}$$

$$M^{n+} + nOH^{-} \rightarrow MO_{\frac{n}{2}} + \frac{n}{2}H_2O(\text{Oxide})$$

$$Zn^{2+} + 2OH^{-} \rightarrow ZnO + H_2O \tag{4.15}$$

4.2.3 *Nucleation and Growth of Oxide Films*

The electrochemical formation of oxide films differs from the corresponding growth process for metallic films. Metal electrodeposition in fact occurs via

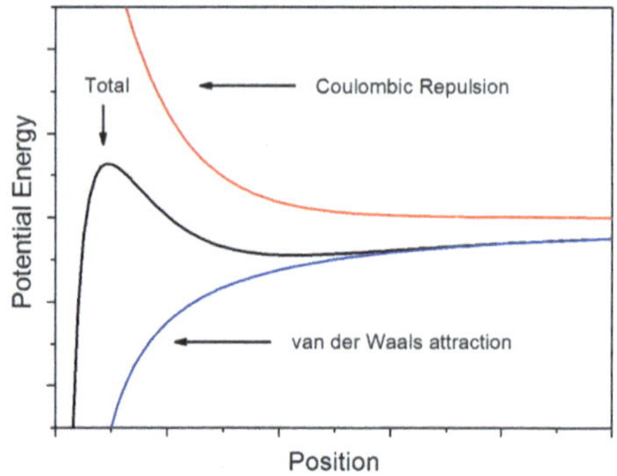

Fig. 4.3 Schematic plot of the repulsive (coulombic) and attractive (van der Waals) terms, as well as their sum, in the DLVO theory predicting oxide coagulation

reduction of metal ions at the electrode surface to form adsorbed atoms, followed by formation of a critical nucleus at the surface, while with oxide films the electrolyte environment close to the electrode surface induces the *homogeneous* nucleation of colloidal particles which coagulate at the electrode to form a solid deposit. Coagulation of the oxide to form the solid film occurs under the conditions described by the Derjaguin–Landau–Verwey–Overbeek (DLVO) theory [43, 44]. In this theory, the interaction between two charged particles may be described by the sum of the repulsive coulombic force associated with the double-layer charge at the particle–solution interface and the attractive van der Waals force. Equations (4.16) and (4.17) describe the repulsive (V_R) and attractive (V_A) potential energies of the interaction, respectively [43, 45].

$$V_R = 2\pi\varepsilon\varepsilon_0 a\psi^2 \ln\left(1 + e^{-\kappa H}\right) \tag{4.16}$$

$$V_A = -\frac{A}{6}\left(\frac{2}{s^2 - 4} + \frac{2}{s^2} + \ln\left(\frac{s^2 - 4}{s^2}\right)\right) \tag{4.17}$$

In the repulsive term, ε is the dielectric constant of the solution, ε_0 is the permittivity of free space, ψ is the surface potential, and H is the interparticle spacing. κ is the inverse of the Debye–Hückel screening length, Eq. (4.18), where the sum is conducted over all the ionic species i. This is a function of N_i and z_i, i.e., the density and valence of the ions in solution contributing to the charge of the double layer, as well as of the temperature T and ε; it also contains the Faraday constant F and Boltzmann's constant k.

$$\kappa = \left(\frac{F^2 \sum_i N_i z_i^2}{\varepsilon\varepsilon_0 kT}\right)^{\frac{1}{2}} \tag{4.18}$$

In the attractive term, A is the Hamaker constant that describes the strength of the interaction, a is the particle radius, and s is defined as: $s = 2 + H/a$.

These potential energies and their sum are plotted in Fig. 4.3. If the coulombic repulsion is large enough, the total potential presents an energy barrier to nucleation via flocculation. When the concentration of the oxide particles in solution is high enough that the barrier is overcome, flocculation occurs and the growth of the film begins. In the presence of cations in solution that decrease the double layer thickness and reduce the effect of coulombic repulsion, the barrier is eliminated leading to a facile coagulation of the colloidal oxide particles [43].

4.2.4 Evolution of Film Structure and the Control Over Morphology

The morphology of oxide films is determined by the particle size and shape of the aggregating particles and by the successive growth of existing films. Particle size and shape can be controlled by the pH and the ionic strength of the electrolyte; in particular, small particles are obtained by precipitation at a pH far from the point of zero charge in solutions of high ionic strength [46]. Control of film morphology instead has been demonstrated by the use of additives in the deposition solution. The evolution of film morphology at low overpotentials, i.e., close to thermodynamic conditions, is determined by the relative surface energies of the various crystallographic facets [47]. Since the crystal seeks to minimize surface energy of the exposed surfaces, high energy surfaces will tend to grow faster and be eliminated as growth proceeds, while low energy surfaces are free to grow larger in area. In the presence of additives that adsorb preferentially to certain facets of the crystal to lower their surface energy, the preferential direction of growth may be altered. Siegfried and Choi for example demonstrated that Cu_2O tends to form cubic crystals in the absence of any additives while the adsorption of sodium dodecyl sulfate stabilizes the {111} facets and promotes the growth of octahedral-shaped crystals [34, 48]. At high overpotentials instead, the diffusion of reactive species to the growth sites is the rate-determining process, and a branching growth is then preferred. The two growth modes can be used in conjunction to finely control the shape of oxide crystals [48]. Similar processes are responsible for the wide variety of structures, from nanorods to nanoplates, observed in ZnO formed by the addition in solution of KCl, NH_4F, NH_4CH_3COO, and ethylenediamine (EDA). Figure 4.4 shows schematically how adsorption of EDA at the side walls of the hexagonal ZnO structure may promote the vertical growth along the (0001) axis, resulting in ZnO nanowires, while preferential adsorption of Cl^- and other polar ions to the (0001) facets results in the formation of hexagonal plates [49]. Figure 4.5 shows actual ZnO materials grown by these methods.

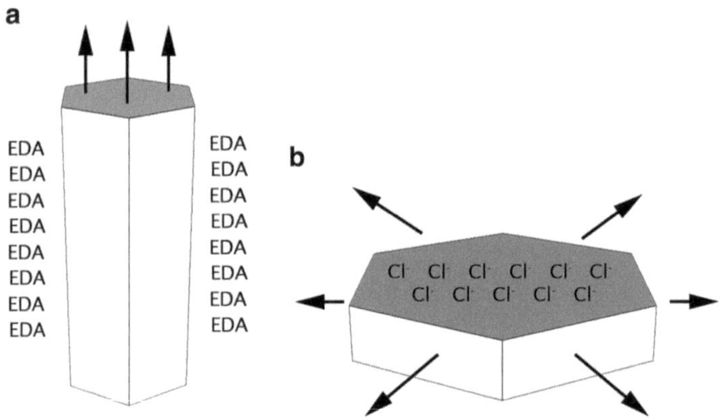

Fig. 4.4 The adsorption of additives can be used to control the growth direction in electrodeposited metal oxides. Shown here for the case of ZnO is (**a**) adsorption of EDA to the side walls promoting vertical growth and (**b**) adsorption of Cl^- to the basal plane promoting lateral growth

Fig. 4.5 (*Left*) Nanowires formed in the presence of EDA and (*right*) hexagonal nano-plates formed in the presence of Cl^-. Reprinted with permission from [49]. Copyright 2005 American Chemical Society

4.3 Notable Examples of Electrochemical Synthesis of Metal Oxides and Their Applications

4.3.1 ZnO

ZnO has been studied over the last several decades as a prototypical metal oxide system that is electrodeposited via the electrogeneration of base; it typically occurs in the hexagonal wurtzite structure, though it can be transformed into the cubic rocksalt structure at high pressure [50] and is a semiconductor with a wide bandgap of 3.37 eV [33]. The formation of ZnO is sensitive to the temperature of electrodeposition. It was found that 34 °C is the minimum temperature needed to form ZnO, since the initially grown $Zn(OH)_2$ forms an insulating layer that prevents further growth [37]. At low temperatures, a mixture of $Zn(OH)_2$ and ZnO forms, while at high temperature only ZnO tends to form owing to the more negative free energy of the dehydration reaction under these conditions [51]. Above 34 °C, $Zn(OH)_2$ is dehydrated to form ZnO seed particles which accelerate nucleation and facilitate the vertical growth [37]. The effect of deposition potential on the nucleation behavior of ZnO was studied by Manzano et al. [33]. They found that at high overpotentials, the large driving force creates large amounts of OH^- and makes the local pH near the electrode highly alkaline. In this condition the intermediate negatively charged $Zn(OH)_4^{2-}$ ions are created that subsequently nucleate as ZnO on the positively charged {0001} facet, resulting in the vertical growth of nanorods. On the other hand at low overpotentials, the smaller amount of OH^- created does not push the pH into as strongly alkaline conditions, and Zn^{2+} or $Zn(OH)_2$ are the predominant species, so nucleation preferentially occurs on the neutral sides, resulting in a plate-like morphology. The effect of the additives discussed above in Sect. 4.2.4 allowed Xu et al. to controllably create structures ranging from hexagonal plates to needles [49].

The energetics of the defect structure in ZnO favors n-type doping in the as-deposited condition, which has made synthesis of p-type ZnO difficult to achieve. Pseudopotential local density approximation calculations produced by Zhang et al. showed that the n-type producing oxygen vacancies and Zn interstitials have low enthalpies of formation while the p-type producing defects of oxygen interstitials and Zn vacancies have high enthalpies of formation [52]. Thus, the intentional introduction of intrinsic p-type dopants is easily compensated by the formation of additional n-type defects, negating p-type doping. Despite the inability to dope ZnO as a p-type semiconductor by using intrinsic defects, the addition of extrinsic defects by introducing different elements has been studied as an alternative. Park et al. studied group I and group V elements with first principles pseudopotential calculations and found that group I elements had lower ionization energies than group V elements, but could be readily compensated by the formation of interstitials [53]. Electrodeposition also provides a method to control the degree and type of doping in ZnO. Rousset et al. showed that the n-type of doping level of ZnO could be modified between 7.4×10^{17} cm^{-3} and 9×10^{19} cm^{-3} by increasing the concentration of Cl^- in the deposition solution [54]. p-type ZnO was also formed by introducing 2 at.% Cu into ZnO by electrodeposition; this process was used to demonstrate the formation of a p–n ZnO homojunction diode [55]. Photoelectrochemical measurements of the open circuit voltage of Ag-doped ZnO formed by electrodeposition showed that modifying the deposition potential below $-0.65\ V_{Ag/AgCl}$ caused a switch from n-type to p-type doping in samples that were annealed after deposition [56].

Considerable research has been devoted to ZnO as an anode material in dye sensitized solar cells, DSSC. DSSCs differ from the conventional p–n or p–i–n junctions used in photovoltaics in that (1) the photosensitive electrode is made from

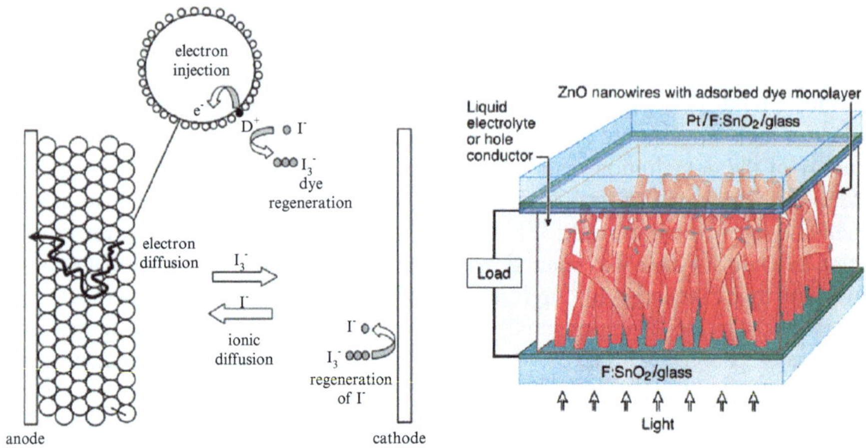

Fig. 4.6 (*Left*) Schematic of a conventional dye sensitized solar cell showing the principle of operation. Reprinted with permission from [58]. Copyright 2007 American Chemical Society. (*Right*) A DSSC device using ZnO nanowires, enabling a high surface area while minimizing the number of grain boundaries that must be crossed by photogenerated charges. Reprinted with permission from [14]. Copyright 2005, American Institute of Physics

porous wide bandgap materials such as ZnO, (2) the absorption of visible light is carried out by a dye adsorbed on the surface of the semiconductor, and (3) current flow within the device occurs via charge transfer to and from an iodine/iodide electrolyte [57, 58]. The device architecture ensures that the charge carriers within the semiconductor electrode are always generated very close to the electrolyte, resulting in high photoconversion efficiency even when using polycrystalline materials; this results in lower production costs [57]. A schematic of the operation of a typical dye sensitized solar cell is shown in Fig. 4.6, left, where the photosensitive electrode consists of nanosized TiO_2 particles. Charges photogenerated in these nanoparticles must undergo a random walk through as many as 10^6 particles for a 10 μm film to reach the conducting back contact [59]; a high surface area electrode is therefore required to generate charge as close as possible to the charge collector, resulting in a more efficient path for charge transport. The ability to fabricate crystalline nanowires of ZnO via electrodeposition without an additional heat treatment step is a promising route to minimizing the losses of the DSSCs due to photogenerated charges crossing many grain boundaries in nanoparticle TiO_2; an alternative ZnO nanowire device (Fig. 4.6, right) would indeed provide direct charge transport paths [14, 60, 61]. An additional advantage is that ZnO exhibits mobility values on the order of 300 cm^2/V-s in the bulk, with mobilities as high as 1,000 cm^2/V-s reported for single crystal nanowires [62, 63]. In comparison, TiO_2 nanotubes were measured with a mobility of 3.6 cm^2/V-s, and nanoparticles of TiO_2 have a mobility of 0.03 cm^2/V-s [64, 65]. Nanowires are also expected to be thick enough to provide a depletion layer with a built-in electric field that facilitates the separation of electron–hole pairs [66]. In principle, the better charge transport properties should make ZnO a much more attractive material for use in DSSCs. However, the efficiency of dye sensitized solar cells made of ZnO is still lower than the 10 % value originally reported by Grätzel for nanoparticles of TiO_2 [67], with recent developments pushing TiO_2-based DSSCs to 11.1 % [68]. Law et al. studied ZnO nanowire DSSC devices and determined that the lower surface area compared to nanoparticle DSSCs contributed to a weak efficiency of 1.2–1.5 % despite faster electron injection and a built in electric field that confines photogenerated charges to the interior of the tubes, enhancing charge transport [15]. ZnO nanowires grown by MOCVD showed efficiencies of 0.5 % [14]. Chen et al. showed that electrodeposited porous ZnO films had an efficiency of 5.08 % by preparing a compact layer of ZnO between the porous structure and the conducting glass substrate [69].

The cause of the differences in the efficiency of TiO_2– and ZnO-based devices come from a variety of sources. Keis et al. found that the formation of inactive Zn^{2+}/dye complexes collecting on ZnO electrodes reduces the efficiency of light absorption, a process which was not seen to occur on TiO_2 [60]. Compared to nanoparticles, the surface area of nanowire ZnO is lower, and a limited amount of light can be collected in this geometry [66]. Guérin and Pauporté carried out a two-step electrodeposition process: ZnO nanowires were formed at high temperature and then the gaps were filled in by a porous ZnO nanocrystalline matrix [70]. The nanocrystalline matrix provides a high surface area, while the nanowires act as conduction pathways for photogenerated charges. The reported efficiency of the DSSCs made with the nanowire–nanocrystalline matrix arrays was 4.08 %, compared to 2.84 % for porous nanocrystalline DSSCs and 0.49 % for nanowires only. Their investigation illustrates the versatility in controlling the anode morphology using electrodeposition. Plank et al. found that a poorly conductive back contact layer is the source of low open circuit voltages that contribute to the weak fill factors [71]. Another cause for low open circuit voltage is that the resistance for electrons injected into ZnO to recombine with the electrolyte is lower for ZnO compared to TiO_2 [72].

4.3.2 Cu_2O

Cu_2O is a semiconductor with a band gap of 2.1 eV and a cubic cuprite structure, which has been investigated mainly for use in photoelectrochemical and photovoltaic devices. Cu_2O can be synthesized as both p-type and n-type depending on the deposition solution's pH [73]. Electrodeposition of p-Cu_2O is performed using an alkaline $CuSO_4$ bath stabilized by the complexing action of a high concentration of lactic acid [16]. Under these conditions {111} oriented p-Cu_2O films could be produced at pH 12. Electrodeposition of n-type Cu_2O occurs at a pH of around 6, in solutions containing copper acetate and sodium acetate [18, 74]. p-type Cu_2O is often deposited in the absence of any additives that would modify the morphology of the growing film, so cubic crystals are obtained. In the presence of acetate, which

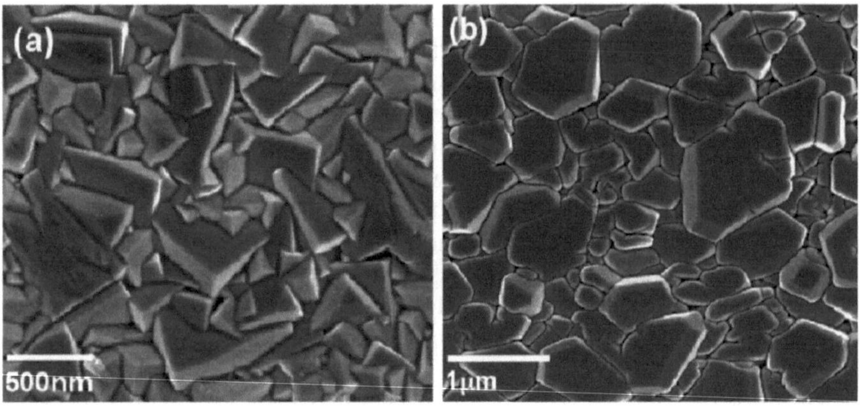

Fig. 4.7 SEM images show the cubic morphology of p-type Cu_2O and the octahedral morphology of n-type Cu_2O with truncated tips. Reprinted with permission from [75]. Copyright 2009 American Chemical Society

adsorbs preferentially onto {111} facets, the Cu_2O forms octahedral crystals with truncated tips [18]. Figure 4.7 shows a typical morphology of cubic p-type Cu_2O and the truncated octahedral structure of n-type Cu_2O [75]. The conductivity of p-type Cu_2O is attributed to Cu vacancies in the lattice [76], while n-type doping is attributed to O vacancies [77]. Wang and Tao have demonstrated that the polarity of doping can be continuously modified by adjusting the pH of the deposition solution and found that Cu_2O is n-type below pH 7.5 and p-type above 9.5 in a lactic acid complexed $CuSO_4$ solution [78]. The doping concentration in Cu_2O films is not homogeneous; solid state conductivity and electrochemical impedance studies by Paracchino et al. determined the bulk doping concentration to be 10^{13} cm^{-3} while the interface exhibited a dopant concentration of 10^{17} cm^{-3} [79]. This would produce a potential barrier in the band structure near the surface.

Photovoltaic applications of Cu_2O have typically coupled p-type Cu_2O with a wide bandgap semiconducting n-type material in a p–n heterojunction configuration. The short diffusion length of photogenerated electron–hole pairs in Cu_2O have represented a barrier to high efficiency devices in planar configurations, but high aspect ratio ZnO nanowires as a substrate for electrodeposition of Cu_2O can collect the photogenerated charges more effectively [80]. Musselman reported using this route to enhance the short circuit current from 3 to 5 mA/cm^2 with a power efficiency of 0.5 % [81]. The ability to plate both p-type and n-type Cu_2O has given rise to investigations of p–n homojunctions, which lack the potential barriers and quantum well that impede the efficient movement of charges that arise when two materials with unequal band gaps are connected. Han and Tao reported a Cu_2O homojunction solar cell with an efficiency of 0.103 % by electrodepositing sequentially in solutions which yield p-type and n-type Cu_2O [82]. It was later identified by McShane et al. that if the orientation of p-type Cu_2O differed from n-type Cu_2O, epitaxial growth of the latter would be inhibited and leave interfacial defects that

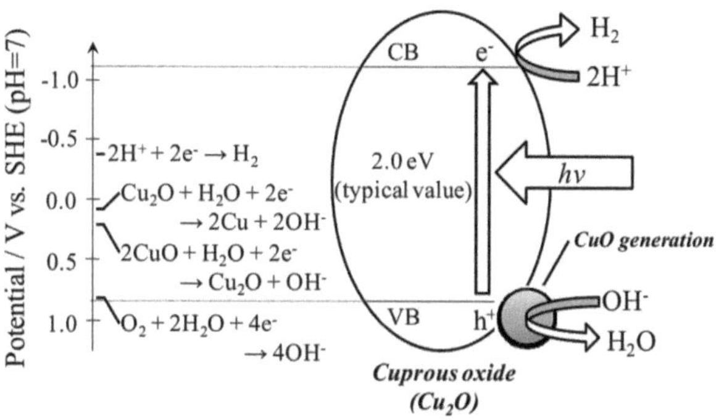

Fig. 4.8 The band structure of Cu_2O shows that for p-type Cu_2O, photoinduced OER is forbidden and HER is permitted, but redox potentials between Cu_2O and CuO are also present within the band gap, leading to potential problems with stability of the oxide. From [92]. Reproduced by permission of ECS—The Electrochemical Society

decrease the efficiency of the system [75]. By matching the orientation so that the ratio of the exposed facets in both p-type and n-type were similar, they achieved an efficiency of 0.2 %.

The utility of Cu_2O for the photoelectrochemical production of hydrogen by water splitting remains controversial. Hara et al. demonstrated water splitting of Cu_2O on timescales of 1,900 h without major degradation [83], but several challenges limit the practical use of Cu_2O as a photoelectrochemical material. The band structure of p-type Cu_2O relative to the solution–electrolyte interface is shown in Fig. 4.8. In the case of p-type Cu_2O, de Jongh et al. show that water reduction by photogenerated charges is thermodynamically disfavored due to the placement of the redox potential for the oxygen evolution reaction (OER) close to the valence band of Cu_2O and to the hydrogen evolution reaction being kinetically unfavorable [84]. Nian et al. have demonstrated that this kinetic limitation for HER can be overcome by coupling Cu_2O with an n-type WO_3 counter electrode and verified using gas chromatography techniques that hydrogen was being produced [85]. Lin et al. deposited a thin layer of NiO_x to the Cu_2O which not only served as a protective layer but also acted to form a p–n junction that enabled the device to produce photocurrents that generate H_2 instead of reducing Cu_2O to copper [86]. However, the presence of redox potentials within the band gap corresponding to Cu_2O oxidation to CuO or reduction to Cu also poses a problem with the inherent instability under illumination, which will be addressed below. Although n-type Cu_2O is expected to have a different band structure relative to the various redox potentials of possible reactions, few investigations have been carried out to determine the feasibility of using this material as a photoanode for water splitting; in particular, limited effort has been devoted to analyze the composition of the gases being formed. McShane and Choi have investigated the photoelectrochemical performance of dendritic n-type Cu_2O, obtaining incident photocurrent efficiency below 1 % in the visible range [18]. In particular, they raised concerns that the efficiency was limited by slow kinetics of the anodic reaction and high resistivity, while durability was limited by photoinduced corrosion. n-type and p-type Cu_2O have been investigated as a visible light sensitizer on TiO_2 nanotubes, with incident photocurrent efficiency of 0.2 % for p-type Cu_2O and 2 % for n-type Cu_2O at 500 nm [87]. An alternative photoelectrochemical application uses Cu_2O to decompose pollutants. Zheng et al. demonstrated that Cu_2O can be used to decompose 20 mg/L of methyl orange dye within 1 h under illumination [88]. Composite systems of TiO_2 and p-type Cu_2O have also been assembled to enhance the degradation of organic pollutant compounds [89–91].

As mentioned above, the susceptibility of Cu_2O to degradation via photocorrosion or reduction to Cu in solution has stimulated several research efforts aimed at stabilization of the material. Cyclic voltammetry studies have shown that Cu_2O tends to transform from a particle-like structure to a flake-like structure, resulting in strongly suppressed photocurrent efficiencies [88, 93]. This process is ascribed to the tendency to preferentially expose lower energy surfaces to the solution, with {111} facets being most stable. Kakuta and Abe found that the transformation of Cu_2O to CuO accompanies hydrogen evolution, limiting photogeneration of H_2 to about 60 h [92]. They also showed that addition of n-propanol that acts as an electron scavenger could maintain the photocurrent and prevent the formation of CuO. An alternate technique that does not require the presence of sacrificial species is to deposit a protective layer on top of Cu_2O to prevent degradation. Paracchino et al. have demonstrated that a layer of amorphous TiO_2 deposited on top of Cu_2O by atomic layer deposition ALD can prevent the gradual performance degradation under light [94, 95]. Cu_2O modified with ALD TiO_2 lasted up to 10 h under 100 mW/cm^2 AM 1.5 simulated sunlight illumination.

4.3.3 *Fe_2O_3 and Iron Oxides*

Hematite (α-Fe_2O_3) has attracted much attention for its use in photoelectrochemical applications because of its abundance in the earth's crust and stability in solution and under illumination [96]. It has the rhombohedral crystal structure of corundum, and its band gap is 2.0 eV, which makes it a good candidate for absorbing visible light [97, 98]. In addition to the obvious formation of Fe_2O_3 via electrodeposition of metallic iron followed by thermal oxidation [99–101], direct electrodeposition of an oxide is a common method to prepare this material. Both cathodic and anodic deposition paths can be followed to make Fe_2O_3, both having in common the formation of an intermediate hydroxide, which is then subjected to a thermal annealing step to dehydrate and crystallize the phase into α-Fe_2O_3. The annealing is typically carried out in a range between 400 and 700 °C; higher temperatures have been associated with better photoelectrochemical performance due to the elimination of defects [102]. Schrebler et al. demonstrated that the reduction of H_2O_2 in a solution also containing Fe^{3+} and F^- ions could be used for the electrogeneration of base that was used to deposit an oxyhydroxide of Fe; successive conversion to Fe_2O_3 was carried out via annealing at 500 °C [103]. The formation of the FeF^{2+} complex shifts the reduction potential of Fe 0.6 V more negative, which allows for the reduction of H_2O_2 to occur preferentially [Eqs. (4.19) and (4.20)] [104].

$$H_2O_2 + 2e^- \rightarrow 2OH^- \tag{4.19}$$

$$FeF^{2+} + 3OH^- \rightarrow FeOOH + F^- + H_2O \tag{4.20}$$

In general, the anodic route has been more widely investigated to prepare Fe_2O_3 [102, 104–108]. Electrodeposition of α-FeOOH from $(NH_4)_2Fe(SO_4)_2$ solutions produces high surface area plate-like structures, which can be readily converted to Fe_2O_3 with an annealing at 400 °C [105, 107]. Spray and Choi produced FeOOH as a precursor by anodic deposition of Fe^{3+} ions [Eqs. (4.21) and (4.22)], in both acidic and alkaline solutions containing $FeCl_2$.

$$Fe^{2+} \rightarrow Fe^{3+} + e^- \tag{4.21}$$

$$Fe^{3+} + H_2O \rightarrow FeOOH + 3H^+ \tag{4.22}$$

The theoretical efficiency for photoelectrochemical water splitting on Fe_2O_3 is 13 % under AM1.5 sunlight [109], but demonstrated efficiencies are much lower, being hampered by low mobility (0.1 cm^2/V-s) and short diffusion lengths (2–4 nm) [110, 111]. These two properties combined limit the collection of charges exclusively to the semiconductor–electrolyte interface's space charge region [112]. Approaches to overcoming these difficulties have included nanostructuring, doping via co-electrodeposition, and modification of Fe_2O_3 with electrodeposited water oxidation catalysts. Spray and Choi [104] compared their work to a previous report by Leibenguth and Cohen with an alkaline solution containing complexed Fe^{2+} ions [106]. Higher surface area per unit thickness was formed in an acidic solution, resulting in an interconnected porous structure, than a tall columnar structure obtained in alkaline conditions. Figure 4.9 compares the morphology of these two systems and their absorbance as well as photocurrent. The high surface area with a shorter path length to the conducting substrate ultimately led to a photocurrent increase by a factor of 2.5×, from 0.2 to 0.5 mA/cm^2 under 3 W/cm^2 Xe illumination.

Numerous dopants have been added to Fe_2O_3 in an effort to increase efficiency; these include Si [113], Sn [114], Zr [115], I [116], Mo [117], Cr [117], Ti [118], Mg [118], and Pt [119]. In particular, doping by co-electrodeposition of 5 % Pt was shown to enhance the photocurrent efficiency at 400 nm from 3 to 12 %[119]. The authors proposed that the increased doping concentration resulted in greater conductivity of Fe_2O_3 and enhanced the electric field at the solution–semiconductor interface that in turn would promote charge separation [119]. Similarly, the increase in conductivity associated with co-electrodeposited Mo and Cr was found to give a 2× enhancement in the photocurrent efficiency of Fe_2O_3 [117].

The water oxidation process is a sluggish reaction requiring the transfer of four holes, so the introduction of oxygen evolution catalysts to accelerate the reaction

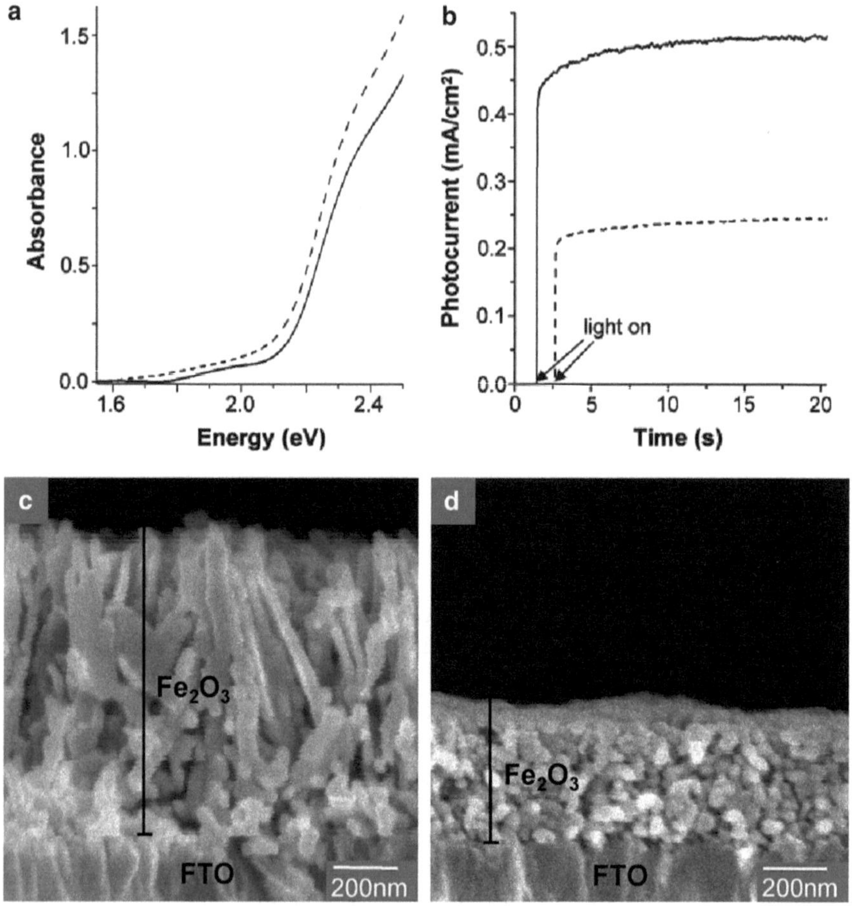

Fig. 4.9 (**a**) Absorbance and (**b**) photocurrent of Fe_2O_3 thin films prepared onto F-doped tin oxide substrates (FTO) by depositing in acidic (*solid lines*) and alkaline (*dashed lines*) conditions. (**c**) and (**d**) show that the morphology of the sample prepared in acidic conditions (**d**) is porous but thin while the tall columnar structures (**c**) provide a high surface area at the expense of a longer average charge transport path to the substrate. Reprinted with permission from [104]. Copyright 2009 American Chemical Society

can decrease the anodic bias to drive the photoelectrochemical reaction. IrO_2 nanoparticles were used in conjunction with nanostructured Fe_2O_3 to produce a 3 mA/cm^2 photocurrent with a cathodic shift of 200 mV versus unmodified Fe_2O_3 [120]. A more widely abundant catalyst compared to IrO_2 is cobalt phosphate (Co-Pi) [121]. Co-Pi prepared by photoelectrodeposition was shown to cathodically shift the oxygen evolution curve by 150–350 mV [122, 123]. Photodeposited Co-Pi/ Fe_2O_3 was shown to yield a 5$\times$ enhancement in photocurrent density compared to bare Fe_2O_3 photoanodes at a potential of 1.0 V vs. RHE [123]. By studying the time constants associated with intermediate steps in the photoelectrochemical reaction,

Peter et al. determined that the mechanism that produces this cathodic shift in Co catalysts is the suppression of surface recombination and is not an acceleration of the oxygen evolution process [112]. Other attempts at minimizing recombination via surface state passivation include ALD deposited Al_2O_3 on top of Fe_2O_3 and a cyclic potential treatment in NaCl that is believed to reduce the number of dangling bonds at the Fe_2O_3 surface [19, 124].

4.3.4 MnO_2

MnO_2 has been used for decades as the positive cathode in both lithium ion and alkaline batteries [24, 125], but recent interest in the electrochemical formation of MnO_2 has shifted towards supercapacitor devices for high current, intermediate density energy storage [26]. Due to its abundance, low cost, and competitive theoretical capacitance (1,370 F/g) [126] with the leading, but expensive RuO_2 (750 F/g) supercapacitors [127], MnO_2 is a promising material for this application. In the presence of cations intercalated into the lattice, MnO_2 stores charge by switching valence between Mn^{3+} (MnOOH) and Mn^{4+}. Early work on MnO_2 focused on the preparation of amorphous oxides. Optimization of electrodeposition conditions (pH, current density, concentration of Mn^{2+} in solution, and temperature) resulted in a maximum of 220 F/g for amorphous MnO_2. Increasing the surface area beyond 200 m^2/g does not yield any further gains in amorphous MnO_2; the ability to select the structure of the tunnels where intercalation occurs instead has made crystalline MnO_2 a more attractive route to increasing the capacitance of MnO_2 supercapacitors [128]. This oxide is notable for the ability to form a number of crystal structures which can be controlled by the synthesis conditions; a non-exhaustive list includes: hollandite (α-MnO_2), pyrolusite (β-MnO_2), ramsdellite, mixed ramsdellite/pyrolusite (γ-MnO_2), the 2D layered birnessite (δ-MnO_2), HCP (ε-MnO_2), spinel (λ-MnO_2), rock salt, and anti-fluorite structures [24, 26, 129]. Figure 4.10 shows a sample of the possible crystal structures for MnO_2 [128]. The MnO_6 octahedra that compose the MnO_2 lattice can be arranged by attachment at different locations. The gaps between these structures permit the infiltration of ions into the lattice, and controlling their size is key to optimizing the supercapacitive behavior of this oxide [26].

The process of anodic electrodeposition of MnO_2 was briefly mentioned above in Sect. 4.2.2, Eq. (4.8). Since it is unlikely that Mn oxidation and the reaction with water occur in a single step, Rodrigues et al. proposed a multistep reaction route where the oxidation from Mn^{2+} to Mn^{4+} precedes the deposition of MnO_2 [130]. Studies of Mn oxidation on Au substrates by Rogulski et al. indicated that the process is a three-step reaction: the initial electrochemical oxidation to Mn^{2+}, the chemical formation of Mn^{3+} and MnOOH intermediates, and the final electrochemical oxidation of MnOOH to MnO_2 [131]. The electrochemical reactions [Eqs. (4.23), (4.24), and (4.25)] that result in the deposition of MnO_2 occur

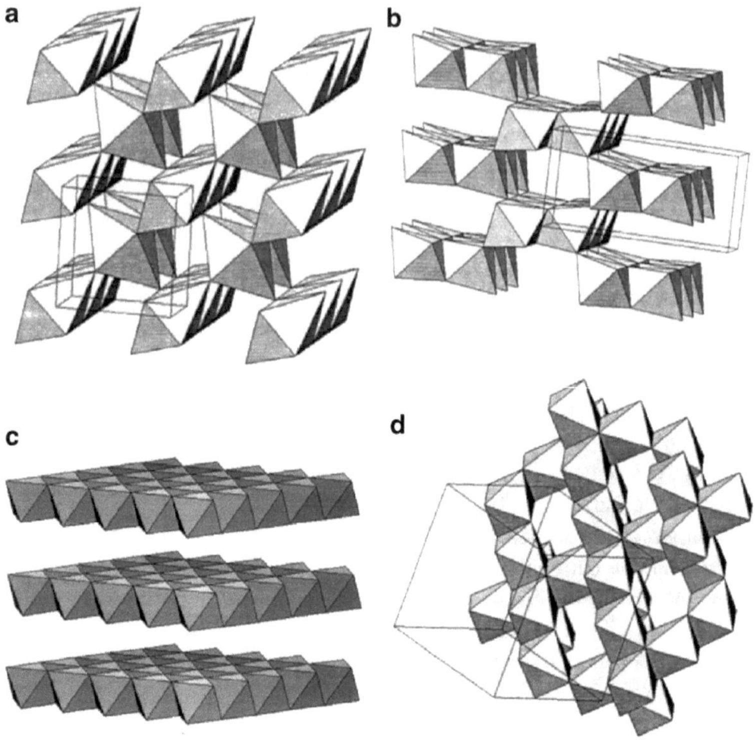

Fig. 4.10 Crystal structures of MnO_2 with different sized gaps for ion intercalation result from the rich variety of possible configurations for the attachment of the MnO_6 octahedral building block. (**a**) pyrolusite (β), (**b**) ramsdellite, (**c**) birnessite (δ), and (**d**) spinel (λ). From [128]. Reproduced by permission of ECS—The Electrochemical Society

concurrently with chemical reactions that convert MnO_2 back to MnOOH, especially under acidic conditions.

$$Mn^{2+} \rightarrow Mn^{3+} + e^- \tag{4.23}$$

$$Mn^{3+} + 2H_2O \rightarrow MnOOH + 3H^+ \tag{4.24}$$

$$MnOOH \rightarrow MnO_2 + e^- + H^+ \tag{4.25}$$

Broughton and Brett also showed that in the presence of a $NaCH_3COO$ additive, the deposition potential could be reduced from 0.8 V vs. AgCl without $NaCH_3COO$ to 0.4 V vs. Ag/AgCl in a 1:1 ratio of Na_2SO_4:$NaCH_3COO$ [132]. The mechanism responsible for this shift in the deposition window was found to be the stabilization of the Mn(III) intermediate by acetate, ultimately accelerating the overall reaction by preventing decomposition of the intermediate. The crystal structure of MnO_2 can be selected by the use of complexing agents; Wei et al. obtained the hexagonal ε-MnO_2 structure in the absence of additives, while a rock salt structure with missing Mn cations was formed in the presence of ethylenediamine tetraacetic acid (EDTA), and an anti-fluorite structure could be deposited from a citrate complexing solution [129]. It was found that among the three, the anti-fluorite structure had the highest specific capacitance owing to the availability of more vacancy sites for cation intercalation. Devaraj demonstrated that surfactants such as sodium lauryl sulfate can increase the specific surface area of δ-MnO_2, leading to capacitances as high as 300 F/g in δ-MnO_2 [133]. The use of sweeping potentials can result in high surface area films with correspondingly higher capacitance, as Prasad and Miura demonstrated [134]. Electrodeposition of γ-MnO_2 using a combined potential sweep between 0.6 V_{SCE} and 0.3 V_{SCE} followed by a potentiostatic deposition at 0.6V_{SCE} could produce high surface area nano-flakes with a specific capacitance of up to 324 F/g [135]. The majority of published work on MnO_2 has involved anodic deposition. Cathodic electrodeposition of MnO_2 with a comparable capacitance of 353 F/g has also been shown from the reduction of Mn^{7+} from $KMnO_4$ solution, with the advantage that using a reduction reaction would not corrode the substrate [136].

Improvements to the capacitance and durability of MnO_2 can be achieved by using mixed oxides or composites with carbon nanomaterials. Mixed oxides of Mn and Mo were synthesized by co-electrodeposition by Nakayama et al.; the increased conductivity contributed by Mo led to capacitances up to 190 F/g compared to 80 F/g for MnO_2 only [40]. Ag doping has shown similar enhancements [137]. Zhao et al. showed that a 7:1 mixed oxide of Mn:Co when formed as nanowires in an anodic alumina template gives a specific capacitance of 396 F/cm^2 [138]. (Ni, Mn) and (Co, Mn) oxides were studied by Prasad and Miura with specific capacitances of 377 and 307 F/g compared with pure MnO_2 systems prepared in similar conditions; they also reported that the mixed oxide system could endure less than 15 % capacity loss after 10,000 cycles [139]. Babakhani attributes this enhancement to the addition of Co to MnO_2, resulting in a decrease in lattice density by 11 %, which makes it easier for the intercalating ions to penetrate [140]. Nanostructured carbon materials including graphene and carbon nanotubes (CNTs) have been used to provide enhancements to the conductivity and increase the longevity of MnO_2 supercapacitors by acting as interspersed low resistance current paths [26, 141]. Reddy et al. produced a conductive radial configuration with a core of CNTs and a shell of MnO_2 and Au, though a low specific capacitance (90 F/g) was reported [142]. Enhanced conductivity was also shown with CNT/MnO_2 systems prepared by both electrodeposition and electrophoretic deposition to produce specific capacitance as high as 356 F/g [143, 144]. Using a fiber-like substrate coated with graphene, Yu et al. showed that 95 % of the capacitance of electrodeposited MnO_2 was maintained up to 5,000 charging and discharging cycles [145].

4.4 Conclusions

The fundamental theory of electrodeposition of metal oxides serves as a guide for the investigation of techniques which may be used to control crystal structure, composition, and morphology. Metal oxides such as ZnO, Cu_2O, and α-Fe_2O_3 are prime examples of earth abundant materials being investigated for the conversion of solar energy directly into electricity or into fuels by photoelectrochemical means. The main research focus for all three materials is an improvement in their efficiency, either by controlling their morphology at the nanoscale or increasing the overall performance by forming composites with dopants or catalysts. In the case of Cu_2O especially, the development of high efficiency devices is also occurring in parallel with investigations as to how to extend the lifetime of these materials, which is essential to any practical application. The case of MnO_2 shows that active research is also ongoing in the field of energy storage with metal oxide systems in batteries and supercapacitors. As with the other materials, approaches of morphology and crystal structure control have been applied, and forming mixed oxides and composites with carbon nanomaterials are exciting routes with high potential to increase the energy storage capacity of these devices. The electrodeposition route continues to be one of the most versatile and cost-efficient for the production of energy materials.

References

1. Heber J (2009) Nature 459:28
2. Liu A, Jones R, Liao L, Samara-Rubio D, Rubin D, Cohen O, Nicolaescu R, Paniccia M (2004) Nature 427:615
3. Szot K, Speier W, Bihlmayer G, Waser R (2006) Nature Mater 5:312
4. Koza JA, He Z, Miller AS, Switzer JA (2011) Chem Mater 23:4105
5. Bae HS, Yoon MH, Kim JH, Im S (2003) Appl Phys Lett 83:5313
6. Nomura K, Ohta H, Takagi A, Kamiya T, Hirano M, Hosono H (2004) Nature 432:488
7. Hussain ST, Siddiqa A, Siddiq M, Ali S (2011) J Nanoparticle Res 13:6517
8. Matsumoto Y (2001) Science 291:854
9. Laurent S, Forge D, Port M, Roch A, Robic C, Elst LV, Muller RN (2008) Chem Rev 108:2064
10. Bessekhouad Y, Robert D, Weber JV (2005) Catal Today 101:315
11. Castillo NC, Ding L, Heel A, Graule T, Pulgarin C (2010) J Photochem Photobio A: Chem 216:221
12. Rühle S, Anderson AY, Barad H, Kupfer B, Bouhadana Y, Rosh-Hodesh E, Zaban A (2012) J Phys Chem Lett 3:3755
13. O'Regan B, Grätzel M (1991) Nature 353:737
14. Baxter JB, Aydil ES (2005) Appl Phys Lett 86:053114
15. Law M, Greene LE, Johnson JC, Saykally R, Yang P (2005) Nature Nano 4:455
16. Golden TD, Shumsky MG, Zhou Y, VanderWerf RA, Van Leeuwen RA, Switzer JA (1996) Chem Mater 8:2499
17. Wijesundera R, Hidaka M, Koga K, Sakai M, Siripala WP (2006) Thin Solid Films 500:241
18. McShane CM, Choi K-S (2009) J Amer Chem Soc 131:2561
19. Le Formal F, Tétreault N, Cornuz M, Moehl T, Grätzel M, Sivula K (2011) Chem Sci 2:737
20. Sivula K, Zboril R, Le Formal F, Weidenkaff A, Tucek J, Frydrych J, Grätzel M (2010) J Amer Chem Soc 132:7436
21. Dall'Antonia LH, De Tacconi NR, Chanmanee W, Timmaji H, Myung N, Rajeshwar, K (2010) Electrochem Solid-State Lett 13:D29
22. Jia Q, Iwashina K, Kudo A (2012) Proc Natl Acad Sci US Amer 109:11564
23. Yu J, Kudo A (2006) Adv Funct Mater 16:2163
24. Thackeray MM (1997) Prog Sol State Chem 25:1

25. Sato Y (1991) J Electrochem Soc 138:L37
26. Xu C, Kang F, Li B, Du H (2011) J Mater Res 25:1421
27. Zhang J, Liu X, Wang L, et al. (2011) Nanotechnology 22:185501
28. Mittiga A, Biccari F, Malerba C (2009) Thin Solid Films 517:2469
29. Cowan AJ, Barnett CJ, Pendlebury SR, Barroso M, Sivula K, Grätzel M, Durrant JR, Klug DR (2011) J Amer Chem Soc 133:10134
30. Carcia PF, McLean RS, Reilly MH, Nunes G (2003) Appl Phys Lett 82:1117
31. Khan SUM, Akikusa J (1999) J Phys Chem B 103:7184
32. Matsumoto Y (2011) MRS Bull 25:47
33. V. Manzano C, Caballero-Calero O, Hormeño S, Penedo M, Luna M, Martín-González MS (2013) J Phys Chem C 117:1502
34. Siegfried MJ, Choi K-S (2004) Adv Mater 16:1743
35. Pauporté T, Lincot D (2000) Electrochim Acta 45:3345
36. Pourbaix M (1974) Atlas of Electrochemical Equilibria in Aqueous Solutions, National Association of Corrosion Engineers, Houston, Texas, USA p. 324–325, 408–409, 495.
37. Goux A, Pauporté T, Chivot J, Lincot D (2005) Electrochim Acta 50:2239
38. Pauporté T, Goux A, Kahn-Harari A, de Tacconi N, Chenthamarakshan CR, Rajeshwar K, Lincot D (2003) J Phys Chem Solids 64:1737
39. Tench D, Warren LF (1983) J Electrochem Soc 130:869
40. Nakayama M, Tanaka A, Sato Y, Tonosaki T, Ogura K (2005) Langmuir 21:5907
41. Zhou Y, Switzer JA (1998) Script Mater 38:1731
42. Therese GHA, Kamath PV (2000) Chem Mater 12:1195
43. Zhitomirsky I (2002) Adv Colloid Interface Sci 97:279
44. Boccaccini AR, Zhitomirsky I (2002) Curr Opin Solid State Mater Sci 6:251
45. Lewis JA (2000) Colloidal Processing of Ceramics. J Am Ceram Soc 83:2341
46. Jolivet J, Froidefond C, Pottier A, Chaneac C, Cassaignon S, Tronc E, Euzen P (2004) J Mater Chem 14:3281
47. Mann S (2000) Angewandte Chemie (Int ed Eng) 39:3392
48. Siegfried MJ, Choi K-S (2005) Angewandte Chemie (Int ed Eng) 44:3218
49. Xu L, Guo Y, Liao Q, Zhang J, Xu D (2005) J Phys Chem B 109:13519
50. Ozgur U, Alivov YI, Liu C, Teke A, Reshchikov MA, Dogan S, Avrutin V, Cho SJ, Morkoc H (2005) J App Phys 98:041301
51. Peulon S, Lincot D (1998) J Electrochem Soc 145:864
52. Zhang S, Wei SH, Zunger A (2001) Phys Rev B 63:075205
53. Park CH, Zhang SB, Wei S-H (2002) Phys Rev B 66:073202
54. Rousset J, Saucedo E, Lincot D (2009) Chem Mater 21:534
55. Marí B, Sahal M, Mollar MA, Cerqueira FM, Samantilleke AP (2012) J Solid State Electrochem 16:2261–2265
56. Thomas MA, Cui JB (2010) J Phys Chem Lett 1:1090
57. Grätzel M (2001) Nature 414:338
58. Peter LM (2007) J Phys Chem C 111:6601
59. Benkstein KD, Kopidakis N, Van de Lagemaat J, Frank AJ (2003) J Phys Chem B 107:7759
60. Keis K, Bauer C, Boschloo G, Hagfeldt A, Westermark K, Rensmo H, Siegbahn H (2002) J Photochem Photobio A: Chem 148:57
61. Karuppuchamy S, Nonomura K, Yoshida T, Sugiura T, Minoura H (2002) Solid State Ionics 151:19
62. Park WI, Kim JS, Yi GC, Bae MH, Lee HJ (2004) Appl Phys Lett 85:5052
63. Albrecht JD, Ruden PP, Limpijumnong S, Lambrecht WRL, Brennan, KF (1999) J App Phys 86:6864
64. Hattori M, Noda K, Nishi T, Kobayashi K, Yamada H, Matsushige K (2013) Appl Phys Lett 102:043105
65. Kroeze JE, Savenije TJ, Warman JM, Tio N (2004) J Amer Chem Soc 126:7608
66. Hochbaum AI, Yang P (2010) Chem Rev 110:527

67. Nazeeruddin MK, Kay A, Rodicio I, Humphry-Baker R, Mueller E, Liska P, Vlachopoulos N, Grätzel M (1993) J Amer Chem Soc 115:6382
68. Chiba Y, Islam A, Watanabe Y, Komiya R, Koide N, Han L (2006) Japan J App Phys 45: L638
69. Chen Z, Tang Y, Zhang L, Luo L (2006) Electrochim Acta 51:5870
70. Guérin V-M, Pauporté T (2011) Energy Env Sci 4:2971
71. Plank NOV, Welland ME, MacManus-Driscoll JL, Schmidt-Mende L (2008) Thin Solid Films 516:7218
72. Ling T, Song J-G, Chen XY, Yang J, Qiao SZ, Du XW (2013) J Alloys Compounds 546:307
73. Le D, Stolbov S, Rahman TS (2009) Surf Sci 603:1637
74. Jayathileke K, Siripala W, Jayanetti J (2010) Sri Lankan J Phys 9:35
75. McShane CM, Siripala WP, Choi K-S (2010) J Phys Chem Lett 1:2666
76. Raebiger H, Lany S, Zunger A (2007) Phys Rev B 76:1
77. Garuthara R, Siripala WP (2006) J Lumin 121:173
78. Wang L, Tao M (2007) Electrochem Solid-State Lett 10:H248
79. Paracchino A, Brauer JC, Moser JE, Thimsen E, Grätzel M (2012) J Phys Chem C 116:7341
80. Musselman KP, Marin A, Schmidt-Mende L, MacManus-Driscoll JL (2012) Adv Funct Mater 22:2202–2208
81. Musselman KP, Wisnet A, Iza DC, Hesse HC, Scheu C, MacManus-Driscoll JL, Schmidt-Mende L (2010) Adv Mater 22:E254
82. Han K, Tao M (2009) Sol Energy Mater Sol Cells 93:153
83. Hara M, Kondo T, Komoda M, Ikeda S, Shinohara K, Tanaka A (1998) Chem Comm 2:357
84. De Jongh PE, Vanmaekelbergh D, Kelly JJ (1999) Chem Comm 12:1069
85. Nian J, Hu C, Teng H (2008) Int J Hydrogen Energy 33:2897
86. Lin C-Y, Lai Y-H, Mersch D, Reisner E (2012) Chem Sci 3:3482
87. Tsui L, Wu L, Swami N, Zangari G (2012) ECS Electrochem Lett 1:D15
88. Zheng Z, Huang B, Wang Z, Guo M, Qin X, Zhang X, Wang P, Dai Y (2009) J Phys Chem C 113:14448
89. Hou Y, Li XY, Zhao QD, Quan X, Chen GH (2010) Adv Funct Mater 20:2165
90. Hou Y, Li X, Zou X, Quan X, Chen GH (2009) Env Sci Tech 43:858
91. Yang L, Luo S, Li Y, Xiao Y, Kang Q, Cai Q (2010) Env Sci Tech 44:7641
92. Kakuta S, Abe T (2009) Electrochem Solid-State Lett 12:P1
93. Wu L, Tsui L, Swami N, Zangari G (2010) J Phys Chem C 114:11551
94. Paracchino A, Laporte V, Sivula K, Grätzel M, Thimsen E (2011) Nature Mater 10:456
95. Paracchino A, Mathews N, Hisatomi T, Stefik M, Tilley SD, Grätzel M (2012) Energy Env Sci 5:8673
96. Bora DK, Braun A, Constable EC (2013) Energy Env Sci 6:407
97. Glasscock JA, Barnes PRF, Plumb IC, Bendavid A, Martin PJ (2008) Thin Solid Films 516:1716
98. Peng H, Lany S (2012) Phys Rev B 85:1
99. Enache CS, Liang YQ, Van de Krol R (2011) Thin Solid Films 520:1034
100. McDonald K, Choi K-S (2010) Chem Mater 23:1686
101. Mohapatra SK, Banerjee S, Misra M (2008) Nanotechnology 19:315601
102. Tamboli SH, Rahman G, Joo O (2012) J Alloys Compounds 520:232
103. Schrebler R, Llewelyn C, Vera F, Cury P, Muñoz E, del Río R, Gómez M, Humberto C, Ricardo D, Enrique AB, Bello K (2006) Electrochem Solid-State Lett 9:D95
104. Spray RL, Choi K-S (2009) Chem Mater 21:3701
105. Jiao S, Xu L, Hu K, Li J, Gao S, Xu D (2010) J Phys Chem C 114:269
106. Leibenguth J-L, Cohen M (1972) J Electrochem Soc 119:987
107. Wu MS, Lee RH, Jow JJ, Yang WD, Hsieh CY, Weng BJ (2009) Electrochem Solid-State Lett 12:A1
108. Wu M-S, Ou Y-H, Lin Y-P (2010) Electrochim Acta 55:3240

109. Murphy A, Barnes P, Randeniya L, Plumb I, Grey I, Horne M, Glasscock J (2006) Int J Hydrogen Energy 31:1999
110. Van Daal H, Bosman A (1967) Phys Rev 158:736
111. Kennedy JH, Frese KW (1978) J Electrochem Soc 125:709
112. Peter LM, Wijayantha KGU, Tahir AA (2012) Faraday Discussions 155:309
113. Cesar I, Sivula K, Kay A, Zboril R, Gra¨tzel M (2009) J Phys Chem C 113:772
114. Ling Y, Wang G, Wheeler DA, Zhang JZ, Li Y (2011) Nano Lett 11:2119
115. Kumar P, Sharma P, Shrivastav R, Dass S, Satsangi VR (2011) Int J Hydrogen Energy 36:2777
116. Khan SUM, Zhou ZY (1993) J Electroanal Chem 357:407
117. Kleiman-Shwarsctein A, Hu Y-S, Forman AJ, Stucky GD, McFarland EW (2008) J Phys Chem C 112:15900
118. Tang H, Yin W, Matin MA, Wang H, Deutsch T, Al-Jassim MM, Turner JA, Yan Y (2012) J App Phys 111:073502
119. Hu Y, Kleiman-Shwarsctein A, Forman AJ, Hazen D, Park JN, McFarland EW (2008) Chem Mater 20:3803
120. Tilley SD, Cornuz M, Sivula K, Grätzel M (2010) Angewandte Chemie (Int ed Eng) 49:6405
121. Kanan MW, Nocera DG (2008) Science 321:1072
122. Zhong DK, Sun J, Inumaru H, Gamelin DR (2009) J Amer Chem Soc 131:6086
123. Zhong DK, Cornuz M, Sivula K, Grätzel M, Gamelin DR. (2011) Energy Env Sci 4:1759
124. Zhang M, Luo W, Zhang N, Li Z, Yu T, Zou Z (2012) Electrochem Comm 23:41
125. Ellis BL, Lee KT, Nazar LF (2010) Chem Mater 22:691
126. Toupin M, Brousse T, Bélanger D (2004) Chem Mater 16:3184
127. Kotz R, Carlen M (2000) Electrochim Acta 45:2483
128. Brousse T, Toupin M, Dugas R, Athouël L, Crosnier O, Bélanger D (2006) J Electrochem Soc 153:A2171
129. Wei W, Cui X, Chen W, Ivey DG (2008) J Phys Chem C 112:15075
130. Rodrigues S, Munichandraiah N, Shukla AK (1998) J App Electrochem 28:1235
131. Rogulski Z, Siwek H, Paleska I, Czerwinski A (2003) J Electroanal Chem 543:175
132. Broughton JN, Brett MJ (2005) Electrochim Acta 50:4814
133. Devaraj S, Munichandraiah N (2005) Electrochem Solid-State Lett 8:A373
134. Prasad KR, Miura N (2004) J Power Sources 135:354
135. Chou S, Cheng F, Chen J (2006) J Power Sources 162:727
136. Wei J, Nagarajan N, Zhitomirsky I (2007) J Mater Proc Tech 186:356
137. Sawangphruk M, Pinitsoontorn S, Limtrakul J (2012) J Solid State Electrochem 16:2623
138. Zhao GY, Xu CL, Li HL (2007) J Power Sources 163:1132
139. Rajendra PK, Miura N (2004) Electrochem Comm 6:1004
140. Babakhani B, Ivey DG (2011) Electrochim Acta 56:4753
141. Zhang H, Cao G, Wang Z, Yang Y, Shi Z, Gu Z (2008) Nano Lett 8:2664
142. Reddy ALM, Shaijumon MM, Gowda SR, Ajayan PM (2010) J Phys Chem C 114:658
143. Li J, Zhitomirsky I (2009) J Mater Proc Tech 209:3452
144. Wang Y, Liu H, Sun X, Zhitomirsky I (2009) Script Mater 61:1079
145. Yu G, Hu L, Vosgueritchian M, Wang H, Xie X, McDonough JR, Cui X, Cui Y, Bao Z (2011) Nano Lett 11:2905

Chapter 5
Luminescence During the Electrochemical Oxidation of Aluminum

Stevan Stojadinović, Rastko Vasilić, Bećko Kasalica, Ivan Belča, and Ljubiša Zeković

5.1 Introduction

Electrochemical oxidation of aluminum has attracted considerable attention because of the widespread use of aluminum oxide layers in different areas of technology and industry. Traditionally, the anodization of aluminum is used to protect metals from corrosion and to increase the abrasive and adsorption properties, hardness, etc. The oxide layers on aluminum are widely used in electronics due to its excellent dielectric properties, perfect planarity, and good reproducibility during production. In recent years, scientists have focused on the formation of self-ordered oxide structures on aluminum in various electrolytes. This is a result of the application of ordered porous oxide layers with pore dimensions ranging from micrometers to nanometers, as the mold in nanotechnology for the synthesis of nanotubes, nanowires, solar cells, micro-optical elements, photonic crystals, etc.

Anodic oxide layers on aluminum show very interesting luminescence properties. There are two types of luminescent phenomena. The first type of luminescence occurs during the electrochemical oxidation of aluminum in some electrolytes, so-called galvanoluminescence. Another type of luminescence is photoluminescence, i.e., occurrence of light emission, mainly in the visible region of the spectrum, the oxide layer upon exposure to UV light. Luminescent techniques are nondestructive and are very useful, and many properties of the oxide layer such as thickness, growth rate, reflection coefficient, porosity, etc. can be determined by these techniques.

There are four types of galvanoluminescence phenomena. The emission of light, mainly in the visible spectrum, during direct electrochemical oxidation of metals is called the anodic galvanoluminescence. Another type of galvanoluminescence is

S. Stojadinović • R. Vasilić (✉) • B. Kasalica • I. Belča • L. Zeković
Faculty of Physics, University of Belgrade, Studentski trg 12-16, 11000 Belgrade, Serbia
e-mail: rastko.vasilic@ff.bg.ac.rs

S.S. Djokić (ed.), *Electrodeposition and Surface Finishing: Fundamentals and Applications*, Modern Aspects of Electrochemistry 57,
DOI 10.1007/978-1-4939-0289-7_5,

cathodic flash that occurs in metals that were previously anodically polarized and in which the cathodic polarization results in the appearance of a short-lasted flash. Repeated anodic polarization gives short-lived anodic flash superimposed on the anodic galvanoluminescence. The fourth type of galvanoluminescence occurs during alternating polarization of metals. Actually, alternative galvanoluminescence is a series of anodic and cathodic flashes emitted in the positive and negative half-periods, correspondingly.

A special type of luminescence occurs during the electrochemical oxidation of aluminum at high voltages, when it comes to dielectric breakdown, followed by microdischarges over the entire surface of the oxide. This type of oxidation is called plasma electrolytic oxidation. Spectral characterization of the microdischarge during plasma electrolytic oxidation process gives very important information about the plasma such as electron concentration, electron temperature, ionization temperature, etc.

In this chapter we will present the results of our research of luminescent phenomena during electrochemical oxidation of aluminum, which were done at the Faculty of Physics University of Belgrade in the last 30 years.

5.2 Anodic Oxide Films on Aluminum

Aluminum is recognized as the third most abundant element (after oxygen and silicon) and the most abundant metal in the Earth's crust. Structural components made from aluminum and its alloys are vital to the aerospace industry and are important in other areas of transportation and structural materials. It is remarkable for the metal's low density and for its ability to resist corrosion due to the phenomenon of passivation. Namely, because of high affinity for oxygen, aluminum is always covered with natural highly resistant oxide film [1, 2].

By putting such metal, naturally covered with oxide, as an anode in electrolyte that doesn't dissolve oxide layer, and enabling an adequate current passing through the system, the film starts growing continuously. Its thickness is determined by electrolysis (anodization) parameters (type of electrolyte, anodization voltage and current, electrolyte temperature). This is the way to obtain the oxide layer thicker and more resistive than natural thin oxide film.

During the anodization process the established electric field in the oxide layer enables directed motion of metal and oxygen ions through the layer enabling its growth. This means that the problem of oxide film growth could be reduced to the ionic conduction in the presence of high electric field, but it is further complicated by the presence of two interfaces: metal–oxide and oxide–electrolyte.

There are two custom regimes for anodic formation of oxide films: constant current regime and constant voltage regime. During the constant current process, every thickening of the oxide layer requires an increase of the voltage to keep

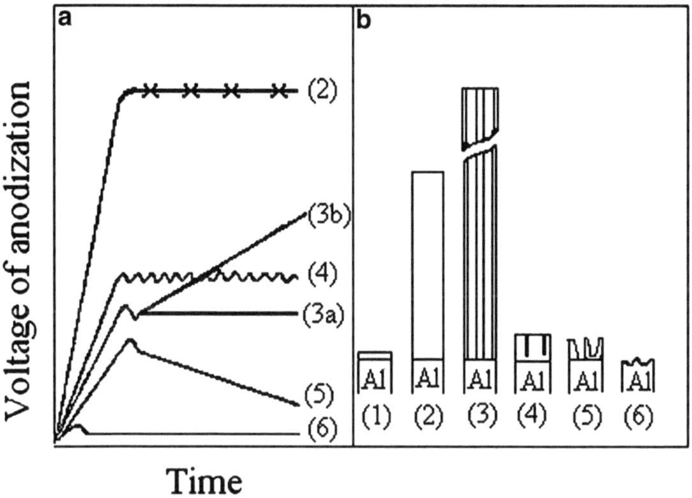

Fig. 5.1 (**a**) Anodization voltage vs. time dependence in various electrolytes for galvanostatic regime; (**b**) corresponding structure of obtained oxide films on aluminum (Reproduced by permission from Elsevier No. 3165800675036)

constant the internal electric field as well as the constant anodization current. In most cases, the electric field doesn't change with an increase in film thickness if anodization current is constant.

Thickness of oxide film depends on several anodization parameters, but the type of the film depends mostly on the type of electrolyte. The growth of various types of films is followed by characteristic voltage/time (constant current regime) or current/time (constant voltage regime) dependence. According to the dependence characteristic and the type of oxide film the general classification of aluminum anodic behavior is suggested in Fig. 5.1 [3]. Figure 5.1a shows the anodization voltage dependence on time in various electrolytes in constant current regimes, while Fig. 5.1b shows corresponding structures of formed oxide layers.

1. Natural oxide layer formed on aluminum sample with thickness of about 2–9 nm.
2. Electrolytes which don't dissolve previously formed oxide film produce so-called barrier oxide films. Their thickness depends on formation voltage (film growth rate is about 1.4 nm/V). Maximum thickness of grown barrier oxide film depends on breakdown voltage and usually reaches 600–700 nm (corresponding voltages are 450–500 V) [2–4]. The anodization of aluminum at voltages over the breakdown voltage leads to intensive sparking covering the whole oxide surface and intensive gas evolution [3, 5]. Most commonly used barrier oxide film forming electrolytes are aquatic solutions of boric acid, ammonium borate, ammonium tartrate, and some organic acids such as citric, malic, and glycolic acid [1, 6–8]. The most important applications of barrier

films obtained by anodization are in production of electrolytic capacitors [9] and thin film transistors [10] and for protection of aluminum films obtained by aluminum evaporation in vacuum.

3. Electrolytes which partially (weakly) dissolve oxide film form porous films. Typical examples of such electrolytes are water solutions of sulfuric, oxalic, chromic, and phosphoric acids [11, 12]. During constant current regime voltage increases almost linearly with time, reaches maximum value, and then decreases until attaining a constant value. The thickness of porous oxide films exceeds that of barrier films several times. Porous oxide films consist of two different layers [13]: internal nonporous barrier layer whose thickness depends on formation voltage and external porous part which is generally much thicker, and its thickness doesn't depend on formation voltage, but depends on other parameters such as anodization current density, time of anodization, electrolyte temperature and concentration, etc. [1, 14].

   Porous oxide layers are utilized in cases where excellent corrosion and abrasion resistivity is needed. For that purpose electrolytically formed porous oxide films are treated in distilled water or water solution of adequate salt on temperature higher than 90 °C. This is a way to close (process of "filling") the porous structure and get nonporous compact layers whose thicknesses exceed thicknesses of barrier films for several orders of magnitude [15–17]. The most probable explanation is the process of bohemite formation by partial hydration of aluminum oxide. Decorative value of these layers increases by their dyeing in chemical or electrochemical processes.
4. Anodic electropolishing occurs during anodization in strong electrolytes (concentrated sulfuric, phosphoric, and perchloric acids and some alkaline electrolytes) and is characterized by initial linear dependence of anodization voltage on time, followed by periodic fluctuation. This process is used for production of highly reflective surfaces [18].
5. The process of anodic etching ("pitting") is characteristic for some monocarboxylic acids (formic, acetic, benzoic, etc.) as well as in some electrolytes with chloride addition. This process is followed by the voltage decrease after the short increase [3].
6. High corrosion of aluminum in some acids, strong alkaline solutions, and halides [3]. The initial voltage is low for a given current and remains constant during the anodization.

5.2.1 Structure of Anodic Oxide Films on Aluminum

Barrier oxide films on aluminum primarily consist of amorphous alumina, but small amounts of crystal phases γ-Al_2O_3 or γ'-Al_2O_3 are also detected by electron diffraction, as well as by XRD [19]. The contribution of crystalline phases in

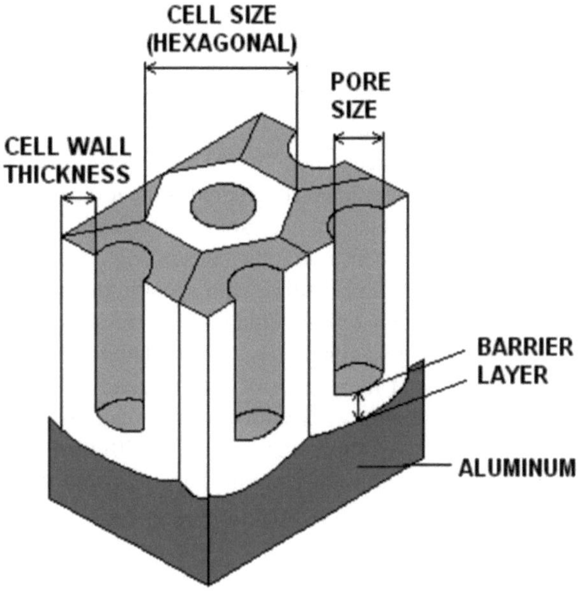

Fig. 5.2 The model of porous anodic oxide film on aluminum

amorphous alumina depends mostly on the anodization voltage [1, 20]. The contribution of γ-Al_2O_3 increases with voltage as well as with electrolyte temperature. There are several phases or oxide types in barrier oxide films. Franklin found at least three: hydrated oxide on oxide/electrolyte interface, irregular islands of γ-Al_2O_3, and amorphous film which occupies the most part of the film [1, 21]. Shimizu et al. found crystal islands of γ-Al_2O_3 immediately above surface irregularities [22]. They also found that volume of such islands increases with anodization voltage. Dorsey supposed that barrier oxide films consist of aluminum trihydrate passing through certain structural changes during conversion into porous type of film [23]. With respect to the anionic species from the electrolyte incorporation which accompany the oxide growth at the oxide/electrolyte interface, the barrier oxide films on aluminum consist of two distinct layers: an outer layer doped with the anionic species of the electrolyte and an inner layer of pure oxide.

The structure of porous film was also investigated by many authors using various analytical techniques. Franklin discovered the structure of hexagonal cells [24]. Their diameter is proportional to anodization voltage. Cell itself consists of several oxide types with varying crystal contents and hydration degrees. Similar results were obtained by Altenpohl [25]. The model of porous film structure suggested by Keller, Hunter, and Robinson seems as the most acceptable till present time [26]. The schematic presentation of their model is showed in Fig. 5.2. The suggested model proposes that every pore is in the center of hexagonal cell. The radius of the pore doesn't depend either on anodization voltage or on anodization time, but on type of the electrolyte.

5.2.2 Kinetics of Barrier Oxide Films Growth

The model of anodic oxide films growth might be reduced to problem of ionic conduction trough the oxide in the presence of a strong electric field obstructed by processes on interfaces (metal/oxide, oxide/electrolyte). Eventually, the first part of the problem that should be considered is the ionic current in some point in the oxide and then the influence of mentioned interfaces. Also the problem of the spatial electric charge influence on ionic current, e.g., on the electric field, should be taken into consideration. For dealing with specific theories for anodic oxide growth it is necessary to recognize two situations where ionic charge transport is possible. In the case of high-field conduction, it is assumed that the electric field strength is high enough to prevent movement of cations against the field direction. In the case of low-field conduction, where movement of cations against the field direction is not negligible, Guntherschultze and Betz [27–30] have shown that under high electric field conditions, the ionic current density j_+ and the electric field strength E are related through the exponential law:

$$j_+ = A_+ \exp(B_+ E), \tag{5.1}$$

where A_+ and B_+ are temperature dependent constants involving parameters of ionic transport. In order to understand the processes that could lead to such an expression for the ionic current, an analogy can be drawn to a simple electrochemical reaction. As in an electrochemical reaction, the rate-determining step for charge transport control is that which has the highest potential energy with respect to the initial state. The other steps are non-rate determining, because they are much faster. There are three possible rate-determining steps to be postulated in the system metal/oxide/electrolyte: (a) ion transfer across the metal/oxide interface, (b) ion transfer through the oxide bulk, and (c) ion transfer across the oxide/electrolyte interface producing a solvated ion. In any reaction these three rate-determining steps are possible and involve the overcoming of energy barriers. The rate-determining process is the step which has the greatest potential energy with respect to the initial state, under a given set of conditions. Theoretical justification of the exponential relationship proposed by Guntherschultze and Betz has been based upon (a), (b), and a combination of (a) and (b) as the rate-determining steps. Several theories are based on above postulations. In Cabrera–Mott theory [31] the rate-determining step is ion transfer across the metal/oxide interface. In Verwey's theory [32] ion movement through the oxide bulk is considered as a rate-determining step. Previous two theories are the boundary cases of more general Young's theory [33].

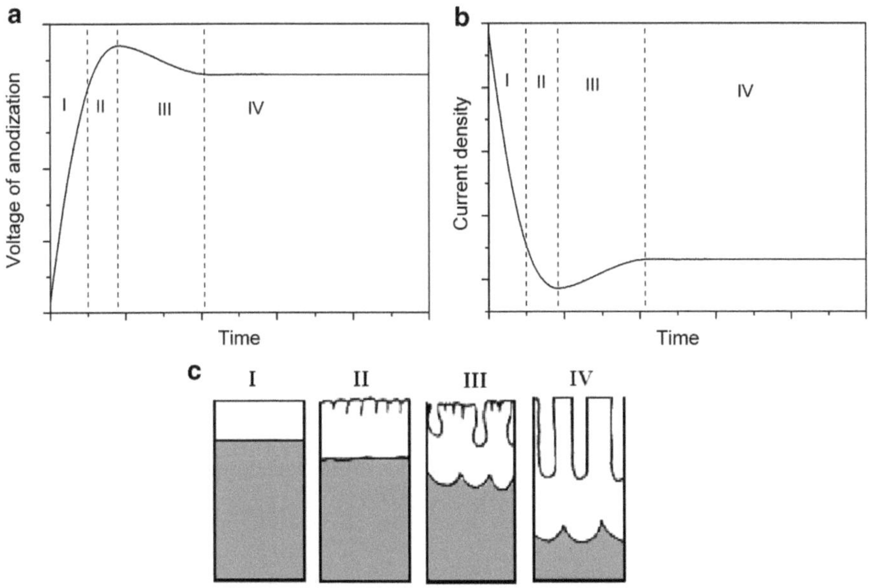

Fig. 5.3 (**a**) Anodization voltage vs. time dependence during growth of porous oxide on aluminum—galvanostatic anodization regime; (**b**) Anodization current vs. time dependence during growth of porous oxide on aluminum—potentiostatic anodization regime; (**c**) corresponding states of porous oxide film

5.2.3 *Kinetics of Porous Oxide Films Growth*

Growth of the porous oxide films passes through several phases that can be easily detected by measuring the anodization voltage (or current) with time or by using microscopy techniques in chosen stages of the anodization [34]. These phases are shown in Fig. 5.3. Figure 5.3a presents the voltage–time dependence during anodization of aluminum in galvanostatic regime which is studied by Kanakala et al. [35]. Figure 5.3b shows the current–time characteristic during anodization of aluminum in potentiostatic regime. Hoar and Yahalom supposed that the relationship between current density and time observed under the constant anodizing potential is a resultant of two overlapping processes: the first one is exponential decrease related to the barrier film formation and the second represents the process of pore formation [36]. The first stage in Fig. 5.3a, b is the growth of the barrier part of the oxide film. Relatively flat oxide films are obtained in this stage (following the morphology of the aluminum substrate), before the onset of the formation of pores in stage II. Further anodization leads to the expansion of pores through barrier film and their bottoms become expanded also. Finally, in stationary stage IV, parallel and regularly arranged pores are obtained. As mentioned before pores are placed in centers of hexagonal cells, with barrier film between their bottoms and aluminum substrate.

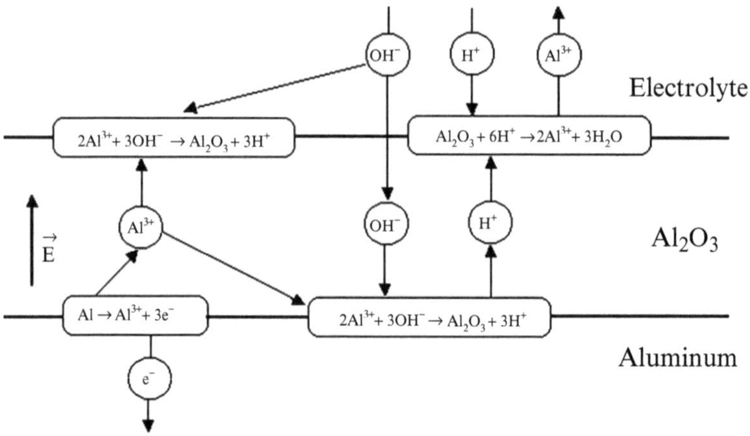

Fig. 5.4 Electrochemical reactions during anodization of aluminum (Reproduced by permission from American Institute of Physics No. 3165811173729)

Basic processes involved in pore formation are shown in Fig. 5.4. In stationary regime pores grow perpendicularly to aluminum substrate when balance between oxide dissolution on electrolyte/oxide interface and oxide layer growth on metal/oxide interface is established. Oxide layer grows due to aluminum cation (Al^{3+}) and anions (O^{2-}, OH^-, and electrolyte anions) migration to opposite sides under the influence of high electric field. Oxide film grows on oxide/electrolyte interface because of aluminum cations migration toward cathode. Al^{3+} cations which are not consumed during migration through the bulk react with O^{2-} and OH^- as well as with electrolyte anions forming the oxide layer. Simultaneously, oxide layer dissolves on the oxide/electrolyte interface at the pore bottom as a result of process of electric field assisted dissolution, while shape of the pores is governed by chemical dissolution in the electrolyte. Electrolyte anions are absorbed on the bottom of the pores as well as in the oxide bulk. Typical levels of incorporation of electrolyte anions are 7.6 % in phosphoric acid, 2.4 % in oxalic acid, 11.1 % in sulfuric acid, and 0.1 % in chromic acid [37].

There are several theoretical models explaining the growth of porous oxide films [6, 38, 39]. Very good agreement with experimental results gives model proposed by Parkhutik and Shershulsky [38]. Their model takes into account the growth at metal/oxide and oxide/electrolyte interfaces, electrochemical and electric field assisted oxide layer dissolution, three-dimensional configuration of the electric field, and current density at the bottom of the semi-spherical surface of the barrier oxide film.

Standard applications of anodic oxide films on aluminum are mostly related to protection of aluminum surface for aerospace industry, other areas of transportation, and structural materials, but new developments in analytical and measurement techniques and in nanoscience and nanotechnology opened the new field of applications for porous oxide films. In that context and in combination with surface engineering, porous oxide films obtained the new role as medium for forming and arranging nanostructures. The main turnover in using porous oxide films for that purpose was made after Masuda et al. reported improving the regularity of the pores in a two-step anodization process and fabrication of the highly ordered, hexagonal close-packed "honeycomb" structure under appropriate processing conditions [40]. In the first anodization step, the pores with almost ideally arranged bottoms are formed. Oxide film is then removed and only hexagonally arranged concave structures on the aluminum surface remain. These structures serve as places of growth of perfectly arranged pores in the second anodization, with same anodization parameters as in the first one.

The two-step process triggered an avalanche of applications as the nanofabrication method using this highly ordered porous anodic alumina template. Two-step process has many advantages and provides a cheap and easy way to fabricate highly ordered arrays of nanostructures made of magnetic and nonmagnetic metals, superconductors, semiconductors, optical materials, etc. [34]. Many authors reported fabrication of nanotubes and nanowires [41–45] using highly ordered porous alumina matrices. These matrices are also used as masks in nanolithography processes such as pattern transferring or forming 2D arrays of nanodots [46, 47]. There are also optical applications of ordered nanopores of porous alumina: filled nanopores with interpore distances from 200 to 350 nm (half the wavelength of light in the visible region) could be used as periodic optical nanostructures called "photonic crystals" [48, 49]. Regardless of applications of ordered porous alumina matrices, the two-step anodization is the most common method for obtaining highly self-ordered pore arrays with regularity along whole pore dimension.

Related methods for obtaining highly ordered arrays of pores are based on artificially pretexturing of aluminum substrate, e.g., forming initiation sites for pore nucleation: nanoimprint method based on SiC molds with an array of convex features, electron beam or focused ion beam lithography, holographic patterning, AFM nanoindentation, etc. [50–56].

5.3 Galvanoluminescence During the Anodization of Aluminum

Phenomenon of the light emission from electrode during the anodization is called galvanoluminescence (GL). Depending on the system under investigation or on assumed emission mechanism, this phenomenon is often termed as electroluminescence, luminescence of anodic oxide films, chemiluminescence, electrochemiluminescence, etc.

First investigation of GL goes back to late nineteenth century [57]. Appearance of the light emission was noticed when electrical current was passing through aluminum/oxide/electrolyte system and aluminum was on a more positive potential

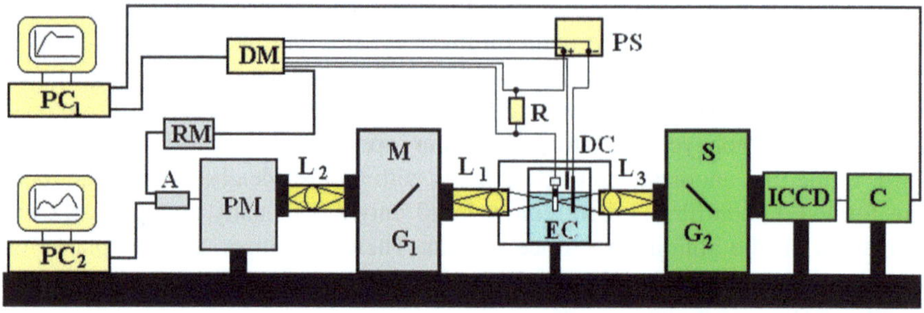

Fig. 5.5 Experimental setup for GL measurements: PS—power supply unit, DC—dark chamber, EC—electrolytical cell with heat exchanger system, L_1—incoming lens system, L_2—outgoing lens system, M—monochromator, G_1—diffraction grating of monochromator, PM—thermo cooled photomultiplier, A—photomultiplier signal amplification system, RM—rate meter, S—spectrograph, L_3—incoming optics, G_2—diffraction grating of spectrograph, ICCD—camera, C—ICCD camera controller, DM—digital multimeter HP 34970A, R—1Ω standard resistor, PC_1—computer 1, PC_2—computer 2 with MCS card

than the platinum electrode. Besides aluminum, similar light emission was observed on metals that form nonconductive anodic oxide coatings by ionic conduction under high electric field (10^6–10^7 V/cm). This group of metals (Ta, Ti, Zr, Zn, Y, W, etc.) is often referred to as "valve metals." Although a vast number of research groups investigated the GL phenomenon [58–105], it has not been completely understood due to the complexity of the system in which GL appears. Metal/oxide/electrolyte system is influenced by many parameters, including the nature of the electrolyte (organic or inorganic), metal surface properties (nature, purity, and pretreatment), anodization conditions (temperature and concentration of the electrolyte, current density, and voltage of anodization), and so on. The influence of these parameters on GL intensity has not been well comprehended and the results differ significantly going from one research group to another. Having in mind that GL appears over the wide range of wavelengths and that its intensity is very low, it is clear that sophisticated optical emission detection systems are required for the investigation of this phenomenon.

5.3.1 Experimental Setup for Galvanoluminescence Measurements

Experimental setup we used for GL measurements is shown in Fig. 5.5. Main parts of the setup are electrolytical cell with heat exchanging system, two independent optical emission detection systems, power supply unit, and data acquisition system. One optical emission detection system uses photomultiplier to detect luminescence, while the other system uses ICCD camera to detect luminescence in the wide range of wavelengths.

Optical emission detection system with photomultiplier consists of incoming lens system (L_1), monochromator (M), outgoing lens system (L_2), and photomultiplier with cooling system (PM). Incoming lens system is made of the objective with 58 mm focal length and 25 mm effective radius that projects the image of the GL emitting sample with 1:1 ratio on the incoming slit of the monochromator. Monochromator SPM-2 (Zeiss) is used to select the GL of desired wavelength and to shine it onto the incoming slit of the photomultiplier tube. This is achieved by the reflection type diffraction grating G_1 (reciprocal dispersion 4 nm/mm, spectral range from 300 to 1,200 nm). The wavelength of the outgoing beam is selected by the rotation of the diffraction grating, which is controlled by step motor and computer PC_2. Both incoming and outgoing monochromator slits are rectangularly shaped, 8 mm high while the width can be changed from 0.1 to 1.5 mm. Outgoing system consists of achromatic 75 mm focal length lens, focusing the outgoing beam on the photomultiplier's cathode.

For the optical emission detection, we used low-noise photomultiplier RCA J1034 CA (PM) with high quantum efficiency in the wavelength range from 200 to 900 nm. Photomultiplier's cathode was cooled to about −40 °C in the custom housing made by Product for Research (TE-222TSRF). By doing so, dark current signal is three orders of magnitude lower compared to the one at room temperature. Signal obtained from photomultiplier is amplified (A) and converted to TTL signal which is sent either directly to MCS card on computer PC_2 or passes through the rate meter (RM) to 20-channel digital multimeter HP 34970A, which is over the RS-232 interface connected to computer PC_1.

Optical emission detection system with ICCD camera consists of incoming lens system (L_3), Hilger spectrograph (S), and ICCD camera with controller (C). Pentax objective (135 mm) projects the image of the anodized sample with 1:1 ratio on the spectrograph slit. Diffraction grating of the spectrograph (300 grooves/mm, blaze on 500 nm, angular dispersion 3.32 nm/rad) selects wavelength in the range of 173 nm and sends it to CCD chip of PIMAX ICCD cooled camera with high quantum efficiency in visible spectrum, made by Princeton Instruments. CCD chip has 1024 × 256 pixels, each of about 26 μm × 26 μm. In order to lower the dark current, CCD chip was cooled at −40 °C.

Electrolytical cell with the heat exchanging system is shown in Fig. 5.6. The main elements of the electrolytical cell are a vessel with flat glass windows (6) dimension of 40 mm × 50 mm × 70 mm, anode (sample) holder (1), two platinum wires (2) as cathode, and temperature sensor embedded in teflon tube (3) that measures the temperature of the electrolyte in proximity to the sample. Two glass tubes passing through the sample holder enable the circulation of the electrolyte through the electrolytical cell/heat exchanger system. Electrolytical cell is placed in the dark chamber that prevents any visible light to hinder GL measurements.

During the anodization, temperature of the electrolyte increases and the evaporation occurs. Therefore, it is necessary to keep temperature of the electrolyte constant in order to maintain same experimental conditions through the whole anodization process. This is achieved by circulating the electrolyte through the

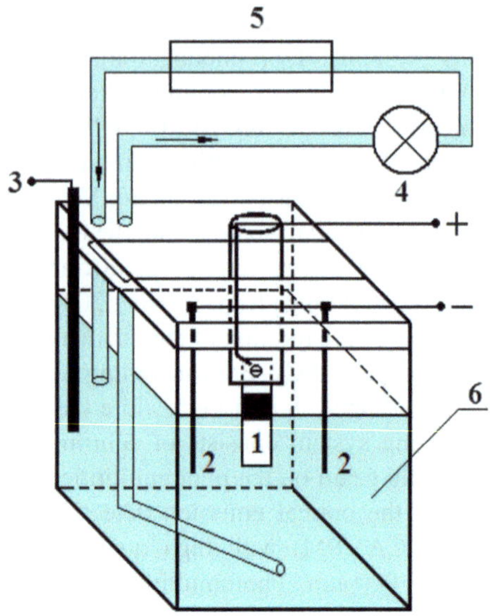

Fig. 5.6 Electrolytical cell: (1)—sample, (2)—platinum wire, (3)—PTC sensor, (4)—peristaltic pump, (5)—heat exchanging system, (6)—vessel with flat glass windows [96] (Reproduced by permission from Elsevier India)

electrolytic cell/heat exchanging system (5). Large volume of the heat exchanging system that is equipped with compressor allows us to maintain the temperature of the electrolyte within 0.1 °C during the experiment.

Before the anodization, pretreatment of the sample surface is required. Most commonly, this is done in three ways [90] (1) electropolishing of the samples—results in flat and shiny surface, (2) chemical cleaning—removes impurities from the surface, and (3) degreasing with acetone, alcohol, and distilled water in ultrasonic bath. In order to define sample's surface and to prevent crawling of the electrolyte, protective insulating mask is deposited on the top part of the sample (on each side).

Systems presented in Fig. 5.5 were checked for their spectral sensitivity before performing actual GL measurements. Spectral sensitivity of the first system (system with monochromator) was inspected by using the standard tungsten lamp (OSRAM Wi-17G) [106]. However, standard tungsten lamps are not applicable for wide wavelength range detection systems as ICCD. Firstly, the stray light cannot be completely eliminated, and secondly, intensity of the lamp radiation varies over the wide range of the wavelengths. Therefore, it is required to have standardized light source that has approximately the same light intensity over the whole range of wavelengths under investigation.

To inspect spectral sensitivity of the detection system with ICCD camera, unique standardized light source with LEDs was engineered, based on three light emitting diodes, each one with different spectral characteristic (Fig. 5.7) [106]. Light emitted from LEDs (a) is mixed using two integration spheres (b), before it reaches outgoing slit (c). System is thermo cooled to prevent the influence of ambient temperature on light intensity.

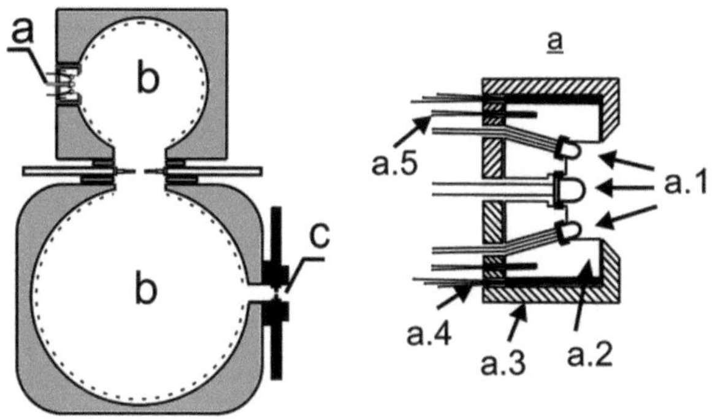

Fig. 5.7 Standardized light source with LEDs: (a) Thermo cooled LEDs: a.1—LEDs, a.2—aluminum block, a.3—thermal insulation, a.4—thermostat system, a.5—temperature sensor; (b) integrating spheres; (c) exit slit [106] (Reproduced by permission from Society of Applied Spectroscopy)

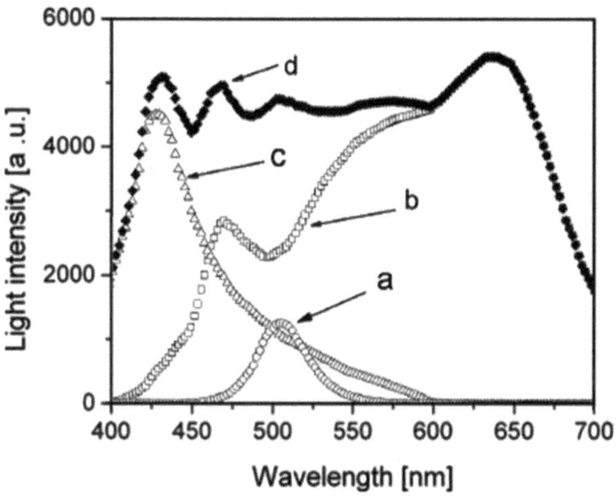

Fig. 5.8 Emission spectrum of standardized light source and individual characteristics of LEDS: (a) turquoise 505 nm; (b) warm white; (c) blue 430 nm; (d) total spectrum [106] (Reproduced by permission from Society of Applied Spectroscopy)

Standardized light source with LEDs was calibrated with respect to the system with monochromator. Yet again, spectral sensitivity of the system with monochromator was inspected by using the standard tungsten lamp [106]. Emission spectrum of the engineered standardized light source is shown in Fig. 5.8. Clearly, light intensity is of the same magnitude over the wavelength range of interest (400–700 nm), allowing us to inspect the spectral sensitivity of the wide wavelength range detection systems as ICCD.

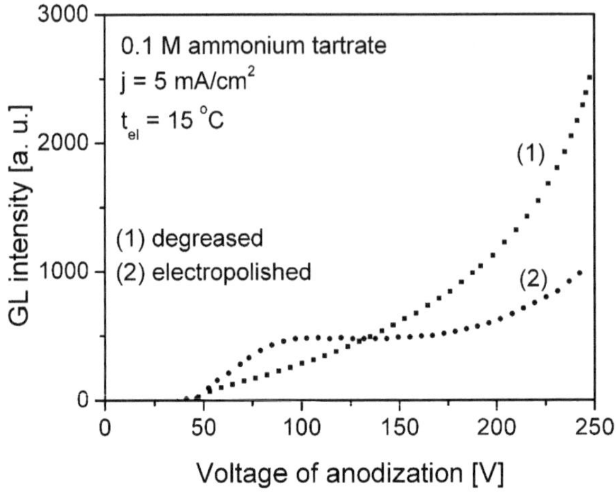

Fig. 5.9 Influence of the aluminum sample's pretreatment on GL intensity [76] (Reproduced by permission from Elsevier No. 3165800868574)

5.3.2 *Galvanoluminescence of Anodic Oxide Films Formed in Organic Electrolytes*

5.3.2.1 GL of Barrier Anodic Oxide Films

Commonly used electrolyte for GL investigation of barrier anodic oxide films on aluminum in organic electrolytes is ammonium tartrate. Detailed investigation of aluminum pretreatment procedure, anodization conditions, and GL mechanism of barrier oxide films on aluminum was conducted by Tajima and Shimizu [76–80].

Influence of the aluminum sample's pretreatment on GL intensity is shown in Fig. 5.9. GL intensity vs. anodization voltage curve for degreased (unpolished) samples mainly follows van Geel's equation [64]:

$$GL = aj[\exp(bU) - 1], \tag{5.2}$$

where a and b are constants, while j is anodization current density. When electropolished samples are used, GL intensity vs. voltage of anodization curve gains in complexity. Zeković et al. showed that interference of light going directly through oxide layer and light reflected from the aluminum surface causes more complex behavior of the curve [83]. In the case of anodization in organic electrolytes, Tajima suggested that carboxylate ions, incorporated into oxide film during the anodization, act as GL centers in such films [76].

Figure 5.9 also shows that GL intensity depends on oxide film thickness, having in mind that relationship between anodization voltage (U) and oxide film thickness (d) follows linear relation $d = \lambda \cdot U$, where $\lambda = 1.4$ nm/V [1]. Shimizu [80] proposed a model that explains dependence of GL on anodization voltage in organic electrolytes (for electropolished samples), based on the following assumptions:

(a) GL appears as a consequence of the excitation of GL centers by their inelastic collisions with electrons from the electronic avalanche, under high electric field 10^6–10^7 V/cm condition, during the anodization process. Carboxylate ions incorporated into oxide film during the anodization act as GL centers.
(b) Electrons injected into conducting band of the oxide film on the oxide/electrolyte interface produce a large number of small nondestructive avalanches by collision ionization. Avalanches are uniformly distributed over the anodic surface, leaving behind slow positive charge, which lowers the effective electric field in the oxide film and prevents the appearance of destructive avalanching.
(c) Number of excited GL centers (and the GL intensity) is proportional to a number of electrons participating in multiplication process.
(d) The maximum of avalanche length (l_{av}) is about 100 nm, corresponding to about 80 V voltage of anodization.

Based on these assumptions, Shimizu explained experimentally obtained GL intensity vs. oxide film thickness curve [80]. When thickness of the oxide film is less than l_{av}, avalanches originating from oxide/electrolyte interface can grow continuously. Therefore, the number of electrons with kinetic energy high enough to excite GL centers sharply increases with the thickening of the oxide film. Under the assumption that the excitation of GL centers is proportional to the number of electrons in the avalanche, then it is clear why GL intensity sharply increases with the increase of the oxide film's thickness. On the other hand, when thickness of the oxide film is higher (but not significantly) than l_{av}, effective electric field beyond l_{av} that causes electron avalanching is very low due to the opposite electric field originating from remaining positively charged ions. Therefore, beyond l_{av} electrons are deficient in kinetic energy required to excite GL centers. Beyond l_{av}, GL intensity remains almost constant until the thickness of the oxide film becomes high enough to allow electrons to gain enough energy for excitation of GL centers and to cause an increase in GL intensity.

Having all this in mind, Shimizu and Tajima [79] have developed theoretical model that explains experimentally obtained dependence of GL intensity on the thickness of oxide film. Their model shows exceptional match with experimental data in the area of electronic avalanche growth, where the influence of positive charge can be neglected. Dependence of GL intensity on oxide film's thickness follows the relation:

$$\mathrm{GL} \cong a[\exp(\alpha b d) - 1], \tag{5.3}$$

where a is a constant related to electron current density on the oxide/electrolyte interface, α is cationic transport number, and b is a constant equal to reciprocal value of mean free path for electron ionization.

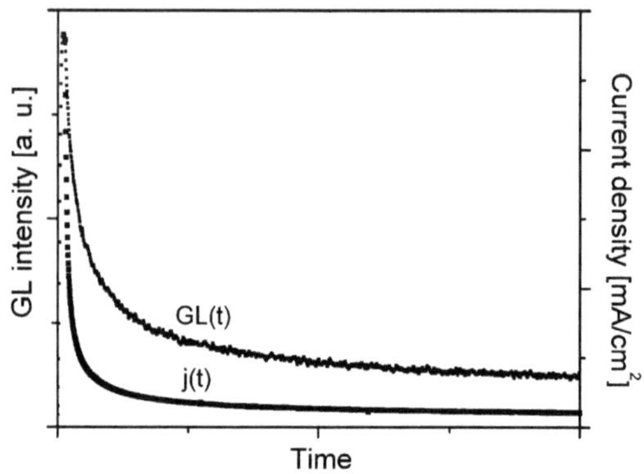

Fig. 5.10 Typical anodization current density and GL intensity vs. time curves under potentiostatic conditions

In their experiments Shimizu and Tajima measured only total GL radiation, but not spectrally resolved GL radiation. For this reason, Shimizu and Tajima were not able to make a correction of GL intensity due to interference effects, leading to less successful interpretation of experimental results.

Belča et al. [89] have developed a theory on GL mechanism that explains the dependence of GL intensity on oxide film thickness, taking into account interference effects, as a consequence of high reflectivity of the aluminum surface [83], and electron capture in traps that are usually defects in the oxide film. This model shows excellent match of experimental and theoretically calculated values for dependence of GL intensity of oxide film thickness. Model also enables the calculation of effective electric field in the cylinder of the electronic avalanche and parameters related to the process of electron multiplication (ionization coefficient α and average lifetime of electrons in oxide film τ).

GL intensity of barrier oxide films increases exponentially with respect to voltage of anodization, during the anodization of aluminum under galvanostatic conditions (voltage of anodization increases linearly with time). This exponential dependence prevents measuring of GL spectra by measuring of GL intensities on a wide range of wavelengths, but requires utilization of time-resolved low-intensity measuring optical emission detection system such as ICCD camera [106]. This obstacle can be overcome by working under potentiostatic conditions. Figure 5.10 presents anodization current density and GL intensity versus time of anodization curves for constant voltage of anodization. Clearly, right after applying a constant voltage U, barrier oxide film begins to grow and anodization current density sharply increases to a certain maximum. During this stage of anodization, anodic current mainly consists of ionic current that forms oxide film. Upon reaching certain oxide film thickness d (corresponding voltage of anodization is U), anodization current density begins to decrease. After some time from the beginning of anodization, oxide film ceases to grow significantly, which corresponds to lower values of anodization current density. This, lower, current density is predominantly electronic current that has no influence on oxide film thickening, but influences GL. Constant values of GL intensity and anodization current density allows

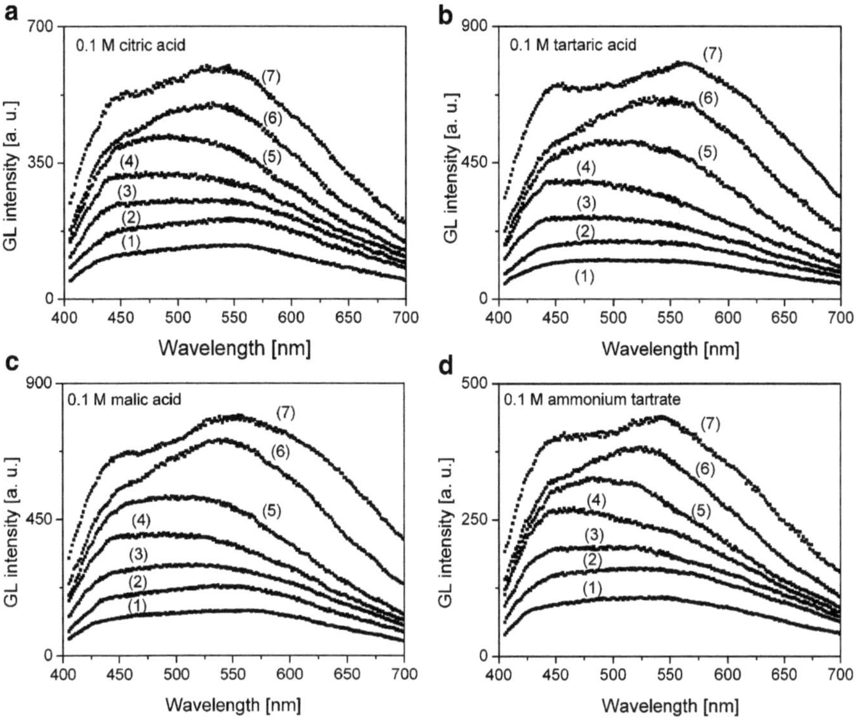

Fig. 5.11 Shape change of the GL spectra of barrier oxide films with respect to voltage of anodization under galvanostatic conditions ($j = 7.5$ mA/cm^2, $t_{el} = 23$ °C) in organic electrolytes [95]: (**a**) 0.1 M citric acid; (**b**) 0.1 M tartaric acid; (**c**) malic acid; (**d**) 0.1 M ammonium tartrate. (1) 75 V; (2) 100 V; (3) 125 V; (4) 150 V; (5) 175 V; (6) 200 V; (7) 225 V (Reproduced by permission from Elsevier No. 3163521207546)

measurement of GL spectra in steady-state regime by simple measurement of GL intensity at various wavelengths.

Figure 5.11 presents the change of the spectra shape with respect to voltage, during the anodization of aluminum in various barrier oxide film forming organic electrolytes (citric acid, tartaric acid, malic acid, and ammonium tartrate). All spectra are obtained under identical anodizing conditions. Used organic electrolytes are characterized by the presence of carboxylate group; hence, GL spectra have the same shape. This is in agreement with the hypothesis that carboxylate ions incorporated in oxide film during the anodization are GL centers.

Influence of the anodization current density and electrolyte temperature on GL intensity and GL spectra shape is presented in Fig. 5.12. GL intensity increases with the increase of anodization current density and with the decrease of electrolyte temperature. This conclusion corresponds to the findings of Mason, who has shown that the incorporation of carboxylate ions in oxide film increases as anodization current density increases, or as electrolyte temperature decreases [107]. It has also

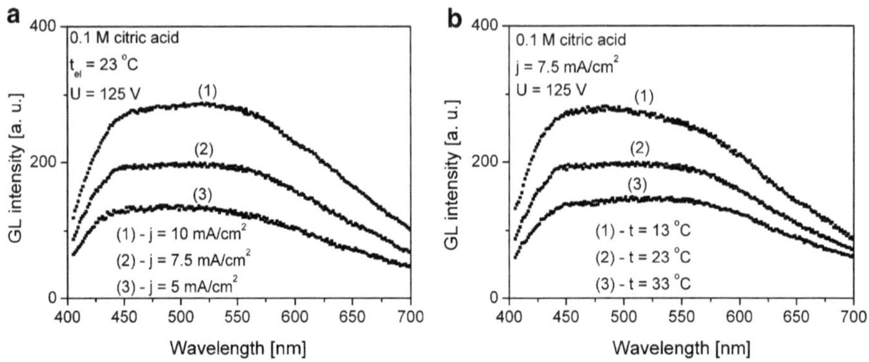

Fig. 5.12 Influence of the anodization parameters on GL spectra shape [95]: (**a**) anodization current density; (**b**) electrolyte temperature (Reproduced by permission from Elsevier No. 3163521207546)

been reported that electrolyte concentration only weakly influences the GL intensity [76]. Figure 5.12 shows that GL spectra shape (for constant voltage of anodization) does not depend on anodization current density, but on electrolyte temperature only [103].

5.3.2.2 GL of Porous Anodic Oxide Films

Malonic and oxalic acids are commonly used for the investigation of GL phenomena of porous anodic oxide films, obtained by aluminum anodization in organic electrolytes [64, 87, 88, 96]. Typical GL intensity vs. time and voltage of anodization vs. time curves for galvanostatic anodization, as well as GL intensity vs. time and anodization current density vs. time curves for potentiostatic anodization, are shown in Fig. 5.13. Obviously, after certain time from the beginning of anodization, voltage of anodization and GL intensity reach constant values in galvanostatic regime. Similar conclusion can be drawn for anodization current density and GL intensity in galvanostatic regime. Constant values of anodization parameters in both cases allow simple measurement of GL spectra when steady-state conditions are reached. This can be achieved by plain measurement of GL intensity values over the selected wavelength range. In the case of porous anodic oxide films, GL radiation is being emitted from the barrier part of the oxide film, due to the fact that in that part of the oxide film electric field reaches maximum.

Figure 5.14a shows the influence of the anodization current density on GL intensity at 450 nm, in the early stage of anodization, on a constant temperature. Van Geel et al. have shown that GL intensity is proportional to current density for constant film thickness (thickness of the barrier layer in porous oxide film is determined by the voltage of anodization) [64]. This finding is in concert with results presented in Fig. 5.14a, i.e., for same voltage of anodization higher anodization current density results in higher GL intensity. Higher anodization current

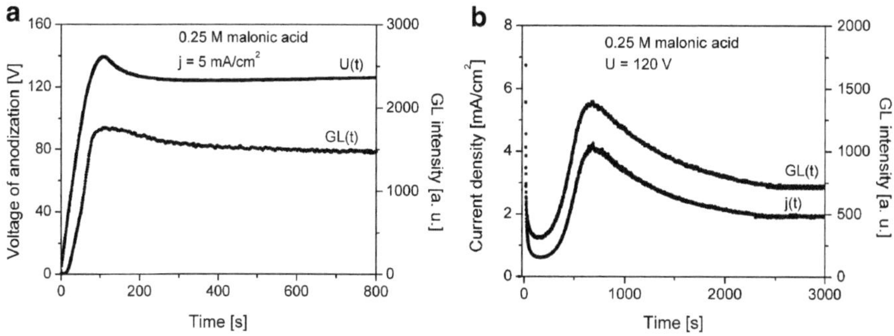

Fig. 5.13 (**a**) GL intensity vs. time and voltage of anodization vs. time curves under galvanostatic conditions; (**b**) GL intensity vs. time and anodization current density vs. time curves under potentiostatic conditions ($t_{el} = 14$ °C, $\lambda = 450$ nm) [96] (Reproduced by permission from Elsevier India)

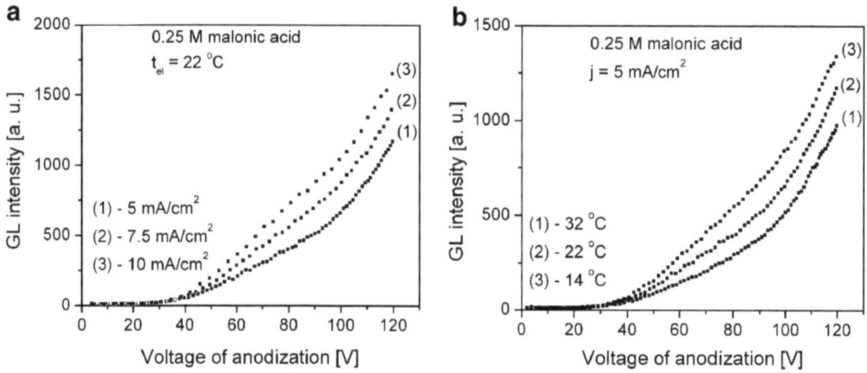

Fig. 5.14 Influence of the anodization parameters on GL intensity [96]: (**a**) anodization current density influence; (**b**) electrolyte temperature influence (Reproduced by permission from Elsevier India)

(ionic + electronic) corresponds to higher number of electrons in conducting zone accelerated by high electric field inside of the barrier layer, resulting in higher GL intensity. GL intensity also depends on temperature of the electrolyte. Figure 5.14b shows the dependence of GL intensity at 450 nm on electrolyte temperature, in the early stage of anodization, for constant anodization current density. Clearly, GL intensity increases for lower temperatures of the electrolyte. Same type of dependence is observed for other organic electrolytes that form both barrier and porous anodic oxide films.

GL spectra of porous oxide films obtained by aluminum anodization in malonic acid with respect to voltage of anodization are presented in Fig. 5.15a. Wide GL bands are present in the spectral range from 400 to 700 nm. Each spectrum is dominated by two observable maxima; first one is centered at about 455 nm, while

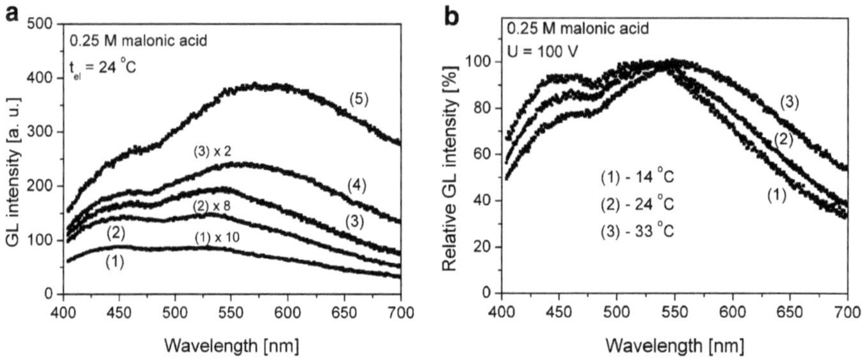

Fig. 5.15 (**a**) GL spectra of porous oxide films for different voltage of anodization in 0.25 M malonic acid [96]: (1) 80 V; (2) 90 V; (3) 100 V; (4) 110 V; (5) 120 V. (**b**) Influence of electrolyte temperature on GL spectra of porous oxide films (Reproduced by permission from Elsevier India)

the second maximum shifts with the voltage of anodization. The relative ratio of the second maximum over the first one increases with the increasing voltage of anodization. At the same time, position of GL maximum and their relative ratio depends on the temperature of the electrolyte, as shown in Fig. 5.15b. When the electrolyte temperature increases, second GL maximum becomes more dominant. Similar dependence of GL spectra shape on anodization conditions was reported for aluminum anodization in oxalic acid [88]. Shift of the spectral maxima suggests that GL maxima have fine structure, i.e., they consist of a number of spectral lines and shifting occurs as a consequence of changing contributions of wide GL band lines. Concurrently, GL spectra shape depends on the electronic current. Electronic component of the anodization current is higher for higher voltage of anodization at constant temperature of the electrolyte, or for higher temperature of the electrolyte for constant voltage of anodization [108].

Similar shape of GL spectra obtained during the anodization of aluminum in various organic electrolytes suggests that centers responsible for the luminescence are of the same type. Oxide films formed by anodization in organic electrolytes are not pure substrate oxides, but are doped with a small amount (a few percents) of carboxylate species [109, 110]. Figure 5.16a shows EDS spectrum of oxide film surface formed by anodization of aluminum in malonic acid. Main constituting elements of the oxide film are aluminum and oxygen, with some carbon. The presence of carbon points to carboxylate species incorporated into oxide film during anodization. ATR FTIR spectrum of oxide film surface formed in the same electrolyte is shown in Fig. 5.16b. Absorption bands with minima at about 1,560 and 1,460 cm^{-1} are typical for oxide films formed in organic electrolytes that comprise functional carboxyl group (–COOH). The band at 1,560 cm^{-1} is related to asymmetric O–C–O stretching vibrations, while the band at 1,460 cm^{-1} is related to the bond between symmetric O–C–O and C–C stretching vibrations [109, 110].

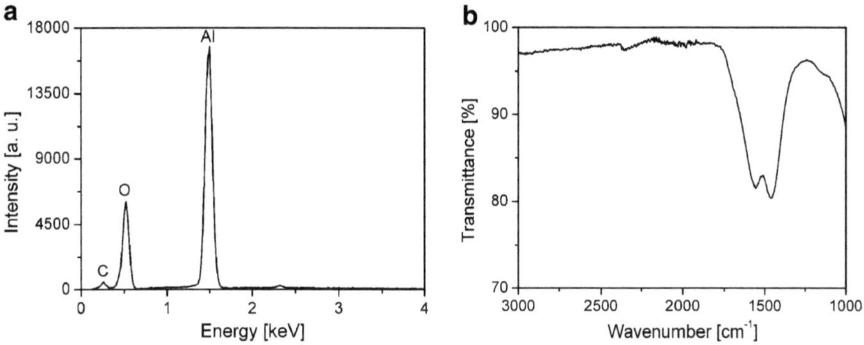

Fig. 5.16 (**a**) EDS spectrum of porous oxide film; (**b**) ATR FTIR spectrum of porous oxide film; (0.25 M malonic acid, $t_{el} = 24$ °C, $U = 100$ V) [96] (Reproduced by permission from Elsevier India)

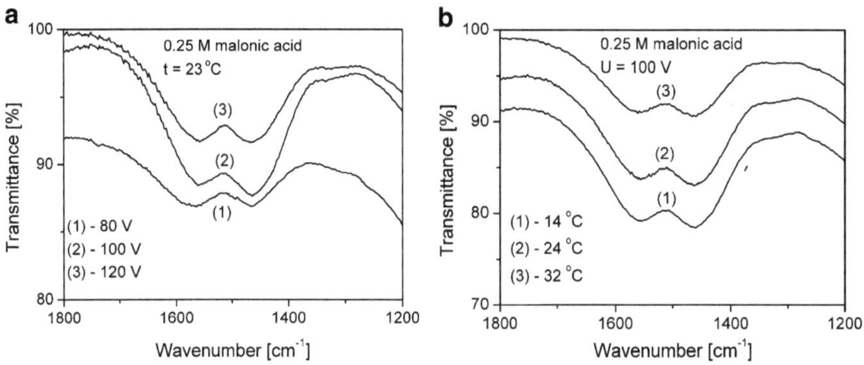

Fig. 5.17 Influence of anodization parameters on ATR FTIR spectra: (**a**) influence of voltage of anodization; (**b**) influence of electrolyte temperature [96] (Reproduced by permission from Elsevier India)

Influence of the voltage of anodization and electrolyte temperature on the shape of ATR FTIR spectra is shown in Fig. 5.17. Position of absorption minimum at about 1,460 cm^{-1} is independent of voltage of anodization value, while the second minimum shifts from about 1,572 cm^{-1} for voltage of anodization of 80 V, to about 1,558 cm^{-1} for voltage of anodization of 120 V (Fig. 5.17a). This shift of the second absorption minima with changing voltage of anodization is comparable to the shift of the second GL maximum in Fig. 5.15. Temperature of the electrolyte has similar influence on the shape of GL spectra (Fig. 5.17b). Position of absorption minima is independent on electrolyte temperature, but their relative ratio increases with the increase of electrolyte temperature. The same influence of anodizing conditions on the shape of GL spectra as well as ATR FTIR spectra also confirms the assumption that carboxylate species incorporated in oxide films act as the GL centers.

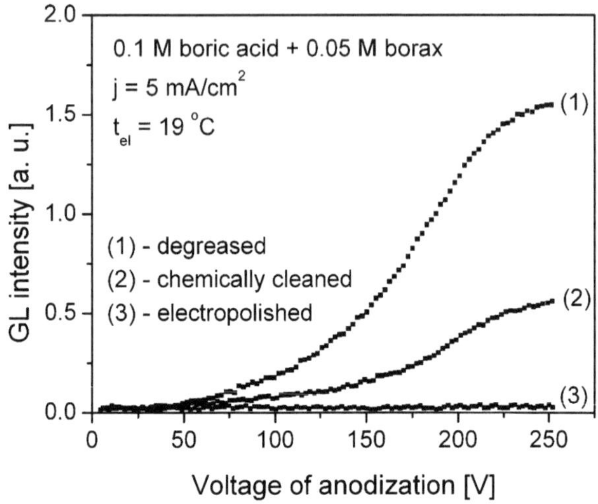

Fig. 5.18 Influence of aluminum surface pretreatment on GL intensity [105]

5.3.3 Galvanoluminescence of Anodic Oxide Films Obtained in Inorganic Electrolytes

5.3.3.1 GL of Barrier Anodic Oxide Films

Detailed investigation on the influence of aluminum surface pretreatment, anodization conditions, and GL mechanism of barrier oxide films in inorganic electrolytes (ammonium borate) was conducted by Tajima and Shimizu with coworkers [76]. Investigation on GL in boric acid/borax solution, conducted within our research group, yields similar results [105]. Main difference between these two sets of results is in experimental method, because Tajima and Shimizu measured total GL intensity, while we measured GL intensity on selected wavelengths in order to eliminate the effect of interference.

Influence of the pretreatment of aluminum surface on GL intensity is shown in Fig. 5.18. The highest GL intensity was obtained for degreased samples; it significantly decreases for chemically cleaned samples and almost disappears for electropolished samples. This suggests that surface properties control GL intensity in inorganic electrolytes, i.e., defects and impurities ("flaws") present in oxide film are responsible for GL [77]. In other words, GL intensity is proportional to the concentration of flaws, as first suggested by Smith [66].

Figure 5.19 shows AFM micrographs of aluminum surface after various pretreatment procedures. Degreased sample has the highest surface roughness, higher than 0.1 μm, which is mostly related to the cold-rolling of aluminum. Chemical cleaning removes chemical contaminants from the aluminum surface, while electropolishing serves to obtain the flattest surface.

Another important factor that influences GL intensity is annealing temperature of the aluminum sample prior to anodization (Fig. 5.20). It has been shown that

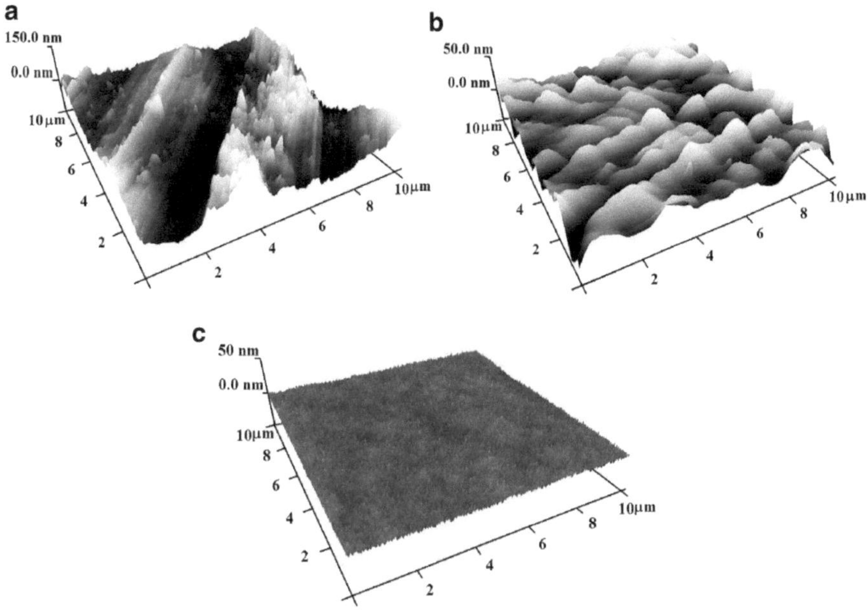

Fig. 5.19 AFM micrographs of aluminum surface after various pretreatment procedures: (**a**) degreasing; (**b**) chemical cleaning; (**c**) electropolishing [100] (Reproduced by permission from Elsevier No. 3163521501120)

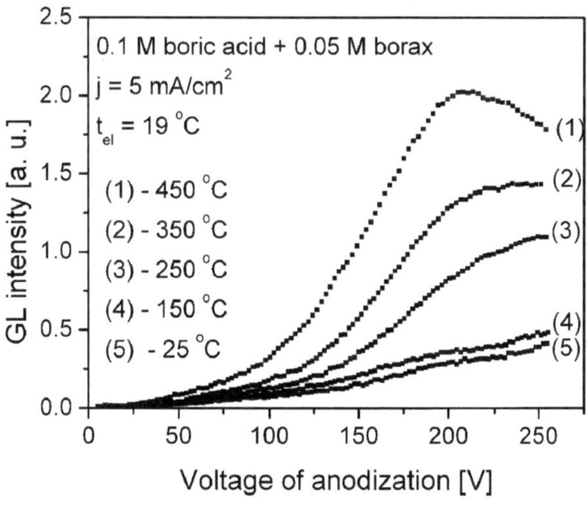

Fig. 5.20 Influence of the annealing temperature on GL intensity [105]

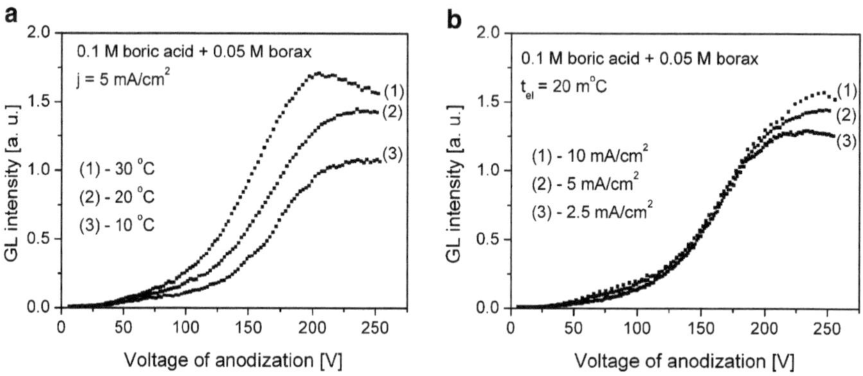

Fig. 5.21 Influence of anodization parameters on GL intensity: (**a**) influence of electrolyte temperature; (**b**) influence of anodization current density [105]

under identical anodizing conditions (anodization current density, electrolyte temperature, and concentration), GL intensity increases with the increase of annealing temperature, as a consequence of increased number of defects on grain boundaries and their orientation [79].

During anodization of aluminum in inorganic electrolytes which form barrier oxide films, when voltage of anodization and thickness of oxide film increase linearly with time, van Geel observed that GL intensity increases exponentially as oxide film thickness (or voltage of anodization) increases [64]. Additional observations can be found in the literature as well. Gardin et al. [111] claim that GL intensity increases linearly as voltage of anodization increases, for all voltages higher than the threshold voltage, while Anderson [59] suggests GL intensity is not directly related to voltage of anodization. Zeković et al. [82] have shown that GL intensity vs. voltage of anodization curve strongly depends on aluminum sample thermal pretreatment. According to [82], exponential and linear increases are related to annealed and non-annealed samples, respectively.

Influence of electrolyte temperature on GL intensity is shown in Fig. 5.21a, while influence of anodization current density on GL intensity is shown in Fig. 5.21b. GL intensity is strongly dependent on electrolyte temperature (GL increases as electrolyte temperature increases), while it shows almost no dependence on anodization current density. This suggests that GL intensity is independent on the amount of electrolyte species incorporated in oxide film. It has been observed that in some electrolytes that contain tungsten species, GL intensity increases with the increase of voltage of anodization, while it is scarcely dependent on electrolyte temperature [101].

Figure 5.22 presents change of the GL spectra shape with respect to voltage of anodization during anodization of aluminum in boric acid + borax electrolyte. Two maxima are superimposed on wide GL bands in the wavelength range from 400 to 700 nm. First maximum is centered at about 430 nm, while the other one shifts from about 600 nm to about 680 nm, when voltage of anodization changes from 200 to 275 V.

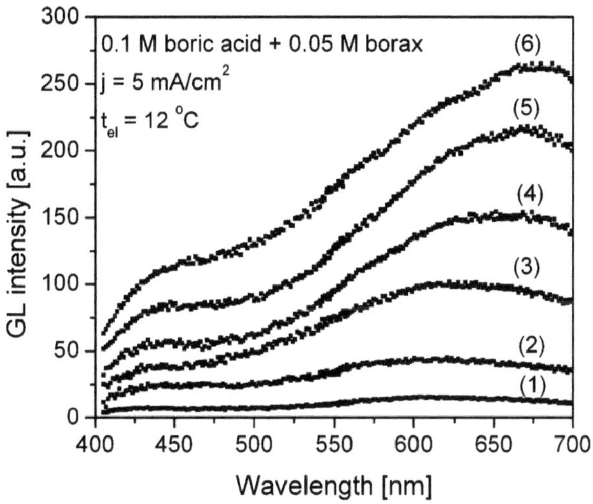

Fig. 5.22 Change in GL spectra of barrier oxide films with respect to voltage of anodization during galvanostatic anodization. (1) 100 V; (2) 150 V; (3) 200 V; (4) 225 V; (5) 250 V; (6) 275 V

Influence of anodization conditions on GL intensity in inorganic electrolytes, such as boric acid + borax electrolyte, is utterly different than in the case of anodization in organic electrolytes. As stated earlier, GL in organic electrolytes gains intensity when electrolyte temperature is lower and/or anodization current density is higher, which is in accordance with the electroluminescence theory in homogeneous oxide film. A conclusion can be drawn that different types of GL occur in organic and inorganic electrolytes: GL in inorganic electrolytes does not appear as a result of collisions, but is a local phenomenon associated with scintillation or some other phenomenon concerning "flaws." "Flaw" is a general term for microfissures, local regions of different composition, and impurities. Confirmation that "flaws" are responsible for GL in inorganic electrolytes is obtained if an electropolished sample is subject to mechanical scratch, chemical etching, or impurity contamination. In that case, "flaws" are artificially introduced on the sample surface, and GL appears only in affected areas [80].

5.3.3.2 GL of Porous Anodic Oxide Films

Commonly used electrolytes for investigation of GL of porous anodic oxide films formed by aluminum anodization in inorganic electrolytes are phosphoric, sulfuric, chromic, and sulfamic acids [90–93, 100]. This paragraph will present the results of GL measurements of porous oxide films obtained by anodization of two types of aluminum samples: ultraclean (99.999 %) aluminum samples (sample A) and 99.5 % pure aluminum samples (sample B). Both sample types were anodized in phosphoric acid. Similar results were obtained for aluminum anodization in sulfuric, chromic, and sulfamic acids, though GL intensity was lower.

Figure 5.23 presents typical voltage of anodization and GL intensity vs. time curves for aluminum sample types A and B, respectively, during anodization in

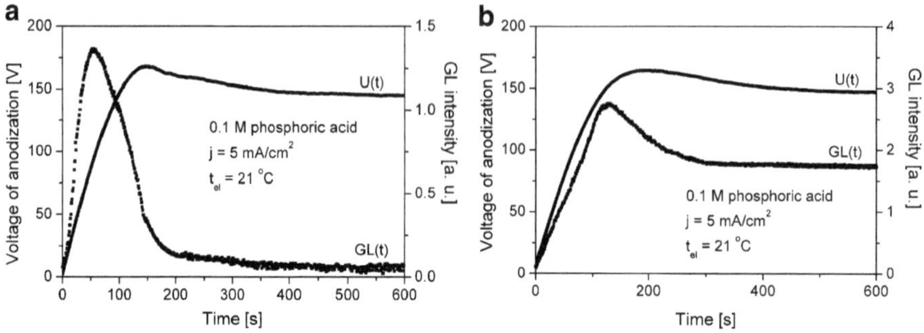

Fig. 5.23 Voltage of anodization and GL intensity vs. time curves [90]: (**a**) sample type A; (**b**) sample type B (Reproduced by permission from Elsevier No. 3163520770775)

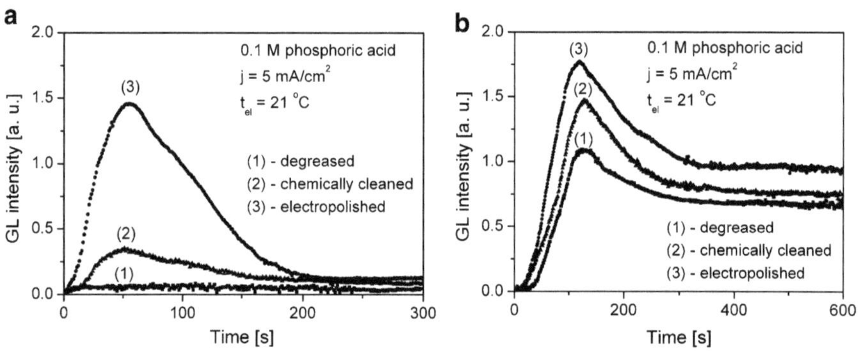

Fig. 5.24 Influence of Al surface pretreatment procedure on GL intensity [90]: (**a**) sample type A; (**b**) sample type B (Reproduced by permission from Elsevier No. 3163520770775)

0.1 M phosphoric acid. Both sample types generate more intense GL in the early phase of anodization (during the formation of barrier oxide layer, i.e., when electric field has highest value). During steady-state regime (thick porous layer) GL almost disappears for sample type A, while for sample type B, GL reaches constant value, which is dependent on voltage of anodization in steady-state regime. Furthermore, maximum GL intensity for sample type A is shifted with respect to voltage of anodization maximum, which corresponds to the onset of pore formation [38]. This shift is insignificant for sample type B.

The difference in GL intensities for sample types A and B points out to surface imperfections ("flaws") as well as to internal impurities as main sources of GL. Amount of internal impurities is determined by the sample purity, while surface imperfections are related to cold-rolling of aluminum samples. Concentration of surface imperfections is independent on aluminum purity and it solely depends on sample's surface treatment.

Influence of aluminum surface pretreatment on GL intensity for sample types A and B is shown in Fig. 5.24. It is to be noted that procedure for pretreatment of the

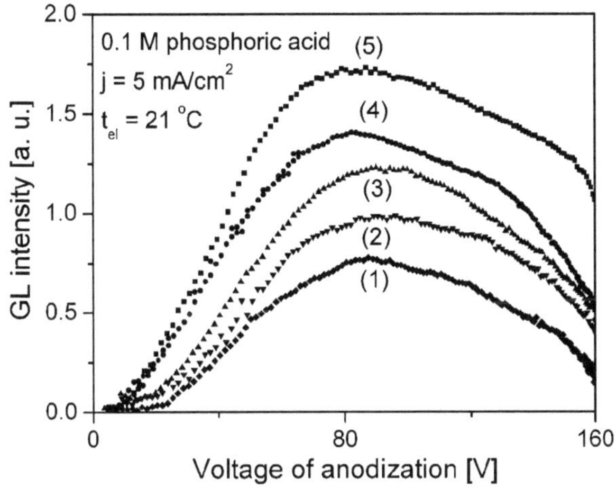

Fig. 5.25 Influence of Al sample annealing temperature on GL intensity for sample type A [90]: (1) 25 °C; (2) 150 °C; (3) 250 °C; (4) 350 °C; (5) 450 °C (Reproduced by permission from Elsevier No. 3163520770775)

sample strongly influences GL intensity for ultraclean samples (type A): degreased samples yield highest GL intensity, chemically cleaned samples have lower GL intensity, and GL of electropolished samples is almost undetectable. These results are similar to results of Tajima and Shimizu, obtained for barrier oxide films in inorganic electrolytes [76]. The maintaining of GL for porous film in B type aluminum can be attributed to a high concentration of impurities. As GL intensity is highest for degreased samples, from now on, only those results will be presented.

Another important aluminum surface pretreatment factor that influences GL intensity is the temperature of annealing. Higher GL intensity is obtained for higher annealing temperature (Fig. 5.25). Increase of annealing temperature increases the number of surface defects on grain boundaries and their orientation, i.e., concentration of "flaws" increases [79]. This finding is valid for samples of type A only, while it does not hold for samples of type B. This can be explained by having in mind that for samples of type B impurities in aluminum act as GL centers, so that the influence of surface defects can be neglected.

Influence of anodizing conditions on GL intensity is presented in Fig. 5.26. According to van Geel [64], GL intensity is proportional to anodization current density for constant oxide film thickness (thickness of the barrier layer of oxide film is determined by voltage of anodization). Curves on Fig. 5.26a confirm this finding, i.e., higher anodization current density results in higher GL intensity, for the same voltage of anodization. Measurement of GL intensity while changing electrolyte temperature shows strong dependence of electrolyte temperature on GL intensity (Fig. 5.26b). Lower electrolyte temperature results in higher GL intensity, which is identical to the relation between these parameters when anodization in organic electrolytes is performed [104]. On the other hand, this is opposite to the relation established between GL intensity and electrolyte temperature when GL of barrier oxide films in inorganic electrolytes is investigated [96]. Literature offers very little data on the influence of electrolyte concentration on GL intensity (Fig. 5.25c).

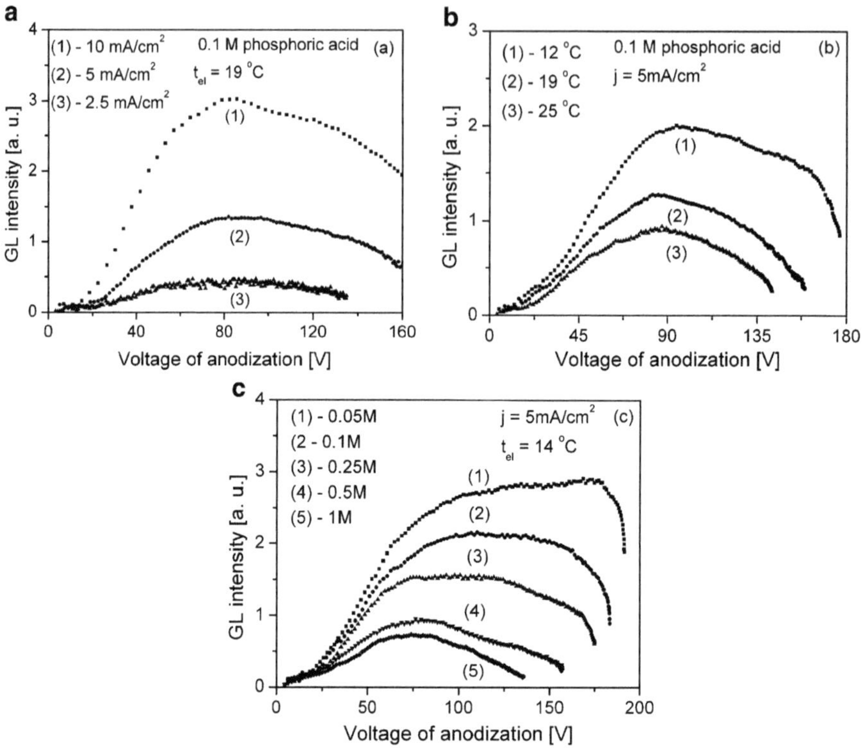

Fig. 5.26 Influence of anodization conditions on GL intensity for sample type A [90]: (**a**) Variation of GL intensity with anodization current density; (**b**) Variation of GL intensity with electrolyte temperature; (**c**) Variation of GL intensity with electrolyte concentration (Reproduced by permission from Elsevier No. 3163520770775)

Our investigations show that higher electrolyte concentrations result in lower GL intensity, although it is not clear how temperature and concentration of phosphoric acid influence GL intensity. Influence of these parameters may be indirect, since electrolyte temperature and concentration affect the distribution of net anodization current. Thus, these parameters influence electronic current, which is responsible for GL.

GL spectra of porous oxide films obtained by aluminum anodization in phosphoric acid (sample type B) were recorded in two regimes: in constant current mode (galvanostatic regime) and in constant voltage mode (potentiostatic regime). GL spectra were recorded in the wavelength range from 400 to 700 nm. GL spectra recording in galvanostatic regime is based on fact that after some time from the beginning of anodization (determined by anodization current density), both voltage of anodization and GL intensity reach constant values (Fig. 5.23b).

Experimental results for influence of anodization conditions on GL spectra in galvanostatic regime suggest that spectra shape is independent of anodization

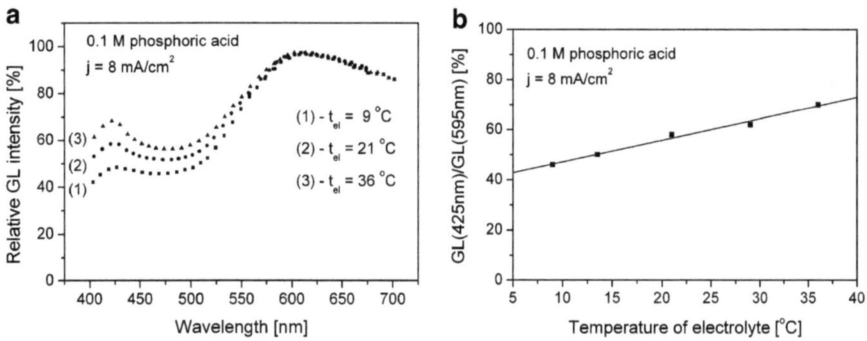

Fig. 5.27 (**a**) Influence of electrolyte temperature on GL intensity in phosphoric acid; (**b**) Influence of electrolyte temperature on the first to second maximum ratio [91] (Reproduced by permission from Elsevier No. 3163520978089)

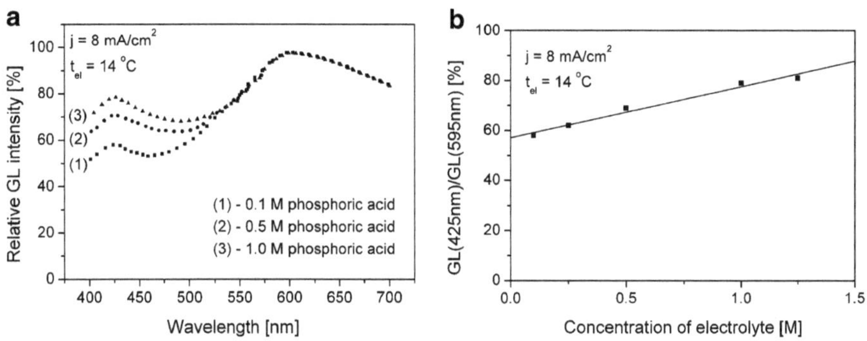

Fig. 5.28 (**a**) Influence of electrolyte concentration on GL spectra shape in phosphoric acid; (**b**) Influence of electrolyte concentration on the first to second maximum ratio [91] (Reproduced by permission from Elsevier No. 3163520978089)

current density, while it depends on electrolyte temperature and concentration. Influence of electrolyte temperature on the shape of GL spectra is presented in Fig. 5.27a. Clearly, each emission GL spectrum is dominated by two maximum at about 425 and 595 nm. With the increase of electrolyte temperature, maximum positioned at about 425 nm becomes more pronounced. Ratio of the first to second maximum shows linear dependence on electrolyte temperature (Fig. 5.27b).

Influence of electrolyte concentration on GL spectra shape is presented in Fig. 5.28a. It is obvious that increase of electrolyte concentration results in distinction of maximum positioned at about 425 nm. Once again, ratio of the first to second maximum shows linear dependence on electrolyte concentration (Fig. 5.28b).

During the early stage of anodization in galvanostatic regime (barrier layer type), GL spectra were recorded both for type A and type B samples. Spectrum was obtained by measuring maximum GL intensity at various wavelengths. An example of the spectrum recorded in this mode is presented in Fig. 5.29. Clearly,

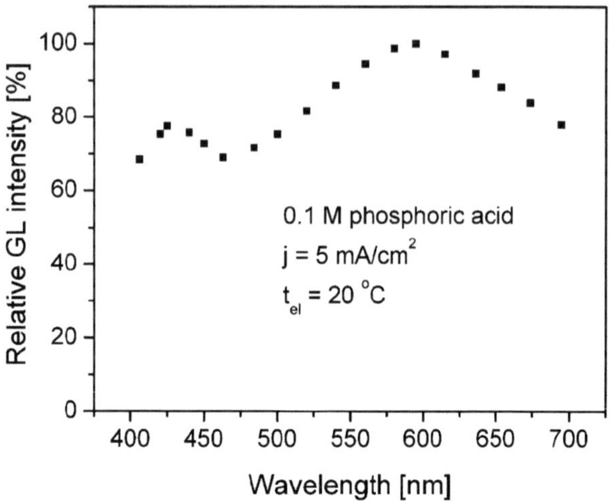

Fig. 5.29 GL spectrum for sample type A obtained during the early stage of anodization (barrier layer type) [90] (Reproduced by permission from Elsevier No. 3163520770775)

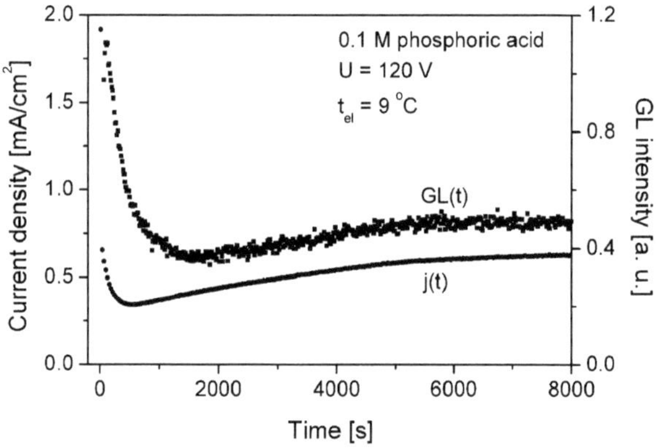

Fig. 5.30 Dependence of anodization current density and GL intensity on time [91] (Reproduced by permission from Elsevier No. 3163520978089)

GL spectrum has the same shape as GL spectrum recorded for thick porous oxide film, which indicates that GL mechanism is the same in both stages of anodization and for both sample types.

Measuring and investigating GL during anodization of aluminum in potentiostatic regime suggests that GL intensity, as well as anodization current density, reaches constant values after 1–2 h after the onset of anodization (Fig. 5.30). This condition is sufficient for the spectrum to remain unchanged during the recording time (about 400 s). In this regime, spectrum was recorded in succession going back and forth several times, for each sample. It is worth mentioning that GL spectrum did not change during the recording, i.e., spectrum shape was the same in every run, for each sample.

Investigation of the influence of anodization conditions on GL spectra in potentiostatic regime shows that spectra shape is not directly related to voltage of anodization, but is closely connected to electrolyte concentration and temperature.

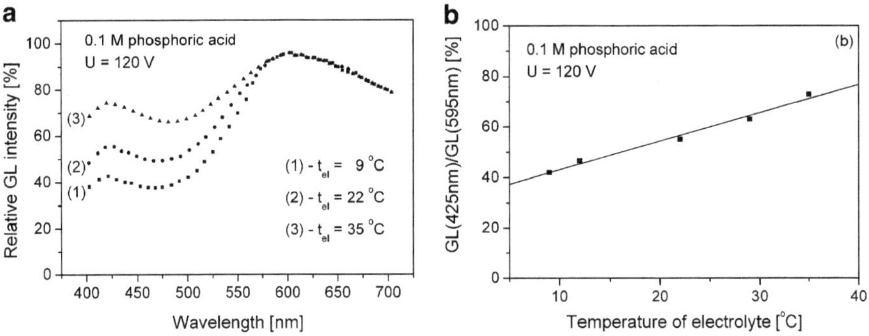

Fig. 5.31 (**a**) Influence of electrolyte temperature on GL spectra shape in phosphoric acid; (**b**) Influence of electrolyte temperature on the first to second maximum ratio [91] (Reproduced by permission from Elsevier No. 3163520978089)

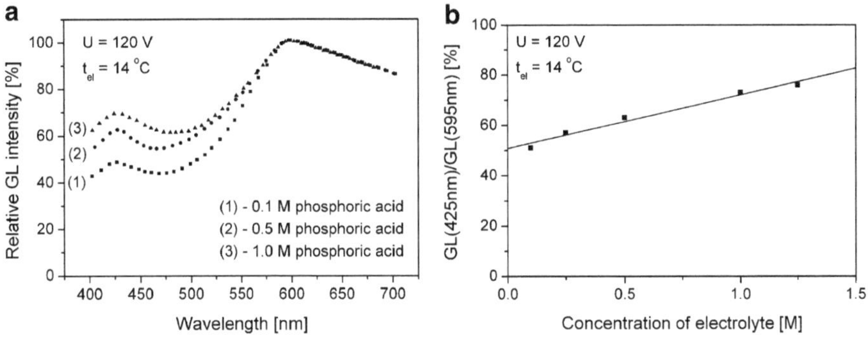

Fig. 5.32 (**a**) Influence of electrolyte concentration on GL spectra shape in phosphoric acid; (**b**) Influence of electrolyte concentration on the first to second maximum ratio [91] (Reproduced by permission from Elsevier No. 3163520978089)

Dependence of GL spectra shape on electrolyte temperature is shown in Fig. 5.31a. As electrolyte temperature increases, spectral maximum at about 425 nm becomes more distinct. Ratio of two observed maxima is a linear function of electrolyte temperature (Fig. 5.31b).

Experimental results showing influence of electrolyte concentration on GL spectra shape are presented in Fig. 5.32a. As expected (based on the results obtained in galvanostatic regime), increase of electrolyte concentration causes maximum at about 425 nm to be more pronounced. Ratio of the first to second maximum also follows linear dependence on electrolyte temperature (Fig. 5.32b).

Again, literature offers insufficient amount of data on GL spectra of porous oxide films obtained by aluminum anodization in phosphoric acid. Ganley [68] conducted an investigation on the influence of impurities on GL spectra obtained by anodization in potentiostatic regime. In this work, two maxima centered at about 430 and 600 nm were observed. Relative ratio of the observed maxima changes depending on anodization regime (anodic or AC) and on impurity concentration. However, data presented by Ganley lack information on electrolyte temperature and electrolyte temperature stability throughout the anodization, sample pretreatment, and details on how the spectra were recorded.

Distribution of anodization current depends on electrolyte temperature and concentration [112], i.e., the sum of dissolution current and electronic current constantly increases with respect to the net current, with an increase in electrolyte concentration and/or temperature. Since relative ratio of the first to second spectral maximum follows the same dependence on electrolyte temperature and concentration, this suggests possible relation between anodization current distribution and ratio of GL maxima obtained during the experiment.

5.3.4 Annealing of Aluminum Above 550 °C and Its Influence on Galvanoluminescence

Annealing of aluminum samples above 550 °C, before anodization, results in increased GL intensity and observable change of GL spectral character. This phenomenon is strongly related to alteration of aluminum structure. Experimental results presented in this chapter are obtained on ultra pure (99.999 % Al) aluminum samples that were annealed on 550 °C for 4 h, prior to anodization.

5.3.4.1 Thermal Treatment Influence on GL of Anodic Oxide Films Obtained in Inorganic Electrolytes

Figure 5.33 presents typical GL intensity vs. time and voltage of anodization vs. time curves obtained during the anodization of aluminum in inorganic electrolytes that form either barrier (a) or porous (b) oxide films. Sharp initial increase of voltage of anodization and GL intensity in presented curves is associated with the existence of oxide layer formed during the annealing of aluminum samples.

During the anodization of aluminum samples annealed at 550 °C prior to anodization in inorganic electrolytes, GL intensity is about a hundred times higher than for aluminum samples annealed below 500 °C. This is related to the formation of regions of high crystallinity, which act as defects in oxide film, therefore causing an increase in GL intensity.

Figure 5.34 shows the influence of aluminum surface pretreatment on GL intensity of porous oxide films obtained by anodization of aluminum in phosphoric acid. The most intense GL is observed for degreased samples, it slightly decreases for chemical treated samples, and it is not present for electropolished samples. The same influence of surface pretreatment procedure on GL intensity is observed for samples anodized in boric acid + borax electrolyte. These results are visibly different than results obtained on samples annealed on temperatures below 500 °C (Figs. 5.18 and 5.24a), where chemical treatment of aluminum samples considerably decreases GL intensity.

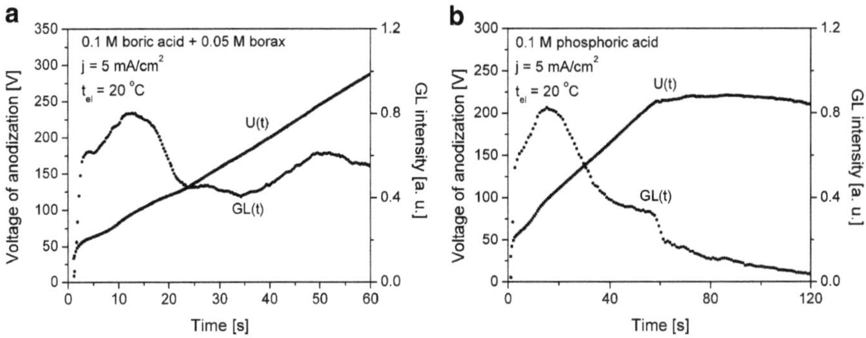

Fig. 5.33 GL intensity and voltage of anodization variation with respect to time of anodization for aluminum samples annealed at 550 °C: (**a**) boric acid + borax; (**b**) phosphoric acid [98] (Reproduced by permission from American Chemical Society)

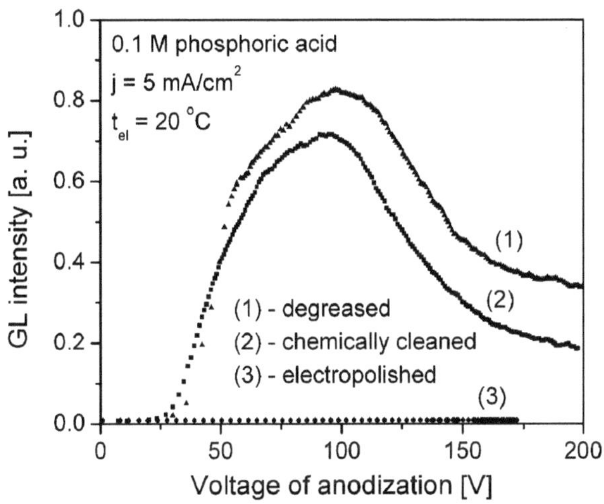

Fig. 5.34 Influence of aluminum surface pretreatment on GL intensity [98] (Reproduced by permission from American Chemical Society)

Chemical treatment serves as a tool to remove amorphous alumina, as well as surface impurities, leaving the regions of crystalline alumina almost intact. On the other hand, electropolishing removes the regions of crystalline alumina from the surface as well. SEM micrograph of chemically treated aluminum sample and corresponding XRD pattern are shown in Fig. 5.35, with clearly visible regions of γ-alumina present of the surface.

Figure 5.36 shows GL spectra change with anodization time for aluminum samples annealed at 550 °C in phosphoric acid. At the beginning of anodization, six broad intensive emission bands can be observed. Emission band maxima are centered at about 430, 483, 544, 575, 601, and 648 nm. Same GL spectra shape is obtained in boric

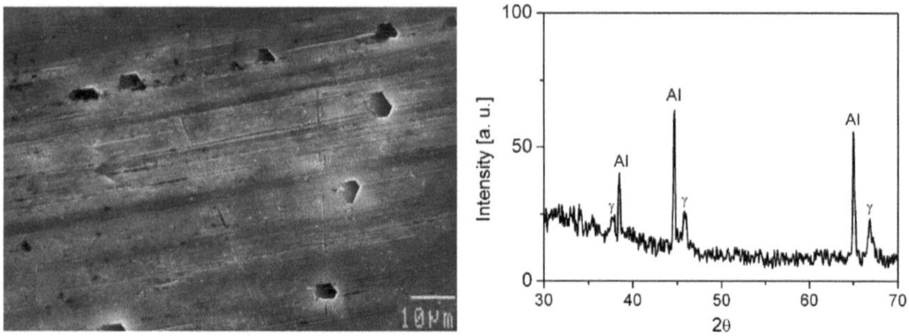

Fig. 5.35 SEM micrograph and XRD pattern of chemically treated aluminum surface annealed at 550 °C [98] (Reproduced by permission from American Chemical Society)

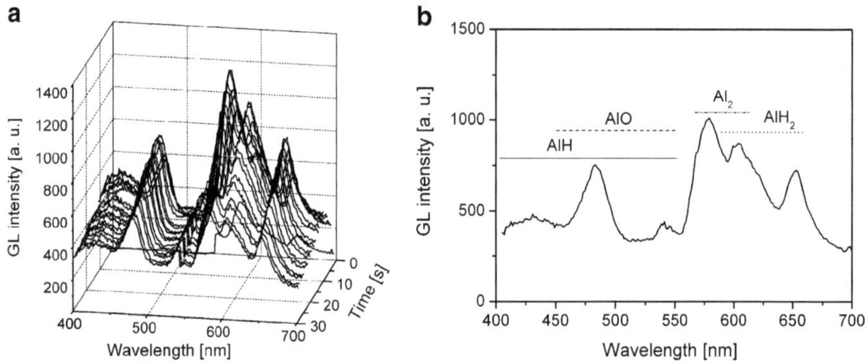

Fig. 5.36 (**a**) GL spectrum shape change during anodization of aluminum annealed at 550 °C in 0.1 M H_3PO_4; (**b**) GL spectrum for 100 V anodization voltage ($j = 10$ mA/cm^2, $t_{el} = 12$ °C) [98] (Reproduced by permission from American Chemical Society)

acid + borax electrolyte. A semiquantitative analysis based on literature data [113] for simple molecular species involving aluminum atom, as well as those atoms whose presence is possible under given experimental conditions (hydrogen, oxygen, etc.), showed that the GL in the range from 400 to 700 nm originated from spectral transitions: $X^1A_1 \leftarrow A_1\Pi$ of AlH, $X^2\Sigma^+ \leftarrow B^2\Sigma^+$ of AlO, $X^3\Sigma_g^- \leftarrow A^3\Sigma_u^-$ of Al_2, and $X^2A_1 \leftarrow A^2A_1(^2\Pi_u)$ of triatomic radical AlH_2. Since broad GL bands are not present for samples annealed below 500 °C, this clearly indicates that their existence is a consequence of annealing on temperatures above 550 °C, i.e., broad GL bands are related to the presence of γ-alumina crystalline regions. Crystalline γ-alumina is porous and consists of spinel structures that form aluminum vacancies in order to gain stability [114]. It is probable that molecules responsible for GL are localized in γ-alumina pores (average pore diameter is 2–5 nm). Another possible explanation of this phenomenon originates from the assumption that regions of crystalline γ-alumina perturbate strong electric field during anodization and generate local dielectric breakdown, similar to DC discharge.

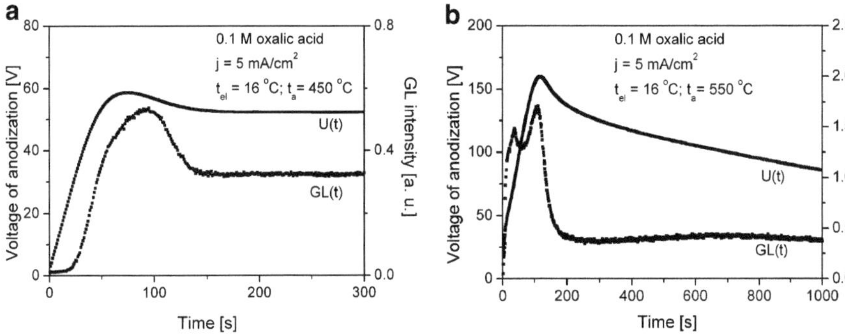

Fig. 5.37 GL intensity and voltage of anodization variation with respect to time of anodization for aluminum samples annealed at (**a**) 450 °C; (**b**) 550 °C [99] (Reproduced by permission from Elsevier No. 3163521319336)

5.3.4.2 Thermal Treatment Influence on GL of Anodic Oxide Films Obtained in Organic Electrolytes

Figure 5.37 presents typical GL intensity and voltage of anodization variations with respect to time of anodization for porous oxide films in oxalic acid. Curves in Fig. 5.37a, b are recorded for aluminum samples annealed at 450 °C and at 550 °C, respectively. Clearly, GL intensity for samples annealed at 550 °C is higher, especially in the early stage of anodization.

Influence of annealing on GL intensity is shown in Fig. 5.38. Annealing on temperatures lower than 500 °C has no significant influence on GL intensity. When annealing temperature is higher than 500 °C, GL intensity increases with an increase of annealing temperature. Oxide films formed on aluminum samples annealed on temperatures higher than 500 °C contain large surface regions of crystalline γ-alumina, while oxide layers formed on aluminum samples annealed on temperatures lower than 500 °C are amorphous.

Influence of pretreatment of aluminum samples annealed at 550 °C on GL intensity is shown in Fig. 5.39. Intense GL is obtained for degreased and chemically cleaned samples, while considerably lower intensities are recorded for electropolished samples.

GL spectrum shape change during the anodization of aluminum annealed at 550 °C in oxalic acid is shown in Fig. 5.40a. In the early stage of anodization GL spectrum has six intense and broad emission maxima centered at about 430, 483, 544, 575, 601, and 648 nm. Just like in the case of aluminum anodization in inorganic electrolytes, identified spectral maxima can be ascribed to transitions $X^1A_1 \leftarrow A_1\Pi$ of AlH, $X^2\Sigma^+ \leftarrow B^2\Sigma^+$ of AlO, $X^3\Sigma_g^- \leftarrow A^3\Sigma_u^-$ of Al_2, and $X^2A_1 \leftarrow A^2A_1(^2\Pi_u)$ AlH_2. During anodization, oxide film grows vertically on the surface of aluminum, balancing between dissolution of oxide film on oxide/electrolyte

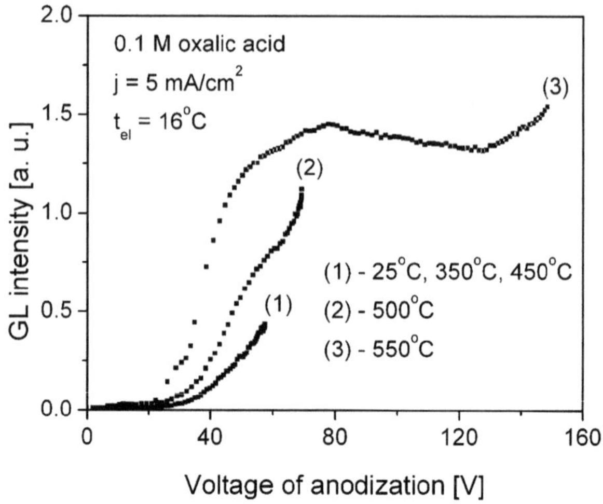

Fig. 5.38 Influence of aluminum annealing temperature on GL intensity [99] (Reproduced by permission from Elsevier No. 3163521319336)

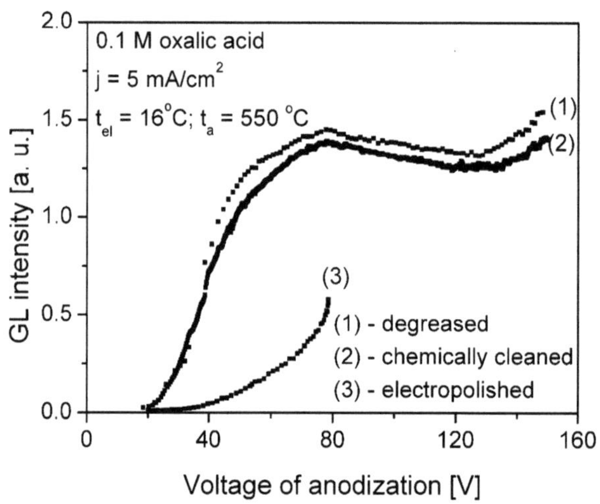

Fig. 5.39 Influence of pretreatment of aluminum samples annealed at 550 °C on GL intensity [99] (Reproduced by permission from Elsevier No. 3163521319336)

interface and oxide film growth on aluminum/oxide and oxide/electrolyte interfaces. In the case of porous oxide films, GL originates only from the barrier layer of the oxide film. Clearly pronounced maxima in GL spectrum can be observed until γ-alumina crystalline grains start moving from barrier to porous layer of the oxide film. From that moment on, GL spectrum has characteristic shape as the GL

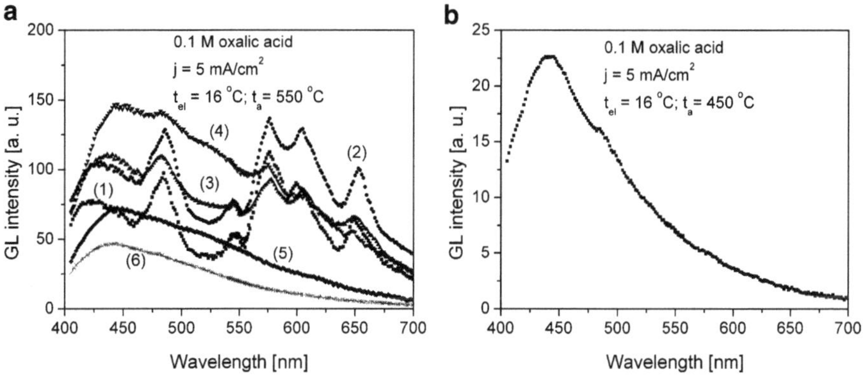

Fig. 5.40 (**a**) GL spectrum shape change during anodization of aluminum annealed at 550 °C: (1) 10 s; (2) 30 s; (3) 60 s; (4) 100 s; (5) 200 s; (6) 600 s; (**b**) GL spectrum of porous oxide film during anodization of aluminum annealed at 450 °C [99] (Reproduced by permission from Elsevier No. 3163521319336)

spectrum of porous oxide films formed by anodization of aluminum samples annealed below 500 °C (Fig. 5.40b) and total emission comes from carboxylate ions incorporated in oxide film during the anodization. Similar spectral features are observed in barrier anodic oxide films [99].

5.3.4.3 GL Spectral Characterization in UV Range

During the anodization of aluminum samples, annealed on temperatures below 500 °C prior to anodization, extremely low GL intensity in UV range is observed. However, upon annealing of aluminum on 550 °C, GL spectrum during anodization features a wide emission band with maximum centered at about 313 nm (Fig. 5.41) [102]. This emission band is observable independently of the electrolyte type. An analysis based on literature data [113], for simple molecules that include aluminum, hydrogen, and oxygen atoms, identifies that observed maximum corresponds to transitions between vibrational levels $C^2\Pi \leftarrow X^2\Sigma^+$ of AlO molecule, which are partially overlapped with transitions from vibrational levels $A^2\Sigma^+ \leftarrow X^2\Pi$ of OH molecule.

5.3.5 Galvanoluminescence Methods as a Tool for Investigation of Anodic Oxide Films

GL of anodic oxide films formed on highly reflective aluminum surfaces in organic electrolytes features clearly pronounced interference maxima that can be used as a tool for determining oxide film thickness and inherent optical parameters.

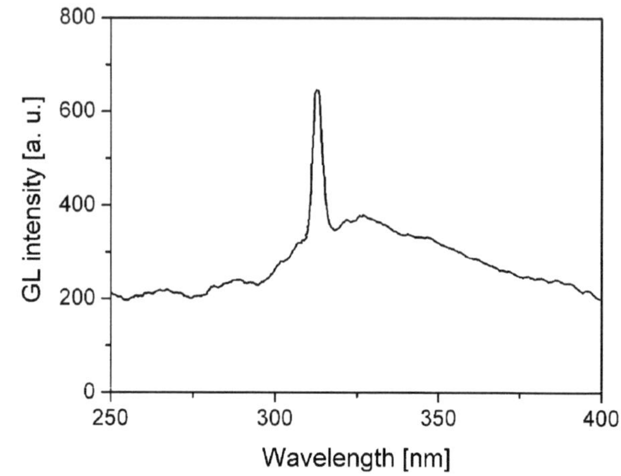

Fig. 5.41 GL spectrum in UV range of anodic oxide film on aluminum annealed at 550 °C [102] (Reproduced by permission from Elsevier No. 3163530121723)

Two methods (based on particular observation angle) for determining such properties are developed:

(a) Measurement at right angle—optical axis is set perpendicular to the oxide film surface,
(b) Measurement at an angle close to total reflection—optical axis is set parallel to the oxide film surface.

5.3.5.1 Investigation of Porous Anodic Oxide Films

Measurement at Right Angle

In this measurement method, GL emission perpendicular to the oxide film surface is detected at a fixed wavelength λ [86]. Figure 5.42 presents detected GL emission pathway (a) and observed GL intensity (b).

Method for determining oxide film thickness is developed based on the assumption that excited GL center (located at distance x from the metal/oxide interface) emits the ray E_{a1} in the optical detection system direction and the ray E_{b1} in the opposite direction. High reflectivity of aluminum surface (obtained by electropolishing) creates multiple reflections on metal/oxide and oxide/electrolyte interfaces, causing interference of rays that have different optical paths. Thickness of obtained oxide films can be determined from these interference maxima as follows:

– Instead of primal ray E_{a1}, under multiple reflections condition, a family of rays is detected

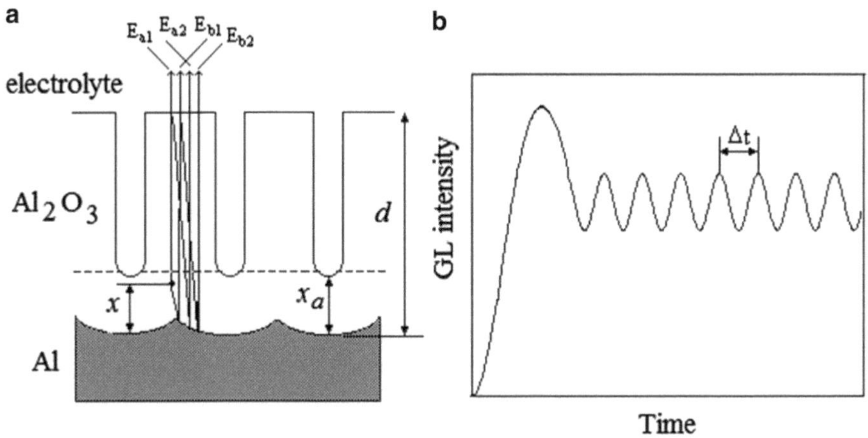

Fig. 5.42 Measurement at right angle: (**a**) detected GL emission pathway and (**b**) observed GL intensity

$$
\begin{aligned}
E_{\mathrm{a}1} &= (1-r)E_{\mathrm{o}}\exp(i\omega t)\\
E_{\mathrm{a}2} &= (1-r)(rR)E_{\mathrm{o}}\exp\left[i\left(\omega t+\pi-\varphi-\frac{4\pi n_2 d}{\lambda}\right)\right]\\
&= rRE_{\mathrm{a}1}\exp\left[i\left(\pi-\varphi-\frac{4\pi n_2 d}{\lambda}\right)\right]\\
E_{\mathrm{a}3} &= (1-r)(rR)^2E_{\mathrm{o}}\exp\left[i\left(\omega t+2(\pi-\varphi)-\frac{8\pi n_2 d}{\lambda}\right)\right]\\
&= (rR)^2E_{\mathrm{a}1}\exp\left[2i\left(\pi-\varphi-\frac{4\pi n_2 d}{\lambda}\right)\right]\\
&\vdots\\
E_{\mathrm{a}(p+1)} &= (1-r)(rR)^pE_{\mathrm{o}}\exp\left[i\left(\omega t+p(\pi-\varphi)-\frac{4\pi n_2 d}{\lambda}p\right)\right]\\
&= (rR)^pE_{\mathrm{a}1}\exp\left[pi\left(\pi-\varphi-\frac{4\pi n_2 d}{\lambda}\right)\right].
\end{aligned}
\tag{5.4}
$$

E_{o} is the amplitude of electric field vector associated with emitted electromagnetic wave, λ is the wavelength of emitted radiation, r is the reflection coefficient on oxide/electrolyte interface, R is the reflection coefficient at metal/oxide interface, and $(\pi - \varphi)$ is the phase shift on metal/oxide interface. Reflection coefficient r is defined by equation:

$$
r = \frac{n_2 - n_1}{n_2 + n_1}, \tag{5.5}
$$

where n_2 is the refraction index of oxide film ($n_2 \cong 1.65$), while n_1 is the refraction index of the electrolyte. Reflection coefficient R and phase shift φ can be calculated from Fresnel's formulae:

$$R = \sqrt{\frac{(n - n_2)^2 - k^2}{(n - n_2)^2 + k^2}} \cong 0.92 - 0.95, \tag{5.6}$$

$$\varphi = \mathrm{arc}tg\left(\frac{2n_2k}{n^2 + k^2 - n_2^2}\right), \tag{5.7}$$

where n and k are real and imaginary part of complex refraction index of aluminum, respectively ($n^* = n - ik$).

Summation of all rays in Eq. (5.4) yields

$$E_\mathrm{a} = \lim_{p\to\infty}\sum_{i=1}^{p+1} E_{\mathrm{a}i} = E_{\mathrm{a}1}\sum_{n=0}^{\infty}\rho^n, \tag{5.8}$$

where

$$\rho = rR\cdot\exp\left[i\left(\pi - \varphi - \frac{4\pi n_2 d}{\lambda}\right)\right]. \tag{5.9}$$

Having in mind that $|\rho| < 1$, it follows that

$$\sum_{n=0}^{\infty}\rho^n = \frac{1}{1-\rho}, \tag{5.10}$$

or, consequently,

$$E_\mathrm{a} = \frac{(1-r)E_\mathrm{o}\exp(i\omega t)}{1 - rR\exp\left[i\left(\pi - \varphi - \frac{4\pi n_2 d}{\lambda}\right)\right]}. \tag{5.11}$$

– Multiple reflections of primal ray $E_{\mathrm{b}1}$ result in formation of rays:

$$E_{\mathrm{b}1} = RE_{\mathrm{a}1}\exp\left[i\left(\pi - \varphi - \frac{2\pi n_2}{\lambda}2x\right)\right]$$

$$E_{\mathrm{b}2} = E_{\mathrm{b}1}(rR)\exp\left[i\left(\pi - \varphi - \frac{4\pi n_2 d}{\lambda}\right)\right]$$

$$E_{\mathrm{b}3} = E_{\mathrm{b}1}(rR)^2\exp\left[2i\left(\pi - \varphi - \frac{4\pi n_2 d}{\lambda}\right)\right]$$

$$\vdots$$

$$E_{\mathrm{b}(p+1)} = E_{\mathrm{b}1}(rR)^p\exp\left[pi\left(\pi - \varphi - \frac{4\pi n_2 d}{\lambda}\right)\right]. \tag{5.12}$$

Upon summation,

$$E_{\mathrm{b}} = \lim_{p\to\infty}\sum_{i=1}^{p+1} E_{\mathrm{b}i} = \frac{(1-r)E_{\mathrm{o}}R\ \exp\left[i\left(\omega t+\pi-\varphi-\frac{4\pi n_2 x}{\lambda}\right)\right]}{1-rR\exp\left[i\left(\pi-\varphi-\frac{4\pi n_2 d}{\lambda}\right)\right]}. \tag{5.13}$$

Total radiation amplitude emitted from the luminescent center is equal to the sum of these two ray families:

$$E_{\mathrm{t}} = E_{\mathrm{a}} + E_{\mathrm{b}} = \frac{(1-r)E_{\mathrm{o}}\exp(i\omega t)\left\{1+R\ \exp\left[i\left(\pi-\varphi-\frac{4\pi n_2 x}{\lambda}\right)\right]\right\}}{1-rR\exp\left[i\left(\pi-\varphi-\frac{4\pi n_2 d}{\lambda}\right)\right]}. \tag{5.14}$$

Corresponding spectral intensity is proportional to the squared radiation amplitude:

$$I_\lambda(x) \cong E_{\mathrm{t}} \cdot E_{\mathrm{t}}^*, \tag{5.15}$$

or

$$I_\lambda(x) \cong \frac{1+R^2-2R\cos\left(\varphi+\frac{4\pi n_2 x}{\lambda}\right)}{1+(rR)^2+2rR\cos\left(\varphi+\frac{4\pi n_2 d}{\lambda}\right)}. \tag{5.16}$$

Total measured intensity of spectral radiation coming from the oxide film can be obtained upon integration of Eq. (5.16):

$$N_\lambda \cong \frac{\int_0^{x_{\mathrm{a}}} I_{\mathrm{o},\lambda}\cdot n_{\mathrm{c}}(x)\left[1+R^2-2R\cos\left(\varphi+\frac{4\pi n_2 x}{\lambda}\right)\right]dx}{1+(rR)^2+2rR\cos\left(\varphi+\frac{4\pi n_2 d}{\lambda}\right)}. \tag{5.17}$$

where x_{a} is the average range of electron avalanches that excite GL centers, $n_{\mathrm{c}}(x)$ is the spatial distribution of GL centers, and $I_{\mathrm{o},\lambda}$ is the initial GL intensity (without interference effect). For the measurement of oxide film thickness only relative intensity of interference maxima as a function of anodization time is taken into consideration. Clearly, Fig. 5.42 shows that interference maxima $N_\lambda(t)$, according to Eq. (5.17), can be associated with the thickness of oxide film d_{m}:

$$\varphi + \frac{4\pi n_2 d_{\mathrm{m}}}{\lambda} = (2m+1)\pi; \quad m = 0, 1, 2\ldots \tag{5.18}$$

For the interference maximum m and $(m + f)$ it can be obtained that

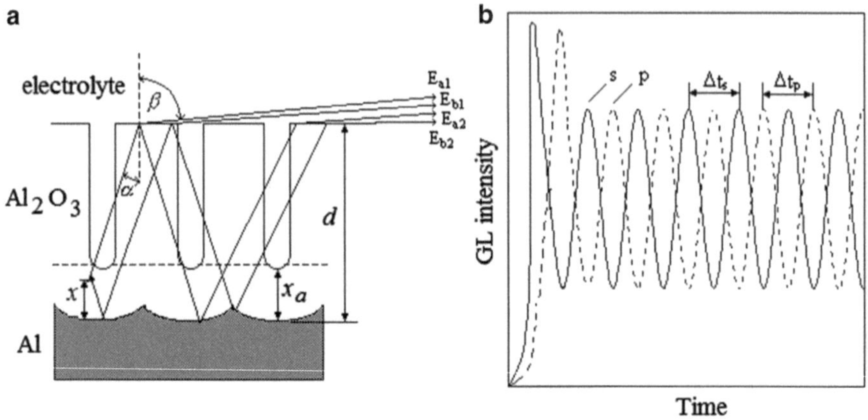

Fig. 5.43 Measurement at angle close to total reflection: (**a**) detected GL emission pathway and (**b**) observed GL intensity

$$d_{\mathrm{m+f}} - d_{\mathrm{m}} = \frac{f\lambda}{2n_2}. \tag{5.19}$$

Taking into consideration the relation between oxide film thickness and anodization time,

$$d = \chi \cdot j \cdot t, \tag{5.20}$$

where χ is the velocity of anodization, and j is the anodization current density. From Eqs. (5.19) and (5.20) the velocity of anodization can be determined:

$$\chi = \frac{\lambda}{2n_2 j} \cdot \frac{f}{t_{\mathrm{m+f}} - t_{\mathrm{m}}} = \frac{\lambda}{2n_2 j \Delta t}. \tag{5.21}$$

Upon determination of the anodization velocity, Eq. (5.20) serves to calculate actual oxide film thickness, for known anodization time t.

Measurement at an Angle Close to Total Reflection

In this method, the system detects only the rays with emission angles α close to the total reflection angle (Fig. 5.43) [86]:

$$n_2 \sin\alpha = n_1 \sin\beta \cong n_1. \tag{5.22}$$

Excited GL center located at distance x from metal/oxide interface emits the ray E_{a1} in the direction of the detection system and the ray E_{b1} in the opposite direction.

Again, multiple reflections on metal/oxide and oxide/electrolyte interfaces occur as a result of high reflectivity of electropolished aluminum surface. Rays following different optical paths interfere and form interference pattern. During multiple reflections, normal and parallel components of the electrical field vector have different phase shifts. On the oxide/electrolyte interface, normal component (s) remains unchanged, because the angle is higher than Brewster's. On the other hand, parallel component (p) shifts its phase for π. On the metal/oxide interface normal component undergoes phase shift of ($\pi - \varphi_s$), while parallel component shifts for ($\pi - \varphi_p$). It is worth mentioning that φ represents the phase shift originating from light absorption on metal surface.

Basic derivation of this method is similar as in the case of measurement under perpendicular angle, but polarized components are treated separately. In the experiment, we select one of the polarized components only by simply placing polarization filter in front of the incoming slit of the detector.

In case of the normal component (s), instead of primal ray E_{a1}, a family of rays is obtained:

$$\begin{aligned}
E_{\mathrm{a1}} &= (1-r_s)E_{\mathrm{o}}\exp(i\omega t) \\
E_{\mathrm{a2}} &= (1-r_s)(r_sR_s)E_{\mathrm{o}}\exp[i(\omega t+\pi-\varphi_s-K_sd)] \\
&= r_sR_sE_{\mathrm{a1}}\exp[i(\pi-\varphi_s-K_sd)] \\
E_{\mathrm{a3}} &= (1-r_s)(r_sR_s)^2E_{\mathrm{o}}\exp[i(\omega t+2(\pi-\varphi_s)-2K_sd)] \\
&= (r_sR_s)^2E_{\mathrm{a1}}\exp[2i(\pi-\varphi_s-K_sd)] \\
&\vdots \\
E_{\mathrm{a}(p+1)} &= (1-r_s)(r_sR_s)^pE_{\mathrm{o}}\exp[i(\omega t+p(\pi-\varphi_s)-pK_sd)] \\
&= (r_sR_s)^pE_{\mathrm{a1}}\exp[pi(\pi-\varphi_s-K_sd)].
\end{aligned} \tag{5.23}$$

Summation of these rays yields

$$E_{\mathrm{a}} = \lim_{p\to\infty}\sum_{i=1}^{p+1}E_{\mathrm{a}i} = E_{\mathrm{a1}}\sum_{n=0}^{\infty}\rho^n, \tag{5.24}$$

where

$$\rho = r_sR_s\cdot\exp[i(\pi-\varphi_s-K_sd)]. \tag{5.25}$$

Again, by applying $|\rho| < 1$, from Eq. (5.10) is obtained:

$$E_{\mathrm{a}} = \frac{(1-r_s)E_{\mathrm{o}}\exp(i\omega t)}{1-r_sR_s\exp[i(\pi-\varphi_s-K_sd)]}. \tag{5.26}$$

Instead of primal ray E_{b1}, under multiple reflections condition, a family of rays is obtained:

$$\begin{aligned}
E_{b1} &= R_s E_{a1} \exp[i(\pi - \varphi_s - K_s x)] \\
E_{b2} &= E_{b1}(r_s R_s)\exp[i(\pi - \varphi_s - K_s d)] \\
E_{b3} &= E_{b1}(r_s R_s)^2 \exp[2i(\pi - \varphi_s - K_s d)] \\
&\vdots \\
E_{b(p+1)} &= E_{b1}(r_s R_s)^p \exp[pi(\pi - \varphi_s - K_s d)].
\end{aligned} \tag{5.27}$$

Summation yields

$$E_b = \lim_{p\to\infty} \sum_{i=1}^{p+1} E_{bi} = \frac{(1 - r_s)E_o R_s \exp[i(\omega t + \pi - \varphi_s - K_s x)]}{1 - r_s R_s \exp[i(\pi - \varphi_s - K_s d)]}. \tag{5.28}$$

Total amplitude of radiation emitted from GL center is

$$E_t = E_a + E_b = \frac{(1 - r_s)E_o \exp(i\omega t)\{1 + R_s \exp[i(\pi - \varphi_s - K_s x)]\}}{1 - r_s R_s \exp[i(\pi - \varphi_s - K_s d)]}. \tag{5.29}$$

Corresponding spectral intensity is proportional to this amplitude squared:

$$I_\lambda(x) \cong E_t \cdot E_t^*, \tag{5.30}$$

or

$$I_{s\lambda}(x) \cong \frac{1 + R_s^2 - 2R_s \cos(\varphi_s + K_s x)}{1 + (r_s R_s)^2 + 2r_s R_s \cos(\varphi_s + K_s d)}. \tag{5.31}$$

Similar analysis gives the expression for spectral intensity of parallel component (p):

$$I_{p\lambda}(x) \cong \frac{1 + R_p^2 - 2R_p \cos(\varphi_p + K_p x)}{1 + (r_p R_p)^2 + 2r_p R_p \cos(\pi + \varphi_p + K_p d)}, \tag{5.32}$$

where

$$K_s = \frac{4\pi}{\lambda}\sqrt{n_{2s}^2 - n_1^2}, \tag{5.33}$$

$$K_p = \frac{4\pi}{\lambda}\sqrt{n_{2p}^2 - n_1^2}. \tag{5.34}$$

Reflection coefficients r_s, r_p, R_s, and R_p and phase shifts φ_s and φ_p can be calculated from Fresnel's formulae:

$$r_s = \frac{n_2\cos(\alpha) - n_1\cos(\beta)}{n_2\cos(\alpha) + n_1\cos(\beta)} = -\frac{\sin(\alpha-\beta)}{\sin(\alpha+\beta)}, \tag{5.35}$$

$$r_p = \frac{n_2\cos(\beta) - n_1\cos(\alpha)}{n_2\cos(\beta) + n_1\cos(\alpha)} = -\frac{tg(\alpha-\beta)}{tg(\alpha+\beta)}, \tag{5.36}$$

$$R_s^2 = \frac{a^2 + b^2 - 2a\cos(\alpha) + \cos^2(\alpha)}{a^2 + b^2 + 2a\cos(\alpha) + \cos^2(\alpha)}, \tag{5.37}$$

$$R_p^2 = R_s^2\frac{a^2 + b^2 - 2a\sin(\alpha)tg(\alpha) + \sin^2(\alpha)tg^2(\alpha)}{a^2 + b^2 + 2a\sin(\alpha)tg(\alpha) + \sin^2(\alpha)tg^2(\alpha)}, \tag{5.38}$$

$$\varphi_s = \mathrm{arc}tg\left[\frac{2b\cos(\alpha)}{a^2 + b^2 - \cos^2(\alpha)}\right], \tag{5.39}$$

$$\varphi_p = \mathrm{arc}tg\left\{\frac{2n_2^2\cos(\alpha)\left[2ank - b\left(n^2 - k^2\right)\right]}{\left(n^2 + k^2\right)^2\cos^2(\alpha) - \left(a^2 + b^2\right)n_2^4}\right\}. \tag{5.40}$$

Parameters a and b are

$$a^2 = \frac{1}{2n_2}\left\{\sqrt{\left[n^2 - k^2 - n_2^2\sin^2(\alpha)\right]^2 + 4n^2k^2} + \left(n^2 - k^2 - n_2^2\sin^2(\alpha)\right)\right\},$$

$$b^2 = \frac{1}{2n_2}\left\{\sqrt{\left[n^2 - k^2 - n_2^2\sin^2(\alpha)\right]^2 + 4n^2k^2} - \left(n^2 - k^2 - n_2^2\sin^2(\alpha)\right)\right\}.$$

Total measured intensity of spectral radiation emitted from the whole oxide film can be obtained by integration of Eqs. (5.31) and (5.32):

$$N_{\lambda,s} \cong \frac{\int_0^{x_a} I_{o,\lambda}n_c(x)\left[1 + R_s^2 - 2R_s\cos(\varphi + K_s x)\right]dx}{1 + (r_sR_s)^2 + 2r_sR_s\cos(\varphi + K_s d)}, \tag{5.41}$$

$$N_{\lambda,p} \cong \frac{\int_0^{x_a} I_{o,\lambda}n_c(x)\left[1 + R_p^2 - 2R_p\cos\left(\varphi + K_p x\right)\right]dx}{1 + \left(r_pR_p\right)^2 + 2r_pR_p\cos\left(\pi + \varphi + K_p d\right)}. \tag{5.42}$$

Regardless of the spatial distribution of luminescent centers, maximal intensity for s component can be calculated from the condition:

$$K_s d_m + \varphi_s = (2m+1)\pi; \quad m = 0, 1, 2 \ldots \tag{5.43}$$

And for the p component from the condition:

$$K_p d_m + \varphi_p + \pi = (2m+1)\pi; \quad m = 0, 1, 2 \ldots \tag{5.44}$$

Utilizing the similar procedure as in the case of measurement at right angle, velocity of anodization can be calculated as

$$\chi = \frac{\lambda}{2j\sqrt{n_{2s}^2 - n_1^2} \cdot \Delta t_s} = \frac{\lambda}{2j\sqrt{n_{2p}^2 - n_1^2} \cdot \Delta t_p}. \tag{5.45}$$

For isotropic oxide films $\Delta t_s = \Delta t_p = \Delta t$, or $n_{2s} = n_{2p} = n_2$ and $K_s = K_p$. Therefore, the velocity of anodization for isotropic oxide films is

$$\chi = \frac{\lambda}{2j\sqrt{n_2^2 - n_1^2} \cdot \Delta t}. \tag{5.46}$$

Overall Equations

By measuring $I_\lambda(t)$ for both geometries (A and B), on the same selected wavelength λ, and for the same anodization current j, Eqs. (5.20), (5.21), and (5.46) yield

$$n_{2,\lambda} = \frac{\Delta t_{\mathrm{B}}}{\sqrt{\Delta t_{\mathrm{B}}^2 - \Delta t_{\mathrm{A}}^2}} \cdot n_1, \tag{5.47}$$

$$\chi = \frac{\lambda \cdot \sqrt{\Delta t_{\mathrm{B}}^2 - \Delta t_{\mathrm{A}}^2}}{2n_1 j(\Delta t_{\mathrm{A}} \cdot \Delta t_{\mathrm{B}})}, \tag{5.48}$$

$$d = \frac{\lambda \cdot \sqrt{\Delta t_{\mathrm{B}}^2 - \Delta t_{\mathrm{A}}^2}}{2n_1(\Delta t_{\mathrm{A}} \cdot \Delta t_{\mathrm{B}})} \cdot t. \tag{5.49}$$

5.3.5.2 Investigation of Barrier Anodic Oxide Films

Measurement at Right Angle

Figure 5.44 shows emitted GL radiation ray pathways (a) and measured GL intensity (b) for constant anodization current and for changing wavelength. Overall GL emission ray E_t is a sum of direct ray E_{a} and reflected ray. If one assumes that multiple reflections can be neglected [83], then

$$E_t = E_{\mathrm{a}} + E_{\mathrm{b}}, \tag{5.50}$$

or, equivalently,

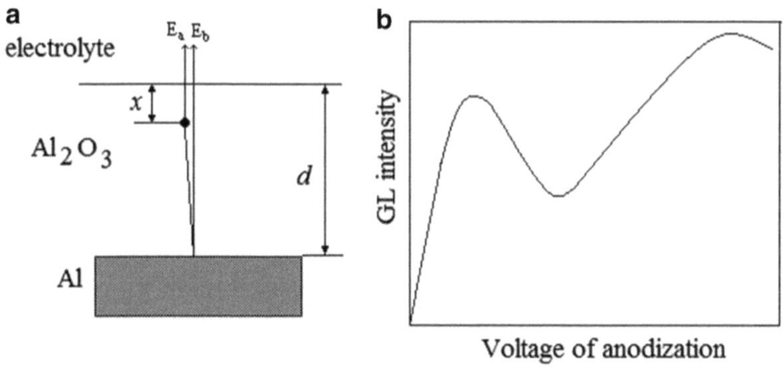

Fig. 5.44 Measurement at right angle: (**a**) detected GL emission pathway and (**b**) observed GL intensity

$$E\exp[i(\omega t + \delta)] = E_o\exp(i\,\omega t) + RE_o\exp[i(\omega t + \pi - \varphi - \phi)]. \qquad (5.51)$$

E_o is the amplitude of electric field vector of the directly emitted electromagnetic radiation, E is the resultant amplitude of the electric field vector, φ is the phase shift related to the absorption on aluminum surface, and ϕ is the phase shift related to the difference in traveling paths of the direct and reflected ray. Phase shift ϕ equals

$$\phi = \frac{4\pi n_2(d - x)}{\lambda}, \qquad (5.52)$$

with values for R and φ given by Fresnel's formulae (5.6) and (5.7). Using the Eq. (5.51) one can obtain

$$E^2 = E_o^2\left[1 + R^2 - 2R\cos\left(\phi + \varphi\right)\right]. \qquad (5.53)$$

Corresponding spectral intensity is proportional to the amplitude squared:

$$I_\lambda(x) \cong E \cdot E^*. \qquad (5.54)$$

Measured GL intensity emitted from GL center on distance x from the oxide/electrolyte interface is

$$I_\lambda(x) = I_{o,\lambda}\left\{1 + R^2 - 2R\cos\left[\frac{4\pi n(d - x)}{\lambda}\right]\right\}, \qquad (5.55)$$

where $I_{o,\lambda}$ represents GL intensity corrected with respect to the interference of emitted rays, i.e., it corresponds to the $R = 0$ ($E_b = 0$, $E_t = E_a$) case. Total measured intensity of spectral radiation from the oxide film can be obtained by simple integration of Eq. (5.55):

$$N_\lambda = \int_0^d I_{\mathrm{o},\lambda}\left\{1 + R^2 - 2R\cos\left[\varphi + \frac{4\pi n(d-x)}{\lambda}\right]\right\}dx. \tag{5.56}$$

In order to calculate spectral intensity, spatial distribution of electrons forming electron avalanche, spatial distribution of GL centers, electronic energy distribution, and excitation probabilities have to be available. It is difficult (if possible) to calculate all these parameters, but good approximation of such system can be obtained under the assumption that spatial distributions of electronic avalanches along mean free path a_v and GL centers throughout the whole oxide film d are uniform and that all electrons in the avalanche have the same probability to excite GL centers. Under these assumptions:

$$I_\lambda(x) = \begin{cases} I_{o,\lambda} & x \in (0, a_v) \\ 0 & x \in (a_v, d) \end{cases}. \tag{5.57}$$

For $d < a_v$, from Eq. (5.56) can be obtained:

$$\begin{aligned} N_\lambda &\cong I_\lambda^{\mathrm{o}}\int_0^d \left\{1 + R^2 - 2R\cos\left[\varphi + \frac{4\pi n_2(d-x)}{\lambda}\right]\right\}dx \\ &= I_\lambda^{\mathrm{o}}\left\{(1+R^2)x + 2R\frac{\lambda}{4\pi n_2}\sin\left[\varphi + \frac{4\pi n_2(d-x)}{\lambda}\right]\right\}_0^d \\ &= I_\lambda^{\mathrm{o}}\left[(1+R^2)d + \frac{R\lambda}{2\pi n_2}\sin(\varphi) - \frac{R\lambda}{2\pi n_2}\sin\left(\varphi + \frac{4\pi n_2 d}{\lambda}\right)\right] \end{aligned} \tag{5.58}$$

For $d > a_v$, from Eq. (5.56) can be obtained:

$$\begin{aligned} N_\lambda &\cong I_\lambda^{\mathrm{o}}\int_0^{a_v} \left\{1 + R^2 - 2R\cos\left[\varphi + \frac{4\pi n_2(d-x)}{\lambda}\right]\right\}dx \\ &= I_\lambda^{\mathrm{o}}\left\{(1+R^2)x + 2R\frac{\lambda}{4\pi n_2}\sin\left[\varphi + \frac{4\pi n_2(d-x)}{\lambda}\right]\right\}_0^{a_v} \\ &= I_\lambda^{\mathrm{o}}\left\{(1+R^2)a_v + \frac{R\lambda}{2\pi n_2}\left[\sin\left(\varphi + \frac{4\pi n_2(d-a_v)}{\lambda}\right) - \sin\left(\varphi + \frac{4\pi n_2 d}{\lambda}\right)\right]\right\} \\ &= I_\lambda^{\mathrm{o}}\left[(1+R^2)a_v - \frac{R\lambda}{\pi n_2}\sin\left(\frac{2\pi n_2}{\lambda}a_v\right)\cdot\cos\left(\frac{4\pi n_2 d}{\lambda} + \varphi - \frac{2\pi n_2 a_v}{\lambda}\right)\right]. \end{aligned} \tag{5.59}$$

From Eqs. (5.58) and (5.59) oxide films' thickness d, velocity of anodization χ, average range of electronic avalanche a_v, and shape of GL spectra corrected with respect to interference $I_{\mathrm{o},\lambda}$ can be calculated. Positions of maxima and minima of the curve $N_\lambda = f(d)$ are obtained by differentiation of Eq. (5.59):

$$\frac{4\pi n_2 d}{\lambda} + \varphi - \frac{2\pi n_2 a_v}{\lambda} = p \cdot \pi \quad p = 0, 1, 2, 3 \ldots \tag{5.60}$$

Further analysis (taking into consideration sign of the second derivative) shows that for

$$a_v < \frac{\lambda}{2n_2}, \tag{5.61}$$

$$\sin\left(\frac{2\pi n_2 a_v}{\lambda}\right) < 0. \tag{5.62}$$

Minima are obtained for $p = 0, 2, 4, \ldots$, while maxima are obtained for $p = 1, 3, 5, \ldots$ For $p = 1$ and $p = 3$, from Eq. (5.60) can be obtained that

$$\frac{4\pi n_2 d_1}{\lambda} + \varphi - \frac{2\pi n_2 a_v}{\lambda} = \pi, \tag{5.63}$$

$$\frac{4\pi n_2 d_2}{\lambda} + \varphi - \frac{2\pi n_2 a_v}{\lambda} = 3\pi, \tag{5.64}$$

where oxide film thickness d_1 and d_2 correspond to first and second maximum of the $N_\lambda = f(d)$ curve. From Eqs. (5.63) and (5.64) $d_2 - d_1$ can be calculated:

$$d_2 - d_1 = \frac{\lambda}{2n_2}. \tag{5.65}$$

Taking into consideration an experimentally derived linear relationship between oxide film thickness d and voltage of anodization U:

$$d = \chi \cdot U, \tag{5.66}$$

velocity of anodization can be calculated as

$$\chi = \frac{\lambda}{2n_2(U_2 - U_1)}, \tag{5.67}$$

where U_1 and U_2 are voltages of anodization corresponding to the first and second maximum, respectively. Oxide film thickness d obtained by anodization up to voltage of anodization U can be calculated from Eq. (5.66), upon determining the velocity of anodization χ from the positions of successive maxima in the $N_\lambda = f(d)$ curve.

Average range of electronic avalanche a_v can be determined from the position of maxima using the following equation:

$$a_v = 2\chi U_1 - \frac{(\pi - \varphi)\lambda}{2\pi n_2}. \tag{5.68}$$

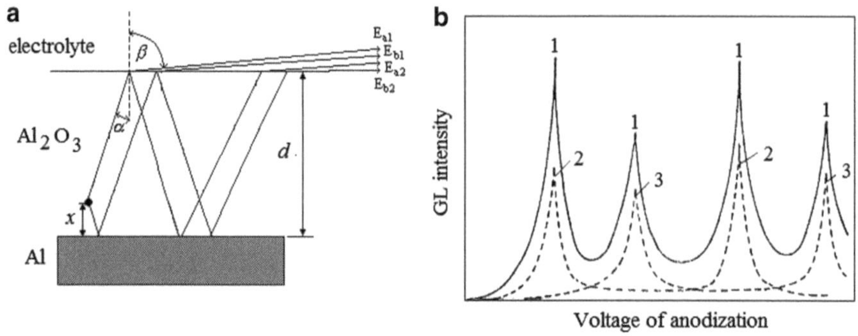

Fig. 5.45 Measurement at angle close to total reflection: (**a**) detected GL emission pathway; (**b**) observed GL intensity: (1) measurements without separation of *s* and *p* polarized components; (2) only the s component; (3) only the p component

Measurement at an Angle Close to Total Reflection

Figure 5.45 shows emitted GL radiation (a) and measured GL intensity for constant anodization current density (b). By applying similar derivation procedure as in the case of porous anodic oxide films [85], position of maxima for normal (*s*) and parallel (*p*) component can be obtained from conditions:

$$Kd_m + \varphi_s = (2m + 1)\pi; \quad m = 0, 1, 2 \ldots \tag{5.69}$$

$$Kd_m + \varphi_p = (2m + 2)\pi; \quad m = 0, 1, 2 \ldots \tag{5.70}$$

If d_1 and d_3 are oxide film thicknesses corresponding to first and third maximum, respectively (Fig. 5.45), then for $m = 0$ and $m = 1$ in Eq. (5.69) one obtains

$$Kd_1 + \varphi_s = \pi, \tag{5.71}$$

$$Kd_3 + \varphi_s = 3\pi, \tag{5.72}$$

or

$$K(d_3 - d_1) = 2\pi. \tag{5.73}$$

Having in mind that parameter K is defined by Eq. (5.33), velocity of anodization χ and oxide film thickness d can be calculated from

$$\chi = \frac{\lambda}{2(U_3 - U_1)\sqrt{n_2^2 - n_1^2}}, \tag{5.74}$$

$$d = \frac{\lambda U}{2(U_3 - U_1)\sqrt{n_2^2 - n_1^2}}, \tag{5.75}$$

where U is the voltage of formation of the oxide film.

Starting from Eq. (5.75) dependence of refraction index of oxide film on wavelength can be calculated as

$$n_2 = \sqrt{n_1^2 + \frac{\lambda^2}{4\chi^2(U_3 - U_1)^2}} = \sqrt{n_1^2 + \frac{A^2(\lambda)}{\chi^2}}, \tag{5.76}$$

where parameter $A(\lambda)$ is equal to

$$A(\lambda) = \frac{\lambda}{2(U_3 - U_1)}. \tag{5.77}$$

At the same time, phase shifts φ_s and φ_p are given by equations:

$$\varphi_s = \pi - Kd_1, \tag{5.78}$$

$$\varphi_p = \pi - Kd_3, \tag{5.79}$$

or, equivalently,

$$\varphi_s = \pi\frac{U_3 - 3U_1}{U_3 - U_1} = \pi\left(1 - \frac{4AU_1}{\lambda}\right), \tag{5.80}$$

$$\varphi_p = 2\pi\frac{U_3 - U_1 - U_2}{U_3 - U_1} = 2\pi\left(1 - \frac{2AU_2}{\lambda}\right). \tag{5.81}$$

Optical constants of aluminum (n and k) can be determined from measured values for phase shifts φ_s and φ_p using graphical method that is based on plotting of $\varphi_s = f(n, k)$ and $\varphi_p = g(n, k)$ [calculated from Fresnel's formulae (5.39) and (5.40)], and then for known values of φ_s and φ_p corresponding pair of n and k can be determined.

5.4 Luminescence of Anodic Oxide Films on Aluminum During the Dielectric Breakdown

Anodization of aluminum in electrolytes that form barrier oxide films on voltages higher than dielectric breakdown voltage is associated with the appearance of sparks (microdischarges) over the whole oxide surface and intensive gas evolution [115, 116]. This phenomenon is often termed plasma electrolytical oxidation (PEO) or micro-arc oxidation (MAO). Understanding microdischarge parameters

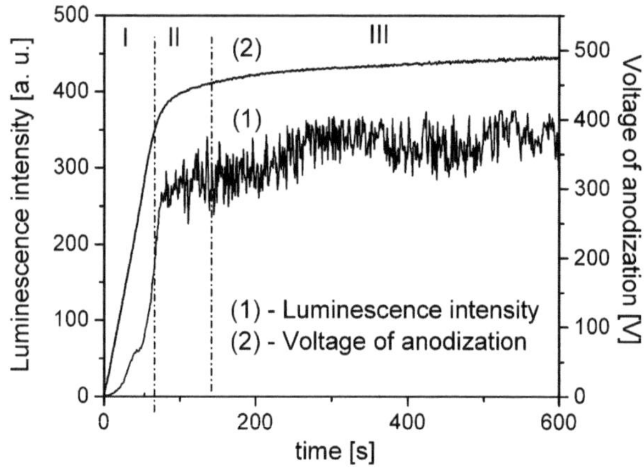

Fig. 5.46 Typical voltage–time and luminescence intensity–time dependence during anodization of aluminum in 0.01 M sodium tungstate ($j = 15$ mA/cm^2, $t_{el} = 20$ °C)

(temperature of plasma in microdischarges, electronic density of plasma, spatial density of microdischarges, active surface covered by microdischarges, and dimensional distribution of microdischarges in various stages of PEO process) is very important for the characterization of PEO process. Distribution and microdischarge type have significant influence on formation mechanism, chemical composition, morphology, and other characteristics of obtained oxide films. There are several available methods used to investigate microdischarges: real-time microdischarge imaging, measuring electrical parameters during the process, optical emission spectroscopy (OES), and surface characterization. Given the liquid environment OES is the best available technique for PEO plasma characterization. The most popular application of OES for PEO diagnostics is spectra characterization and observation of temporal evolution of spectral lines in the visible and near UV spectral region [117–125]. The main difficulty in an application of OES for PEO characterization comes from space and time inhomogeneity of microdischarges appearing randomly across the anode surface. Results presented in this chapter are obtained by PEO process on aluminum in citric acid and in sodium tungstate.

5.4.1 Time Variation of Voltage and Luminescence Intensity During Anodization

Figure 5.46 shows typical voltage vs. time and luminescence intensity vs. time characteristics during anodization of aluminum samples in 0.01 M sodium tungstate at current density of 15 mA/cm^2. Depending on the increasing trend of the voltage and luminescence intensity during anodization, three regions can be clearly identified. From the beginning on anodization, the voltage increases approximately linearly with time to about 380 V with average slope of 6 V/s, resulting in the constant rate of increase of the oxide film thickness (stage I). Simultaneously, low anodic luminescence (galvanoluminescence—GL) is observed. This stage of

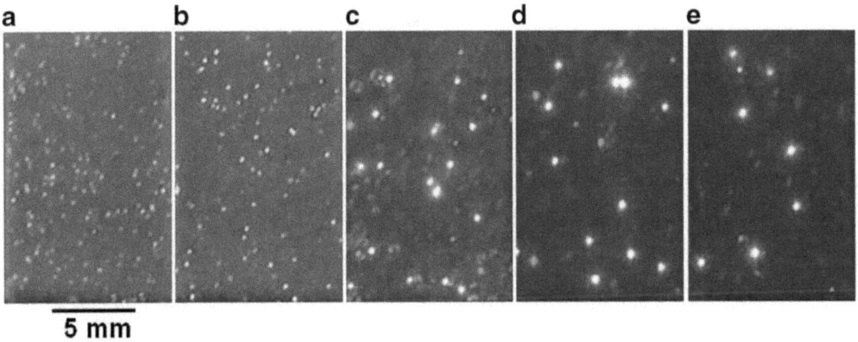

Fig. 5.47 Appearance of microdischarges at various stages of PEO process: (**a**) 2 min; (**b**) 5 min; (**c**) 15 min; (**d**) 30 min; (**e**) 45 min

anodization is followed by apparent deflection from linearity in voltage–time curve, starting from so-called breakdown voltage (stage II). After the breakdown, voltage continually increases, but the voltage–time slope decreases and a large number of small size microdischarges appear, evenly distributed over the whole sample surface. Sparking luminescence combines with the anodic luminescence and, as a result, the total luminescence intensity increases. Further anodization results in relatively stable value of the voltage of anodization (stage III).

In the anodization process total current density is the sum of ionic current density and of electron current density [126]. In the stage I, the electric field strength for a given current density remains constant during the anodic growth and the ionic current is two to three orders of magnitude larger than the electronic component. During anodization electrons are injected into the conduction band of the anodic oxide and accelerated by the electric field producing avalanches by an impact ionization mechanism [126]. When the avalanche electronic current reaches critical value the breakdown occurs [127]. In stage II a relatively low voltage is required to maintain the same total current density (compared with the stage I), due to the independence of electron current density with anodic oxide film thickness. In stage III, the fraction of electron current density in total current density becomes the dominant one. In this stage, the total current density is almost independent of the anodic oxide film thickness and the voltage–time slope is close to zero.

Figure 5.47 shows the appearance of microdischarges at various stages of PEO process. Microdischarges are visible after about 1.5 min. The size of microdischarges becomes larger, while spatial density of microdischarges becomes lower, with increasing time of PEO. It can be seen that relatively small microdischarges, with average cross-sectional area ~0.03 mm^2, are dominant in the early stage of PEO process. The population of small microdischarges decreases during PEO. On the other hand, large microdischarges become noticeable with extended PEO time and after about 30 min large microdischarges with average cross-sectional area ~0.35 mm^2 are dominant. In view of the fact that microdischarges are generated by dielectric breakdown through weak sites in the oxide film, the number of weak sites is reduced with increasing time

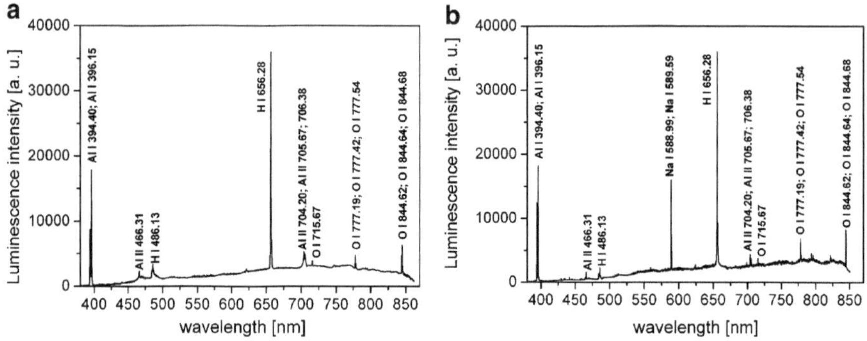

Fig. 5.48 Optical emission spectrum of PEO process on aluminum in (**a**) 0.01 M citric acid; (**b**) 0.01 M sodium tungstate [123] (Reproduced by permission from Elsevier No. 3163530232265)

of anodization, i.e., with increasing thickness of the oxide films. The increased size of microdischarges with increasing time of PEO is ascribed to reduced number of discharging sites through which higher anodic current is able to pass [128].

5.4.2 *OES of PEO Process on Aluminum*

Typical optical emission spectra of microdischarges during PEO process of aluminum in citric acid and sodium tungstate are presented in Fig. 5.48. The species that are identified originate either from aluminum electrode or from the electrolyte. Two strong emission lines at 394.40 and 396.15 nm belong to Al I. Also, four lines of Al II at 466.31, 704.20, 705.67, and 706.38 nm are observed. The spectral lines of oxygen, hydrogen, tungsten, and sodium from electrolyte are observed, too. The strongest lines belong to Na I at 588.99 and 589.59 nm, Balmer lines H_α (656.27 nm) and H_β (486.13 nm), and three lines of O I at 844.62, 844.64, and 844.68 nm. Also, line of O I at 715.67 nm, three lines at 777.19, 777.42, and 777.54 nm, and many lines of O II and W I are detected. In aforementioned results notations I and II refer to neutral atoms and the single ionized atoms, respectively. Apart from atomic and ionic lines, strong AlO band at 484.23 nm and broad AlO bands in the range between 500 and 556 nm are presented, corresponding to the transition of the AlO from the excited state $B^2\Sigma^+$ into ground state $X^2\Sigma^+$ [129]. The continuum emission between 380 and 850 nm results from collision—radiative recombination of electrons [101] and bremsstrahlung radiation [117].

Determination of PEO electron number density (N_e) is possible using broadened profiles of the hydrogen lines H_α and H_β (Fig. 5.49) and the Al II 704.2 nm line (Fig. 5.50). Analysis of the H_α line shape in Fig. 5.49a shows that this line can be properly fitted only if two Lorentz profiles having the Full Width at Half Maximum (FWHM) of 0.1 and 0.6 nm are applied. These Lorentz profiles correspond to

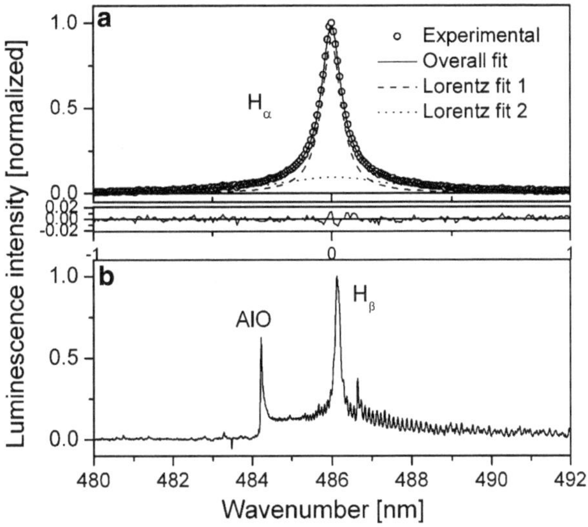

Fig. 5.49 (**a**) The H_α line experimental profile fitted with two Lorentzians and residue plot; (**b**) The H_β line profile [123] (Reproduced by permission from Elsevier No. 3163530232265)

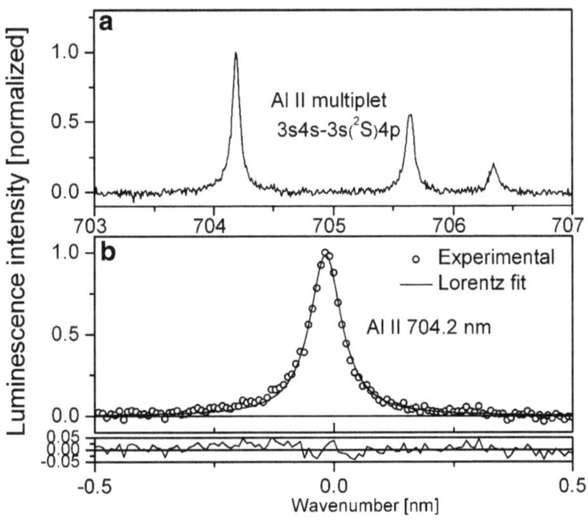

Fig. 5.50 (**a**) The recording of Al II 3s4s-3s(^{2}S)4p multiplet 3; (**b**) Experimental profile and best fit of the Al II 704.2 nm line [123] (Reproduced by permission from Elsevier No. 3163530232265)

electron number density $N_e = 0.7 \times 10^{16}$ cm^{-3} and $N_e = 1.0 \times 10^{17}$ cm^{-3}, respectively [130]. Although the fit quality is exceptionally good, these results are not considered confident. The H_α line is very strong in PEO and self-absorption may broaden the line considerably, while the line shape may be still well fitted with Lorentz profiles [131]. The presence of the H_α self-absorption can be proven after analysis of the H_β line profile (Fig. 5.49b). In spite of interference with AlO molecular band, the upper part of the profile allows the estimation of FWHM ~0.17 nm which corresponds to $N_e \approx 0.8 \times 10^{15}$ cm^{-3} (estimated error ± 20 %) [132]. This indicates that the H_α line is probably self-absorbed.

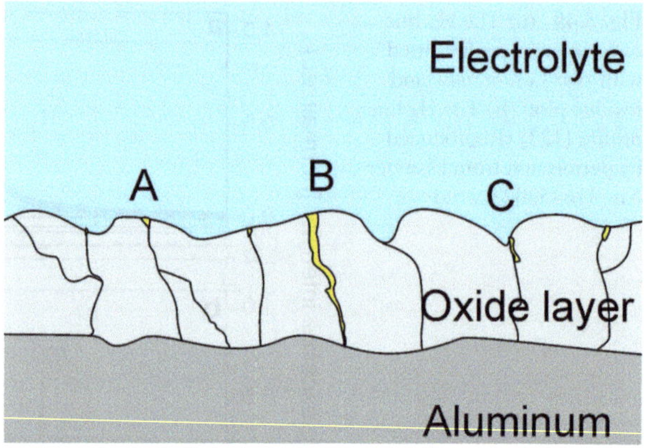

Fig. 5.51 Model of PEO process on aluminum

Analysis of hydrogen lines H_α and H_β during PEO process on tantalum [133], titanium [134], and magnesium [124] shows that H_α line also can be properly fitted only with two Lorentz profiles. FWHM of these Lorentz profiles correspond to similar values for the electron number density as for aluminum. However, H_β lines of these metals do not overlap with the AlO band. Analysis of the H_β line profile shows that this line can be properly fitted only by using two Lorentz profiles with FWHMs corresponding to the electron number density of $N_e \cong 0.9 \times 10^{15}$ cm^{-3} and $N_e \cong 2.2 \times 10^{16}$ cm^{-3}. This shows that H_α line is self-absorbed during the PEO process and cannot be used to determine the electron number density.

The electron number density can be calculated by using the shape of Al II line at 704.2 nm, belonging to multiplet no. 3 (3s4s-3s(^{2}S)4p transition) (Fig. 5.50). FWHM of this line corresponds to electron number density of $N_e \cong 6.0 \times 10^{16}$ cm^{-1} with the estimated error of 20 % [123]. Relative intensities of the Al II lines from multiplets are used to check for the presence of any self-absorption. Ratio of the intensity of the strongest and the medium lines is equal to the theoretical relative intensities 5:3, which confirms that there is no self-absorption during the PEO process.

Three different electron number densities obtained by analyzing hydrogen H_α and H_β lines and Al II line indicate three different types of microdischarges. Hussein et al. have proposed three types of plasma discharge models (Fig. 5.51): metal/oxide interface discharge type (B) and oxide/electrolyte interface discharge types within the coating upper layer (A) and at the coating top layer (C) [121]. Two electron number densities obtained from the hydrogen lines are probably related to the discharge of types A and C, while the electron number density obtained from Al II lines corresponds to discharge type B. Investigation of PEO process of different metals has shown that type of discharge with evaporation of anodic material (type B) always occurs during PEO of aluminum and magnesium (metals with a low melting temperature), regardless of the type of electrolyte [123, 124], whereas in metals with high melting temperature (tantalum, titanium, and zirconium), this type of discharge is strongly dependent on the type of electrolyte [104, 133–135].

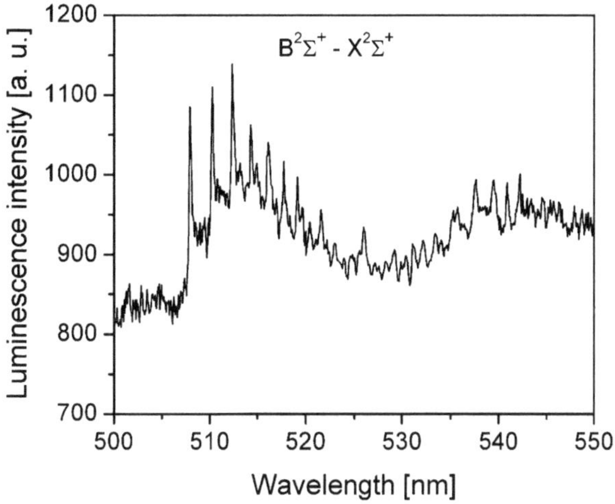

Fig. 5.52 Optical emission spectrum of PEO process on aluminum in the range from 500 to 550 nm. The peaks are assigned to bands of the $B^2\Sigma^+ - X^2\Sigma^+$ system of AlO

Spectral lines used for calculation of electron temperature must belong to the same atomic or ionic species and are emitted in the same ionization stage. Hussein et al. [121] used the relative intensities of Al I at 396.2 and 309.1 nm in order to estimate the electron temperature. Their results pinpoint the electron temperature range from (4,500 ± 450) K and (10,000 ± 1,000) K, with the assumption that it was optically thin plasma in partial thermodynamic equilibrium. The relative intensities of O II and W I lines can also be used to determine the electron temperature using Boltzmann plot technique (BP) [123]. For the application of this assumption one must be sure that energy levels used for BP are populated in accordance with Boltzmann equilibrium distribution, i.e., that the upper energy level of the spectral line used for BP is above the lowest level determined by partial local thermal equilibrium condition. The electron temperature of about 40,000 K is obtained from relative intensity of O II lines and about 3,300 K from relative intensity of W I [123].

Temperature of microdischarge during PEO process on aluminum can be estimated from the $B^2\Sigma^+ - X^2\Sigma^+$ emission transition of AlO molecule [129]. The luminescent spectrum in the wavelength range from 500 to 550 nm is shown in Fig. 5.52. Measured luminescence intensities in combination with quantum–chemical calculations allow calculation of relative populations of vibrational levels of $B^2\Sigma^+$ electronic state and hence the estimation of plasma temperature. Estimated plasma temperature is (8,000 ± 2,000) K [129]. Similar results are obtained from the intensity of vibrational bands $B^1\Sigma^+ - X^1\Sigma^+$ of MgO molecules during the PEO process on magnesium alloy [136, 137].

5.5 Conclusions

This chapter presents the main findings of the research focused on galvanoluminescence during the electrochemical oxidation of aluminum. It is shown that type of oxide films that forms on aluminum surface is electrolyte dependent (organic or inorganic). Carboxylate ions are identified as responsible for appearance of GL in organic electrolytes, while surface pretreatment of samples (surface preparation and annealing) influences GL in inorganic electrolytes. The pretreatment of samples governs the concentration of "flaws" in oxide films, which are related to the GL mechanism. The annealing temperature of samples is another pretreatment factor that affects the GL intensity. Higher annealing temperature results in higher GL intensity. Annealing at different temperatures influences state of the sample's surface, number of defects, crystal grains, and their orientation, in other words, on the concentration of "flaws." GL of oxide films formed by anodization of aluminum samples annealed at temperatures above 500 °C showed that the sudden rise in the formation of gamma-alumina crystalline regions is strongly related to the appearance of GL and its intensity. An analysis on simple molecular species involving the Al atom, as well as those atoms whose presence was possible under given experimental conditions (hydrogen, oxygen, etc.), showed that the sources of GL are molecules AlH, AlO, Al_2, and AlH_2. In the case of organic electrolytes GL is agitated by collision of electrons, injected into the oxide film at the electrolyte–oxide interface and accelerated by high electric field (nearly 10^7 V/cm), with luminescence centers (carboxylate ions) inside the oxide film.

GL of anodic oxide films formed on highly reflective aluminum surfaces in organic electrolytes features clearly pronounced interference maxima that can be used as a tool for determining oxide film thickness and inherent optical parameters. Following this finding, two methods (based on particular observation angle) for determining such properties are developed and presented in this work.

Anodization of aluminum above the breakdown voltage leads to a formation of plasma, as indicated by the presence of sparks on the metal surface, accompanied by gas evolution. Sparking luminescence combines with GL, and as a result, total luminescence intensity increases. Spectroscopy characterization of plasma allows the determination of electron temperature and electron number density.

Acknowledgments The authors would like to express their gratitude to the Ministry of Education, Science, and Technology of the Republic of Serbia for long-term funding of the research presented herein.

References

1. Diggle JW, Downie TC, Goulding CW (1969) Chem Rev 69: 365
2. Norman JE (1977) Corr Sci 17: 39
3. Tajima S (1977) Electrochim Acta 22: 995

4. Li Y, Shimada H, Sakairi M, Shigyo K, Takahashi H, Seo M (1997) J Electrochem Soc 144: 866
5. Yerokhin AL, Nie X, Leyland A, Matthews A, Dowey SJ (1999) Surf Coat Tech 122: 73.
6. Despic A, Parkhutik V (1989) Electrochemistry of Aluminium in Aqueous Solutions and Physics of Its Anodic Oxides, Chapter 6: In Bockris JOM, White RE and Conway BE (eds), Modern Aspects of Electrochemistry, No.22, Plenum Press, New York, p.401
7. Thompson GE, Wood GC (1983) Anodic Films on Aluminium, Chapter 5: In Scully, J C (ed) Treatise on Materials Science and Technology, No. 23, Academic Press Inc.: New York, p. 205
8. Takahashi H, Fujimoto K, Nagayama M (1988) J Electrochem Soc 135: 1349
9. Jackson NF, Campbell DS (1976) Thin Solid Films 36: 331
10. Liang CW, Luo TC, Feng M.S, Cheng HC, Sue D (1996) Mater Chem Phys 43: 166
11. Nielsch K, Choi J, Schwirn K, Wehrspohn RB, Gosele U (2002) Nano Lett 2: 677
12. O'Sullivan JP, Wood GC (1970) P Roy Soc A-Math Phy 317: 511
13. Setoh S, Miyata A (1932) Sci Paper Inst Phys Chem Res 19: 273
14. Wernick S, Pinner R, Sheasby PG (1987) The Surface Treatment and Finishing of Aluminium and its Alloys, Fifth edition, Vol. 1, ASM International Finishing Publications, Metals Park, OH
15. Machkova M, Girginov A, Klein E, Ikonopisov S (1981) Surf Tech 14: 241
16. Machkova M, Klein E, Girginov A, Ikonopisov S (1984) Surf Tech 22: 21
17. Girginov A, Zahariev A, Machkova M (2002) Mater Chem Phys 76: 274
18. Neufeld P, Southall DM (1975) Electrodepos Surf Treat 3: 159
19. Klein IE, Yaniv AE (1971) Metallography 4: 403
20. Altenpohl D (1954) Convention Record, I.R.E. Nat. Conv. 3: 35
21. Franklin RW (1961) Proceedings of the Conference on Anodising Aluminium (Nottingham), (Aluminium Development Assoc., London, 1962) 96
22. Shimizu K, Tajima S, Thompson GE, Wood GC (1981) Electrochim Acta 25: 1481
23. Dorsey Jr GA (1966) J Electrochem Soc 113: 169
24. Franklin R.W. (1957) Nature 180: 1470
25. Altenpohl D, Post W (1961) J Electrochem Soc 108: 628
26. Keller F, Hunter MS, Robinson DL (1953) J Electrochem Soc 100: 411
27. Guntherschultze A, Betz H (1931) Z Phys 68: 145
28. Guntherschultze A, Betz H (1931) Z Phys 71: 106
29. Guntherschultze A, Betz H (1934) Z Phys 91: 70
30. Guntherschultze A, Betz H (1934) Z Phys 92: 367
31. Cabrera N, Mott NF (1948-49) Rep Prog Phys 12: 133
32. Verwey EJ (1935) Physica 2: 1059
33. Young L (1959) Can J Chem 37: 276
34. Sulka G (2008) Highly Ordered Anodic Porous Alumina Formation by Self-Organized Anodizing, in Nanostructured Materials in Electrochemistry, Edited by Ali Eftekhari WILEY-VCH Verlag GmbH & Co. KGaA, Weinheim
35. Kanakala R, Singaraju PV, Venkat R, Das B (2005) J Electrochem Soc 152: J1
36. Hoar TP, Yahalom J (1963) J Electrochem Soc 110: 614
37. Thompson GE (1997) Thin Solid Films 297: 192
38. Parkhutik VP, Shershulsky VI (1992) J Phys D Appl Phys 25: 1258
39. Thamida SK, Chang HC (2002) Chaos 12: 240
40. Masuda H, Fukuda K (1995) Science 268: 1466
41. Li J, Papadopulos C, Xu JM (1999) Appl Phys Lett75: 367
42. Suh JS, Lee JS (1999) Appl Phys Lett75: 2047
43. Pradhan BK, Kyotani T, Tomita A (1999) Chem Commun 14: 1317
44. Gao T, Meng G, Wang Y, Sun S, Zhang L (2002) J Phys-Condens Mat 14: 355
45. Choi J, Sauer G, Nielsch K, Wehrsphn RB, Gosele U (2003) Chem Mater 15: 776
46. Crousea D, Lo YH, Miller AE, Crouse M (2000) Appl Phys Lett 76: 49
47. Masuda H, Satoh M (1996) Jpn J Appl Phys 35: L126

48. Miskulas I, Juodkazis S, Jagminas A, Meskinis S, Dumas JG, Vaitkus J, Tomasiunas R (2001) Opt Mater 17: 343
49. Choi J, Luo Y, Wehrsphn RB, Hillebrand R, Schilling J, Gosele U (2003) J Appl Phys 94: 4757
50. Choi J, Nielsch K, Reiche M, Wehrsphn RB, Gosele U, (2003) J Vac Sci Technol B 21: 763
51. Li AP, Muller F, Gosele U (2000) Electrochem Solid St 3: 131
52. Liu CY, Datta A, Wang YL (2001) Appl Phys Lett 78: 120
53. Sun Z, Kim HK (2002) Appl Phys Lett 81: 3458
54. Shingubara S, Murakami Y, Morimoto K, Takahagi T (2003) Surf Sci 532-535: 317
55. Masuda H, Kanezawa K, Nishio K (2002) Chem Lett 31: 1218
56. Belča ID, Petković M, Stojadinović S, Kasalica B, Belča JS, Zeković LjD (2011) Appl Phys A-Mater 104: 295
57. Braun F (1898) Ann Chim Phys 65: 361
58. Guntherschulze A (1906) Ann Phys 21: 929
59. Anderson S (1943) J Appl Phys 14: 601
60. Lewowski T (1961) Acta Phys Pol 20: 161
61. Lewowski T (1963) Acta Phys Pol **28** 215
62. Van Geel WCh, Bouma BC (1951) Philips Res Rep 6: 401
63. Van Geel WCh, Pistorius CA, Bouma BC (1956) Philips Res Rep 11: 471
64. Van Geel WCh, Pistorius CA, Bouma BC (1957) Philips Res Rep 12: 465
65. Smith AW (1956) Can J Phys 35: 1151
66. Smith AW (1959) Can J Phys 37: 591
67. Ganley WP, Mooney PM, Humnik D (1969) Thin Solid Films 3: 377
68. Ganley WP (1972) Thin Solid Films 11: 91
69. Ikonopisov S, Nankov N (1967) Phys Status Solidi A 23: K61
70. Ikonopisov S (1975) Electrochim Acta 20: 783
71. Ikonopisov S (1975) Electrochim Acta 20: 795
72. Ikonopisov S, Elenkov N, Klein E, Andreeva L (1978) Electrochim Acta 23: 1209
73. Ikonopisov S, Girginov A, Machkova M (1989) Electrochim Acta 34: 631
74. Girginov A, Machkova M, Ikonopisov S (1990) Electrochim Acta 35: 825
75. Tajima S (1977) Electrochim Acta 22: 995
76. Tajima S, Shimizu K, Baba N, Matsuzawa S (1977) Electrochim Acta 22: 845
77. Shimizu K, Tajima S (1977) Electrochim Acta 25: 259
78. Shimizu K, Tajima S, Baba N, Matsuzawa S (1977) Thin Solid Films 41: L35
79. Tajima S, Shimizu K, Baba N, Matsuzawa S (1977) Electrochim Acta 22: 851
80. Shimizu K, (1978) Electrochim Acta 23: 295
81. Shimizu K, Tajima S (1979) Electrochim Acta 24: 309
82. Zeković LjD, Urošević VV, Panić BM (1980) Surf Sci 101: 310
83. Zeković LjD, Urošević VV (1981) Thin Solid Films 78: 279
84. Zeković LjD, Urošević VV (1981) Thin Solid Films 86: 347
85. Zeković LjD, Urošević VV, Jovanić BR (1982) Appl Surf Sci 11/12: 90
86. Zeković LjD, Urošević VV, Jovanić BR, Panić B, Žikić A (1988) Fizika 20: 441
87. Kasalica B, Belča I, Zeković Lj, Jovanic B (1998) Solid State Phenom 61–62: 325
88. Belča I, Kasalica B, Zeković Lj, Jovanić B, Vasilić R (1999) Electrochim Acta 45: 993
89. Belča ID, Zeković LjD, Jovanić B, Ristovski G, Ristovski Lj (2000) Electrochim Acta 45: 4095
90. Stojadinović S, Zeković Lj, Belča I, Kasalica B (2004) Electrochem Commun 6: 427
91. Stojadinović S, Zeković Lj, Belča I, Kasalica B, Nikolić D (2004) Electrochem Commun 6: 708
92. Stojadinović S, Belča I, Zeković Lj, Kasalica B, Nikolić D (2004) Electrochem Commun 6: 1016
93. Kasalica B, Stojadinović S, Zeković Lj, Belča I, Nikolić D (2005) Electrochem Commun 7: 735

94. Stojadinović S, Belča I, Kasalica B, Zeković Lj, Tadić M (2006) Electrochem Commun 8: 1621
95. Stojadinović S, Tadić M, Belča I, Kasalica B, Zeković Lj (2007) Electrochim Acta 52: 7166
96. Stojadinović S, Belča I, Tadić M, Kasalica B, Nedić Z, Zeković Lj (2008) J Electroanal Chem 619/620: 125
97. Stojadinović S, Vasilić R, Petković M, Nedić Z, Kasalica B, Belča I, Zeković Lj (2010) Electrochim Acta 55: 3857
98. Kasalica B, Belca I, Stojadinović S, Sarvan M, Perić M, Zeković Lj (2007) J Phys Chem C 111: 12315
99. Sarvan M, Stojadinović S, Kasalica B, Belča I, Zeković Lj (2008) Electrochim Acta 53: 2183
100. Stojadinović S, Vasilic R, Belča I, Tadić M, Kasalica B, Zeković Lj (2008) Appl Surf Sci 255: 2845
101. Stojadinović S, Vasilic R, Petkovic M, Nedić Z, Kasalica B, Belča I, Zeković Lj (2010) Electrochim Acta 55: 3857
102. Sarvan M, Perić M, Zeković Lj, Stojadinović S, Belča I, Petković M, Kasalica B (2011) Spectrochim Acta A 81: 672
103. Stojadinović S, Vasilić R, Petković M, Belča I, Kasalica B, Perić M, Zeković Lj (2012) Electrochim Acta 59: 354
104. Stojadinović S, Vasilić R, Petković M, Belča I, Kasalica B, Perić M, Zeković Lj (2012) Electrochim Acta 79: 133
105. Stojadinović S, Zeković Lj, Belča I, Kasalica B, Nikolić D (2004) 11th Congress of Physicists of Serbia and Montenegro, Petrovac na Moru 4/179
106. Kasalica B, Belča I, Stojadinović S, Zeković Lj, Nikolić D (2006) Appl Spectrosc 60: 1090
107. Mason RB (1955) J Electrochem Soc 102: 671
108. Charlesby A (1953) Proceedings of the Physical Society LXVI 7-B: 533
109. Fan DH, Ding GQ, Shen WZ, Zheng MJ (2007) Micropor Mesopor Mat 100: 154
110. Yamamoto Y, Baba N (1983) Thin Solid Films 101: 329
111. Gardin YuE, Odynets LL, Tukmakov BC (1970) Elektrohimija 6: 1562
112. Bocchetta P, Sunseri C, Chiavarotti G, Di Quarto F (2003) Electrochim Acta 48: 3175
113. Herzberg G, Molecular Spectra and Molecular Structure III. Electronic Spectra of Polyatomic Molecules, Van Nostrand, New York, 1966
114. Nabatame T, Yasuda T, Nishizawa M, Ikeda M, Horikawa T, Toriumi A (2003) Jpn J Appl Phys 42: 7205
115. Petković M, Stojadinović S, Vasilić R, Belča I, Nedić Z, Kasalica B, Mioč UB (2011) Appl Surf Sci 257: 1995
116. Stojadinović S, Vasilić R, Belča I, Petković M, Kasalica B, Nedić Z, Zeković Lj (2010) Corr Sci 52: 3258
117. Klapkiv M, Nykyforchyn H, Posuvailo V (1994) Mater Sci **30:** 333
118. Mécuson F, Czerwiec T, Belmonte T, Dujardin L, Viola A, Henrion G (2005) Surf Coat Tech 200: 804
119. Dunleavy CS, Golosnoy IO, Curran JA, Clyne TW (2009) Surf Coat Tech 203: 3410
120. Kasalica B, Petković M, Belča I, Stojadinović S, Zeković Lj (2009) Surf Coat Tech 203: 3000
121. Hussein RO, Nie X, Northwood DO, Yerokhin A, Matthews A (2010) J Phys D Appl Phys 43: 105203
122. Hussein RO, Nie X, Northwood DO (2010) Surf Coat Tech 205: 1659
123. Jovović J, Stojadinović S, Šišović NM, Konjević N (2011) Surf Coat Tech 206: 24
124. Jovović J, Stojadinović S, Šišović NM, Konjević N (2012) J Quant Spectrosc Radiat Transf 113: 1928
125. Kasalica B, Stojadinović S, Belča I, Sarvan M, Zeković Lj, Radić-Perić J (2013) J Anal Atom Spectrom 28: 92
126. Albella JM, Montero I, Martinez-Duart JM (1987) Electrochim Acta 32: 255
127. Ikonopisov S (1977) Electrochim Acta 22: 1077
128. Sundararajan G, Rama Krishna L (2003) Surf Coat Tech 167: 269

129. Stojadinović S, Perić M, Petković M, Vasilić R, Kasalica B, Belča I, Radić-Perić J (2011) Electrochim Acta 56: 10122
130. Gigosos MA, Gonzalez MA, Cardenoso V (2003) Spectrochim Acta B 58: 1489
131. Konjević N, Roberts JR (1976) J Phys Chem Ref Data 5: 209
132. Ivković M, Jovićević S, Konjević N (2004) Spectrochim Acta B 59: 591
133. Stojadinović S, Jovović J, Petković M, Vasilić R, Konjević N (2011) Surf Coat Tech 205: 5406
134. Stojadinović S, Vasilić R, Petković M, Zeković Lj (2011) Surf Coat Tech 206: 575
135. Stojadinović S, Vasilić R, Petković M, Kasalica B, Belča I, Žekić A, Zeković Lj (2013) Appl Surf Sci 265: 226
136. Stojadinović S, Perić M, Radić-Perić J, Vasilić R, Petković M, Zeković Lj (2011) Surf Coat Tech 206: 2905
137. Ranković R, Stojadinović S, Sarvan M, Kasalica B, Krmar M, Radić-Perić J, Perić M (2012) J Serb Chem Soc 77: 1483

Chapter 6
Electrochemical Aspects of Chemical Mechanical Polishing

K. Cadien, L. Nolan, H. Pirayesh, K. Dawkins, and Z. Xu

6.1 Introduction

Metal and glass surfaces have been polished to high levels of planarity and tight tolerances for hundreds of years. The telescope of Galileo, for instance, was enabled by accurate glass lens polishing. Chemical Mechanical Polishing, also known as chemical mechanical planarization or CMP, was developed specifically for the semiconductor industry from the same historical principles, beginning in 1983 at IBM [1] with the successful planarization of reflown glass "bulges" on a wafer surface. Although reflown glass was never adopted as a commercial technique, and despite initial reluctance to incorporate the use of small particles into cleanroom technology, the use of CMP has flourished. Today, a typical integrated circuit is polished dozens of times during manufacture; the planarization of each device layer permits new layers to be built upon it, leading to devices with eight or more separate layers of metallization. CMP is thus industrially important as an enabler of new geometries, as well as a manufacturing technique in its own right.

CMP technology has evolved considerably since its inception and is now well established for metals such as tungsten, aluminum, and copper. Techniques for polishing semiconductors such as single crystal and polycrystalline silicon were also developed. Dielectric materials such as silicon dioxide and tantalum nitride have also been thoroughly investigated. Future challenges for CMP have been

K. Cadien (✉) • H. Pirayesh • K. Dawkins • Z. Xu
Department of Chemical and Materials Engineering, University of Alberta,
9107 – 116 Street, Edmonton, AB, Canada T6G 2 V4
e-mail: kcadien@ualberta.ca

L. Nolan
Portland Technology Development, Intel Corporation, Hillsboro, OR, USA

S.S. Djokić (ed.), *Electrodeposition and Surface Finishing: Fundamentals and Applications*, Modern Aspects of Electrochemistry 57,
DOI 10.1007/978-1-4939-0289-7_6, © Springer Science+Business Media New York 2014

identified in the International Technology Roadmap for Semiconductors and they include [2]:

- Increasing within-wafer removal rate uniformity and decreasing dishing for shallow-trench isolation (STI)
- Reducing scratching and other defects
- Scaling to 450 mm wafers

CMP should not be confused with ECMP, or Electrochemical Mechanical Polishing, a separate but related polishing technique in which a voltage is applied to a conducting surface during polishing, making it the working electrode in a three-electrode electrochemical cell. In ECMP, polishing is achieved by electrochemical dissolution of the working electrode rather than by mechanical means. This allows low polishing pressures and very low abrasive particle concentrations to be used, but does not typically achieve the same surface planarity as CMP [3]. This technique is not discussed here.

6.1.1 CMP Tools

The typical CMP tool consists of a carrier which supports the wafer and presses it down against the polishing pad, while both the pad and carrier (and hence wafer) rotate. A polishing slurry, containing water, chemical components, and fine abrasives, is introduced onto the polishing pad so that it flows between the pad and wafer. The pad is usually significantly larger than the wafer and is mounted on a platen that resists the load from the wafer. Polishing pads are polymeric and may have grooves inscribed on the surface to assist in slurry distribution. They may also contain microscopic pores, creating asperities, or peaks, on the pad surface. The asperities gradually wear away with use and the pad becomes "glazed," resulting in a drop in polishing rate; to counteract this, a diamond-grit disc conditioner passes across the pad during polishing to maintain a consistent surface. Schematic representations of these components are shown in Figs. 6.1 and 6.2.

Basic CMP tools used for research have been assembled from these components by using, for example, a modified drill press as the wafer carrier and a metallurgical polishing wheel as the platen. More sophisticated research-grade CMP tools have these components preassembled and may additionally contain vacuum chucks to grip the wafer, and some robotic parts to automate wafer loading and unloading. Specialist metrology instruments may also be installed to monitor friction, pad temperature, acoustic emissions, and other process parameters. Industrial CMP tools are self-contained "dry-in, dry-out" units which process cassettes of wafers on multiple platens and incorporate wafer cleaning and drying. These typically also contain dedicated metrology instruments. Regardless of tool type, the user controls the process by selecting the polishing pad and slurry, the rate at which slurry flows on to the pad, the speed of wafer and platen rotation, and the pressure applied to the system.

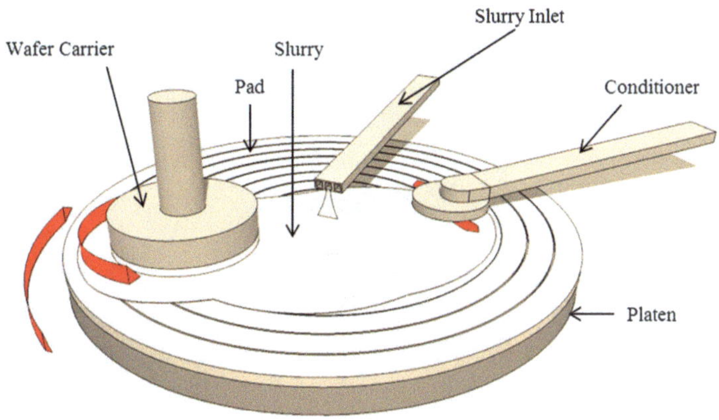

Fig. 6.1 Schematic diagram showing the typical polishing components

Fig. 6.2 Cross section through wafer carrier, showing position of wafer

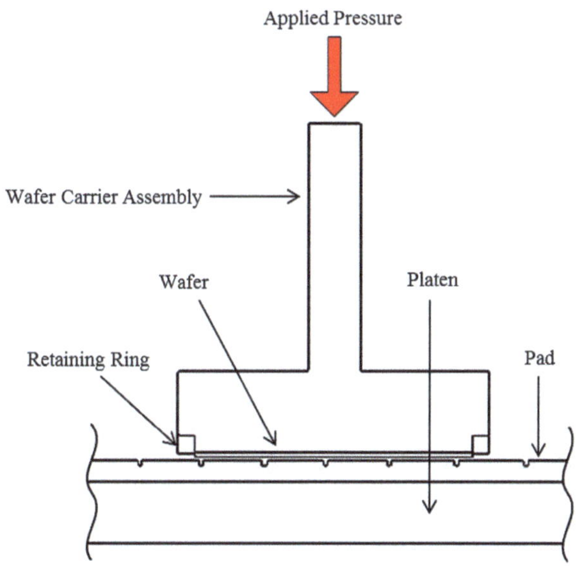

Unlike many other nanofabrication techniques, CMP is carried out in ambient conditions and does not require special conditions beyond a clean environment to prevent unwanted particulates from contaminating the process.

The polishing slurry, polishing pad, and conditioning disk are consumed during processing and must be periodically replaced. These components determine polishing outcomes and make up a significant portion of the tools' cost of ownership, so they should be selected carefully. The typical features and components of these consumables are described below.

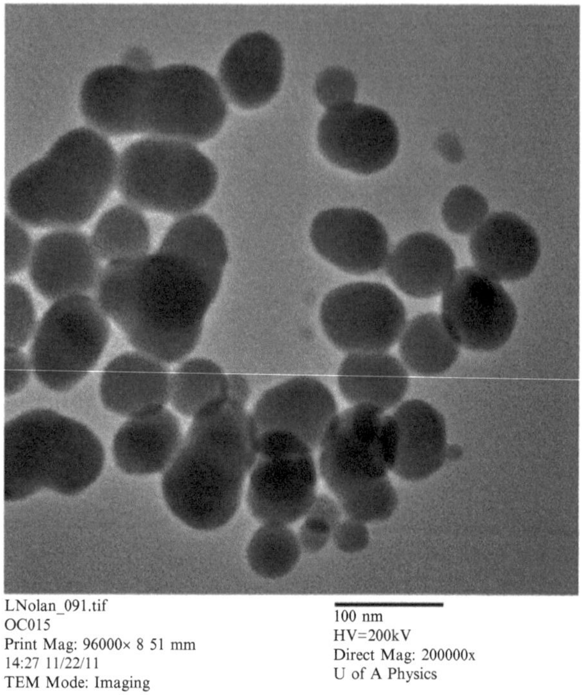

Fig. 6.3 TEM micrograph of colloidal silica abrasives (bright field image, direct magnification = ×200,000)

6.1.2 Slurry

CMP slurries are aqueous and contain components to bring about both chemical and mechanical effects on the wafer. Mechanical work is brought about by abrasive particles of between 20 nm and several microns in size. These are typically hard oxides such as silica, alumina, and ceria, although softer polymeric abrasives have been investigated in order to reduce defects [4]. Mixtures of different types of oxides have also been used in order to exploit the properties of multiple abrasive types, such as the high removal rates associated with ceria and the low cost and good size control of silica [5, 6]. Small abrasives are generally fabricated through a colloidal process and hence are approximately spherical in shape. Larger abrasives may be crushed and are jagged. Specialty abrasive shapes, such as cubes, prolate spheroids, and flakes, have also been used experimentally. A TEM image of colloidal silica particles is shown in Fig. 6.3, while mixed-oxide particles of unconventional shape are shown in Fig. 6.4.

Increasing polishing rates with decreasing particle size have been demonstrated when polishing tungsten with alumina particles, down to a particle size of 250 nm [7]. Some dependence of polishing rate on particle size distribution has also been observed [8]. However, a clear decrease in polish rate is observed with very small particles, suggesting that the abrasives must be larger than a critical size to be effective [5]. An example of this is the work of Lin et al. [5]. The study

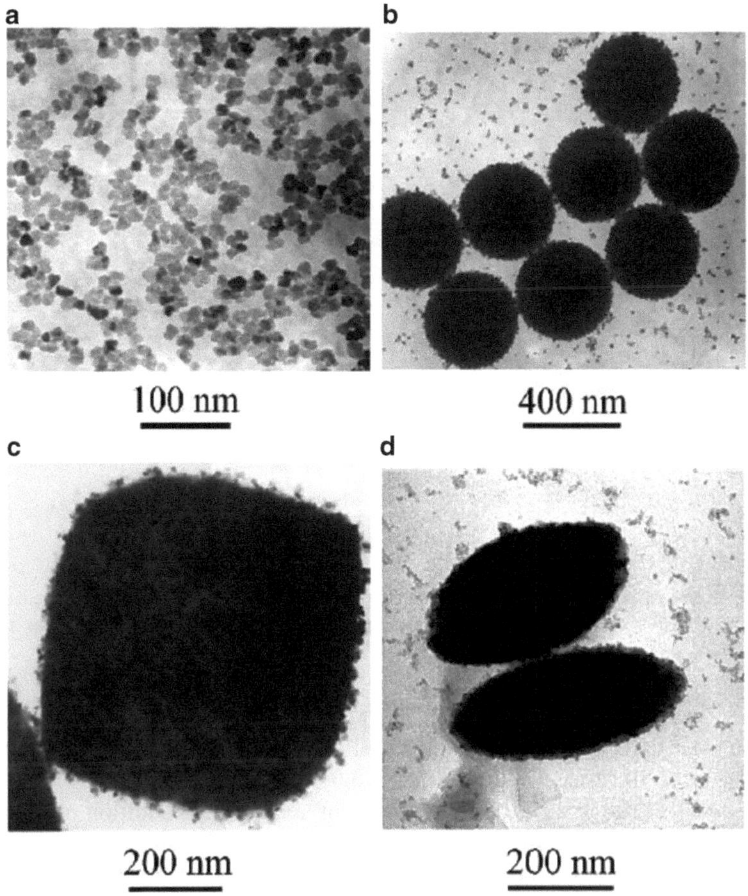

Fig. 6.4 TEM micrographs of (**a**) ceria particles, mean diameter of 20 nm; (**b**) 400 nm colloidal silica particles mixed with the ceria particles shown in (**a**)—electrostatic attraction ensures adhesion of ceria to silica; (**c**) 700 nm cubic hematite coated in 35 nm of silica, mixed with the ceria particles shown in (**a**); 400 nm ellipsoidal hematite coated in 35 nm of silica, mixed with the ceria particles shown in (**a**) (Reproduced from [6])

demonstrated a 20-fold increase in oxide polish rates when using composite particles of a silica to ceria weight ratio of 0.1 (5 wt% silica particles and 0.5 wt% ceria particles), compared to polishing with either silica or ceria alone. This is due to the fact that ceria particles are very small and acting alone as polishing slurry cannot effectively transfer the shear forces from the pad to the wafer.

The concentration of particles in the slurry also affects the rate of polishing. Numerous studies have indicated an increase in polishing rate with increasing abrasive concentration, up to a critical concentration [7, 9]. If the amount of abrasives is increased beyond this point, no further increase in polishing rate is

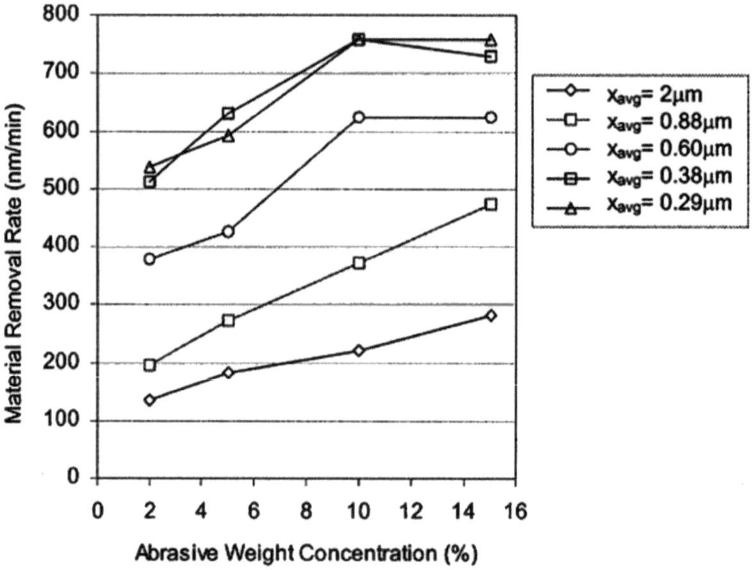

Fig. 6.5 Removal rate of Tungsten by silica particles of the sizes indicated increases with increasing abrasive concentration, until a constant rate is reached. Further increases in abrasive concentration do not result in any further polish rate increases (Reproduced from [8])

observed, and in fact the polishing rate may decrease. This has been observed in both tungsten and copper CMP, using abrasives ranging from around 85 nm to 2 μm in diameter, as illustrated in Figs. 6.5 and 6.6.

Chemical work during polishing is brought about by the addition of oxidizers, corrosion inhibitors, chelating agents, pH adjusters, buffers, and surfactants. These vary considerably, depending on the surface to be polished, and will be discussed in the appropriate part of Sect. 6.3. It should be noted that the active components make up only a small fraction (typically 10 wt% or less) of the slurry, with water making up the remainder.

6.1.3 Pads and Conditioners

Polishing pads are polymeric and relatively soft, with a durometer hardness between 55 Shore A and 60 Shore D, depending on their application [10]. They are often made of a polyurethane blend and contain microscopic pores. A cross-sectional SEM image of a typical pad is shown in Fig. 6.7. The surface of the pad is swept by a diamond-grit conditioner during polishing. This continually refreshes the surface and opens pores, resulting an approximately constant microscopic surface roughness. The sizes of asperities on the pad have, like many rough surfaces, been shown to follow an approximately Gaussian distribution [11, 12].

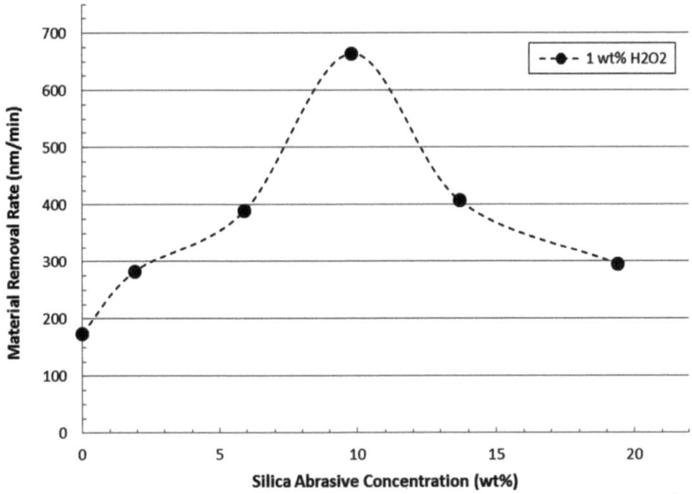

Fig. 6.6 Removal rate of Copper by silica particles with mean diameter of 84 nm indicates that removal rate increases with increasing abrasive concentration, up to a maximum value. Further increases in abrasive concentration decrease the polishing rate (Plotted from data in [9])

Fig. 6.7 Cross-sectional SEM of a typical polishing pad. Numerous pores and the cross section of a groove are visible

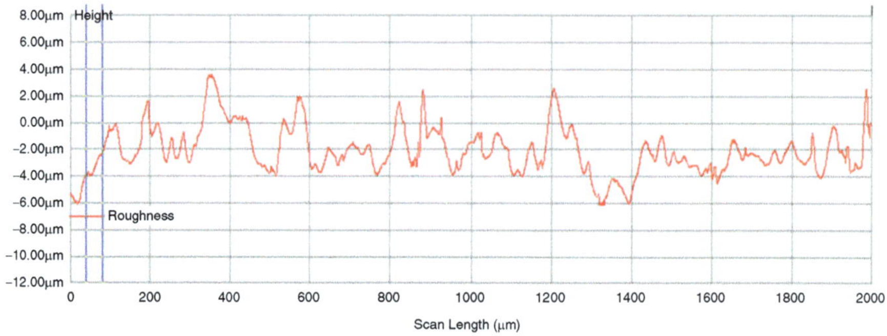

Fig. 6.8 Profilometer scan of pad surface (Scan length 2 mm, total *y*-axis height 20 µm)

A profilometry scan of the surface is shown in Fig. 6.8. Typical values for the average pad roughness are between 1 and 20 µm.

Polishing pads are often grooved to facilitate slurry transport across the surface. This is important as the slurry must travel from near the center of the pad to a radial distance equal to the outer edge of the wafer, a length of 30 cm or more, under the centrifugal action of the rotating pad alone. Common pad groove patterns include spirals, radial and/or concentric lines, square grids, and logarithmical or “k” grooves. Despite such patterning, utilization of slurry in the CMP process is thought to be fairly low, with published estimates ranging from 90 % to as low as 2 %, depending on process conditions [13, 14]. Given that slurry costs make up a significant portion of CMP cost of ownership and that treating slurry waste after polishing is a significant burden on fabrication facilities, increasing slurry utilization is an important driver in CMP research.

6.1.4 Pressure and Velocity: The Preston Equation

Aside from the choice of consumables, the two main parameters controlled by the user are the rotational speed of the wafer and pad and the pressure or “downforce” exerted on the wafer. The relationship between pressure, velocity, and the overall rate of material removal has therefore been the subject of considerable research. One of the earliest and most widely known of these relationships is the Preston equation, which was developed in 1927 to describe plate glass polishing [15]. Equation (6.1) states that the removal rate (MRR) is proportional to the product of the velocity (V) and applied pressure (P). The proportionality constant K_P is known as the Preston coefficient and is experimentally determined for a given system.

$$\mathrm{MRR} = K_{\mathrm{P}} \times PV \tag{6.1}$$

The term “system” in this context refers to a tool type and its consumables, such as the surface being polished, the polishing pad and slurry, and the conditioner type and regime. This highlights the limitation of the Preston equation: while useful within a system, the data obtained from one tool or set of consumables cannot be transferred to another. Additionally, some experimental work is required to determine the value of the constant before the relationship can be used.

A number of models which attempt to overcome these limitations have been developed in response to industry calls for a priori design capabilities. These are based on various understandings of the mechanisms at work during CMP, such as the rates of chemical interactions [16, 17] or the force between the abrasives and the wafer [18–20], and incorporate techniques such as fluid dynamics and finite element analysis [21–23]. Many also require some experimentally derived quantities as model inputs. Pressure and velocity are usually incorporated by assessing their impact on the frequency of the modeled polishing event. The output of the model, typically the overall polish rate, is obtained by summing these events over a specified time interval.

The models referenced here are just some of a plethora of such models, and the interested reader is directed to the CMP literature for a more complete list. The lack of wide acceptance for any one of these models over the others reflects the ambiguity surrounding the fundamental polishing actions, and the difficulty of directly characterizing the CMP process. However, good progress has been made in in situ measurement techniques, and models are becoming more accurate and more widely used.

6.2 CMP Fabrication Techniques

There are many fabrication techniques in which CMP plays a crucial role, and two of the most widely used are STI and copper dual damascene.

The dual damascene method is used to fabricate copper-filled vias and trenches to connect chip components. In this method, the vias and trenches are etched through multiple layers of low-k dielectric and silicon nitride capping material. The internal surface of the etched parts is then coated in a very thin layer of a barrier material, such as tantalum, to prevent diffusion of copper from the interconnect into the device and to act as a seed for electrochemical deposition of the fill metal. Copper is then deposited into the vias and trenches; however, in order to fill the trench completely, and an excess is deposited in non-patterned areas. CMP is then done afterward to remove the excess copper and the diffusion barrier from the surface of the chip, leaving a flush connecting structure. The process is shown schematically in Fig. 6.9. This technique can be performed by etching either the

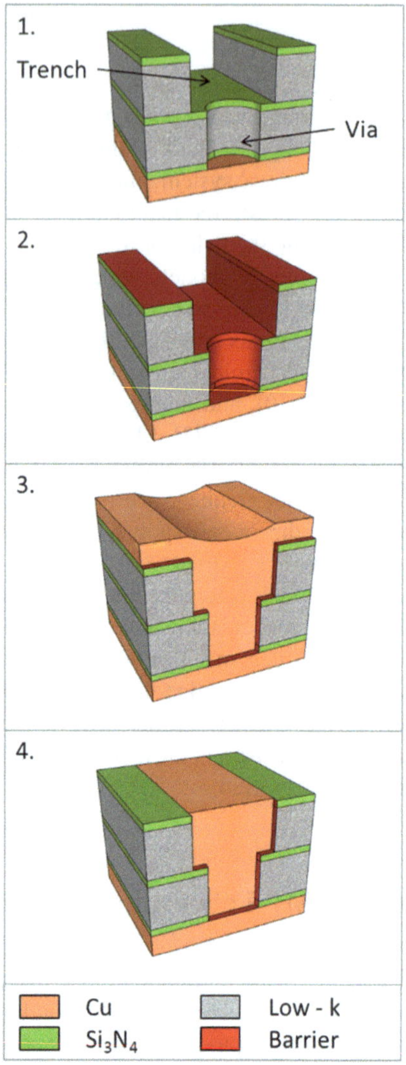

Fig. 6.9 Schematic of the dual damascene process: Trenches and vias are etched through device layers (1) and then a diffusion barrier is deposited (2). An excess of copper is deposited to ensure that the interconnect structure is completely filled (3). The excess copper and the diffusion barrier are then removed using CMP (4), leaving a planar surface

trench first ("trench-first dual damascene") or the via first ("via-first dual damascene"). Where only one structure is created, filled, and polished at a time, the technique is referred to simply as "damascene."

The STI method is used to electrically isolate transistors embedded in the silicon substrate from one another and is also enabled by CMP. In this technique, illustrated in Fig. 6.10, a trench is cut into the silicon substrate using a hard silicon nitride layer as mask. The trench is then overfilled with CVD silicon oxide to ensure complete filling. CMP is used to remove excess oxide and to planarize the top of the filled trench. The nitride mask is then stripped away, leaving an isolating structure behind.

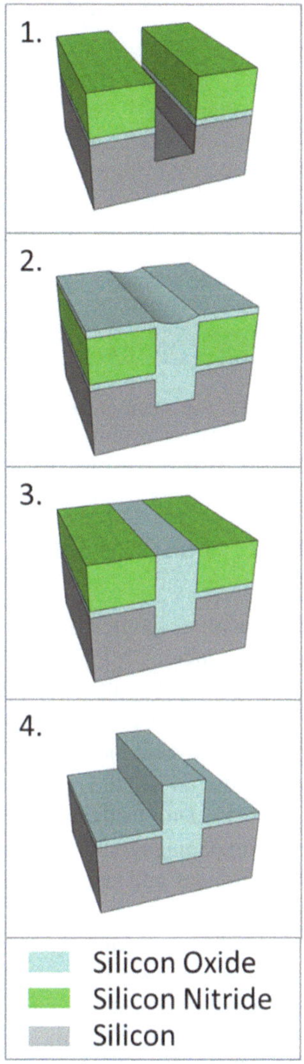

Fig. 6.10 Schematic of the STI process: a trench is cut through the silicon nitride cap layer into the silicon substrate (1) and excess silicon oxide is deposited (2). CMP is then used to planarize the surface, removing the excess oxide (3). Finally, the nitride cap is removed (4), leaving a trench structure in the silicon that isolates one region from the next

6.3 Electrochemical Phenomena in CMP

In the previous section, CMP techniques used for both conducting and dielectric materials were described. While there are many practical similarities between polishing the two material types, the underlying mechanisms for polishing conductors such as metals are based on electrochemical principles. The principles of the polishing of dielectrics are non-electrochemical by nature and as such will not be discussed further. The remainder of this chapter therefore focuses on the polishing of metals and other conducting materials.

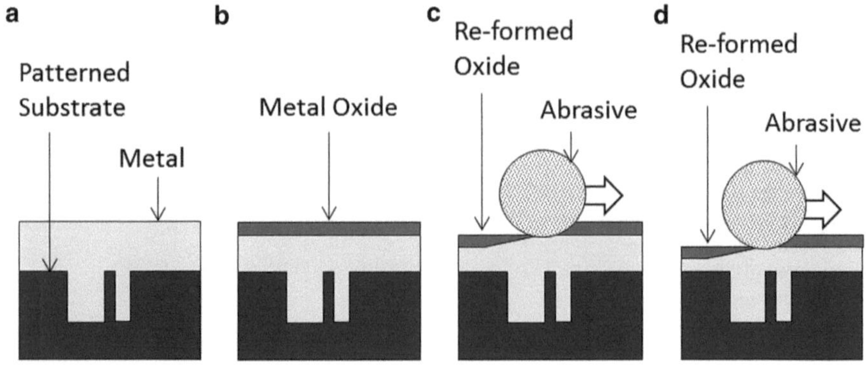

Fig. 6.11 Schematic of the metals CMP mechanism: the patterned metal surface (**a**) is oxidized by the chemical components of the slurry (**b**). The oxide layer is then removed by abrasion (**c**). This process is repeated (**d**) until the metal is removed and the metal-filled patterns remain

6.3.1 Tungsten CMP

Chemical mechanical polish was developed initially to process vitreous materials. The first metal CMP process was introduced some 15 years later as a more reliable and reproducible alternative to plasma etching for the formation of tungsten plugs at the sub-0.35 μm node [24]. Rather than the "chemical tooth" identified as the primary removal mechanism in oxide polishing [25], W-CMP relies on the sequential oxidation and then removal of the metal surface. This is shown schematically in Fig. 6.11. Due to their ductility, metals are typically difficult to remove without causing damage to the surface. By forming a hydrated oxide as the abradable layer, the hardness and ductility of the polishing surface are reduced, while the metal underneath is protected from ongoing, uncontrolled etching. Both an oxidizer and an abrasive are required for polishing to occur [9].

The key chemical requirements of the metal CMP slurry include both oxidation and passivation of the surface. An initial design point for such a slurry is the potential pH, or Pourbaix, diagram, which is calculated from thermodynamic data and shows the equilibrium relationship between pH, potential, and oxidation products formed. The Pourbaix diagram for tungsten in water is shown in Fig. 6.12. From this figure, it is apparent that passivation of tungsten occurs at all positive potentials at a pH of 4 or lower. Consequently, polishing slurries for tungsten contain an oxidizer, such as hydrogen peroxide, nitric acid, or ferric nitrate, and have a pH of less than 4.

Characterization of the electrochemical behavior of tungsten can be a useful tool in assessing new slurry chemistries. For example, Lim et al. [27] have used techniques such as cyclic voltammetry and potentiodynamic polarization to characterize the formation of tungsten oxidation products. They demonstrated that a continuous, passivating oxide can be induced on the metal surface of tungsten and that the formation of such an oxide leads to improved CMP outcomes.

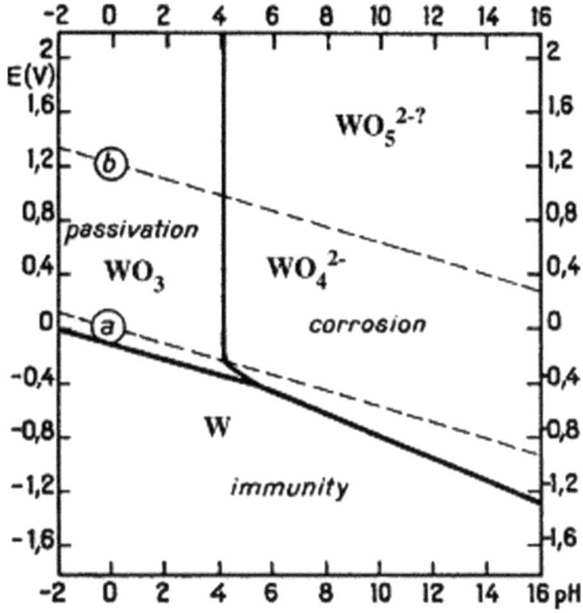

Fig. 6.12 Pourbaix diagram for tungsten in water [26]

6.3.2 *Copper CMP*

The process of polishing copper is somewhat more complicated than polishing tungsten or other self-passivating metals such as aluminum. Unlike these metals, copper does not form a stable oxide layer by itself and uncontrolled oxidation can lead to pitting and etching of the copper surface. To counter this, the chemical components of the polishing slurry are manipulated to produce controlled electrochemical effects by including both an oxidizer to form a reacted oxide surface layer and a corrosion inhibitor to stabilize the oxide layer and prevent unwanted, excessive corrosion. Each of these components has a specific electrochemical function which will be discussed in the following sections.

6.3.2.1 Oxidizers

In the presence of water at standard conditions, neither Cu(I) nor Cu(II) oxides form spontaneously. This is confirmed by the negative standard potentials for the formation of CuO and Cu_2O:

$$Cu \rightarrow Cu^{2+} + 2e^- \quad E^\circ = +0.3419\ \text{V} \tag{6.2}$$

$$2Cu \rightarrow 2Cu^+ + 2e^- \quad E^\circ = +1.042\ \text{V} \tag{6.3}$$

$$\tfrac{1}{2}O_2 + H_2O + 2e^- \rightarrow 2OH^- \quad E^\circ = +0.2005\ \text{V} \tag{6.4a}$$

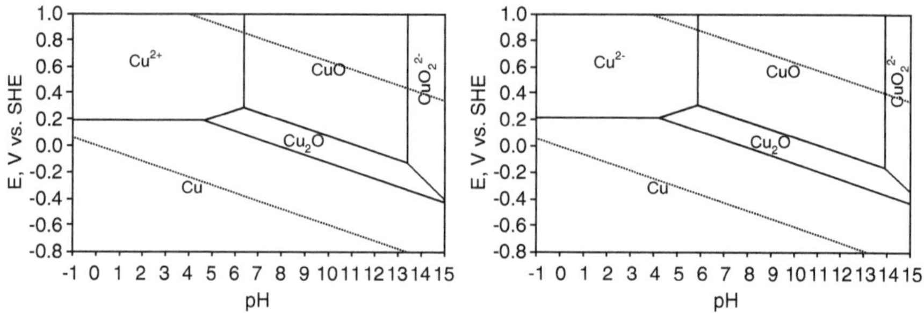

Fig. 6.13 Pourbaix diagrams for the copper–water system at 25 °C. The diagram on the left is for a total dissolved copper activity of 10^{-5}, while the one on the right is for $a_{CuT} = 10^{-4}$ [29]

$$Cu + \tfrac{1}{2}O_2 + H_2O \rightarrow CuO + 2OH^- \quad E^{\circ}{}_{cell} = -0.1414\ V \tag{6.4b}$$

$$2Cu + \tfrac{1}{2}O_2 + H_2O \rightarrow Cu_2O + 2OH^- \quad E^{\circ}{}_{cell} = -0.8415\ V \tag{6.4c}$$

To induce oxidation of copper, an oxidizer is added to the slurry. Commonly used oxidizers are nitric acid (HNO_3), ammonium hydroxide (NH_4OH), and hydrogen peroxide (H_2O_2). These are discussed individually below.

Nitric Acid

Nitric acid was one of the first oxidizers selected for copper CMP because nitrate ions promote the formation of Cu(II) ions, rather than Cu(I) ions, and hence CuO rather than Cu_2O, which has been associated with more robust polishing [28]. However, the high pH of strong acids such as nitric acid encourages the dissolution of copper, as indicated by the Pourbaix diagram in Fig. 6.13. Potentiodynamic scans of copper in nitric acid, such as that shown in Fig. 6.14, show that this dissolution is more prominent than oxidation, and nitric acid does not effectively passivate the copper surface. While removal rates for nitric acid are fast, the etching action also results in poor final planarization. In one study, Carpio et al. [30] demonstrated that polishing with nitric acid actually increased the roughness of the copper surface by a factor of 15, with the surface roughness increasing from 10 to 150 Å after polishing.

Ammonium Hydroxide

Ammonium hydroxide has also been examined as an oxidizer in copper CMP because of its ability to passivate copper in potentiodynamic studies. An example of such a study is shown in Fig. 6.15.

This effect is thought to be at least partially due to the production of hydroxide ions by NH_4OH in solution, raising the pH to the point where oxide formation is thermodynamically favorable according to the Pourbaix diagram in Fig. 6.13.

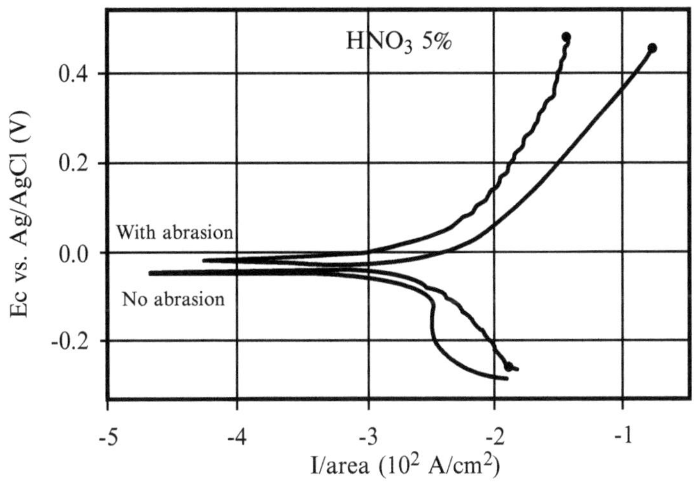

Fig. 6.14 Potentiodynamic profiles for copper in a solution of 5 % HNO_3 in air-saturated water, with and without abrasion during testing. The scan rate was 5 mV/s, starting at 0.25 V below the open circuit potential [28]

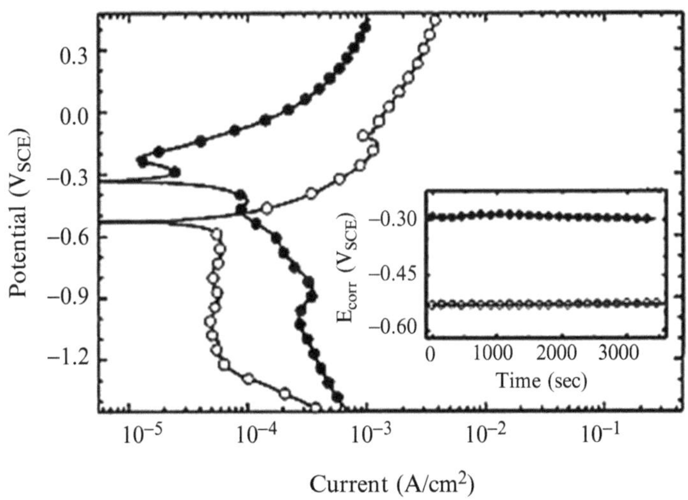

Fig. 6.15 Potentiodynamic profiles of copper in solutions with low (2.35 g/L, *filled circle*) and high (30 g/L, *open circle*) concentrations of ammonium hydroxide. Here the scan rate was 1 mV/s. Prior to the commencement of testing, the system was allowed to reach steady state for an hour. The recorded open circuit potential E_{corr} over this hour is shown in the inset [31]

However, ammonia ions also chelate the cupric oxide formed on the copper surface, encouraging dissolution. The overall effect on the passive film is that it is weakened, and pronounced intergranular corrosion may occur. Examples of this type of uncontrolled corrosion are shown in Fig. 6.16 in scanning electron micrographs of copper surfaces polished in different concentrations of ammonium hydroxide for

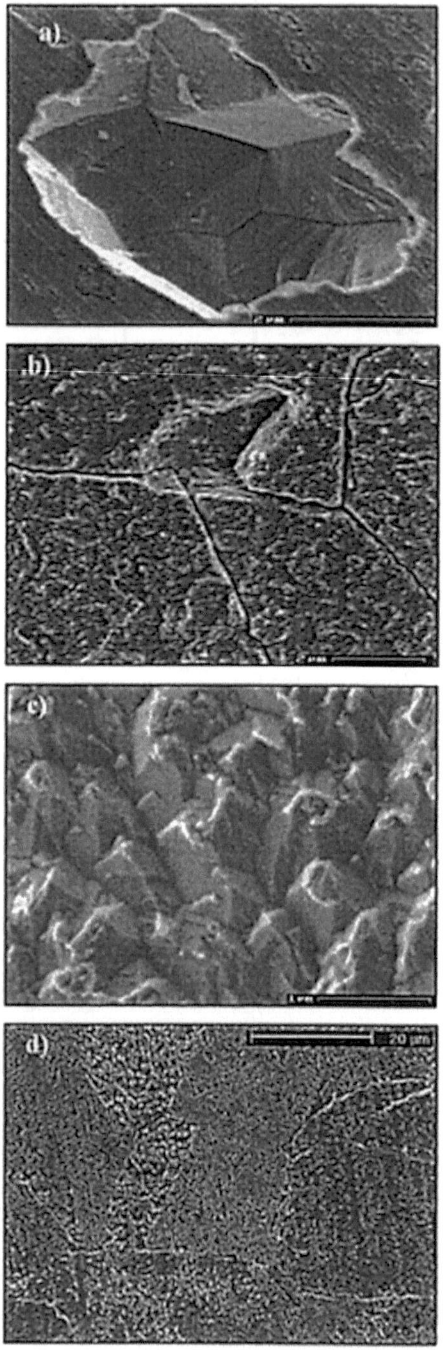

Fig. 6.16 SEM images of copper surfaces exposed to varying concentrations of NH_4OH at the open circuit potential for varying lengths of time: weak solution (2.35 g/L) for (**a**) 1 min and (**b**) 1 h, and strong solution (30 g/L) for (**c**) 1 min and (**d**) 1 h [29]

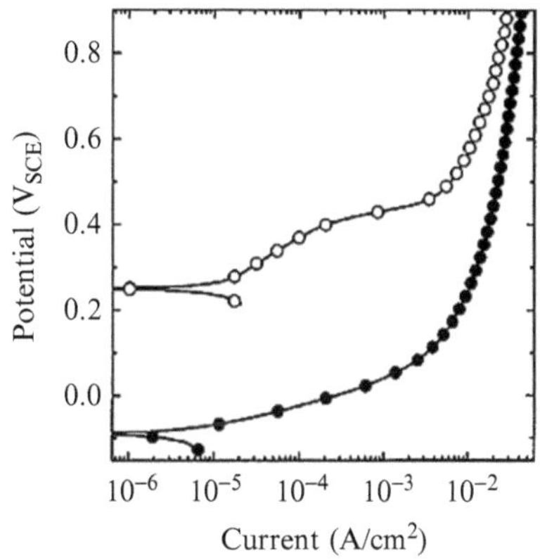

Fig. 6.17 Potentiodynamic profiles for copper electrodes without H_2O_2 (*filled circle*) and with 3 vol.% H_2O_2 (*open circle*). The solution is buffered at pH 4 and also contains 10 g/L Na_2SO_4 to increase the solution conductivity. The scan rate is 5 mV/s and the increase in E_{corr} observed is approximately 0.35 V [32]

varying lengths of time. The copper corrosion occurs primarily between the grains and in some cases is severe enough to excise an entire grain, as demonstrated in Fig. 6.16a.

Ammonium hydroxide also has relatively slow corrosion rates, with studies indicating that i_{corr} for copper in a 5 % solution of NH_4OH is just 0.3304 mA/cm^2, more than 30 times less than i_{corr} for copper in a 5 % solution of nitric acid [31]. Although the chemical reaction rate does not correlate linearly with the overall material removal rate [9], slow corrosion rates in potentiodynamic studies can indicate that the studied oxidizer is not appropriate.

Hydrogen Peroxide

Hydrogen peroxide has become the oxidizer of choice for copper CMP, both as pure H_2O_2 or when present in polishing slurry as a dissolution product of chemicals such as sodium percarbonate. Hydrogen peroxide is a strong oxidizer in and of itself, and can additionally decompose into hydroxyl radicals. This decomposition is catalyzed by transition metal ions, such as Cu^{2+}, which are thought to be abundant in copper CMP slurries during processing. This process is known as the Fenton cycle and occurs by the following reactions:

$$Cu^+ + H_2O_2 \rightarrow Cu^{2+} + OH^- + {}^*OH \tag{6.5}$$

$$Cu^{2+} + H_2O_2 \rightarrow Cu^+ + {}^*OOH + H^+ \tag{6.6}$$

Hydrogen peroxide in copper polishing slurries therefore leads to a sharp increase in oxidation potential E_{corr}, as demonstrated in the potentiodynamic curves shown in Fig. 6.17. However, strong oxidation is not the only criterion in

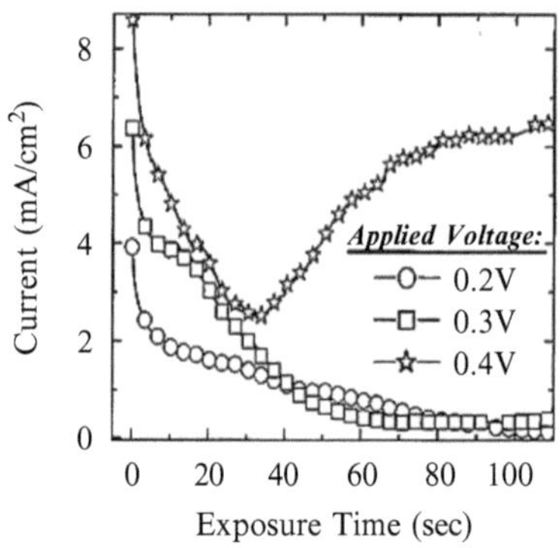

Fig. 6.18 Chronoam perometric profiles of copper in a solution of 3 vol.% H_2O_2 with 10 g/L of Na_2SO_4, buffered at pH 4. Measurements are taken at potentials of 0.2, 0.3, and 0.4 V. The current stabilizes at a low value for potentials at or below 0.3 V, indicating the development of an oxide film on the copper surface [30]

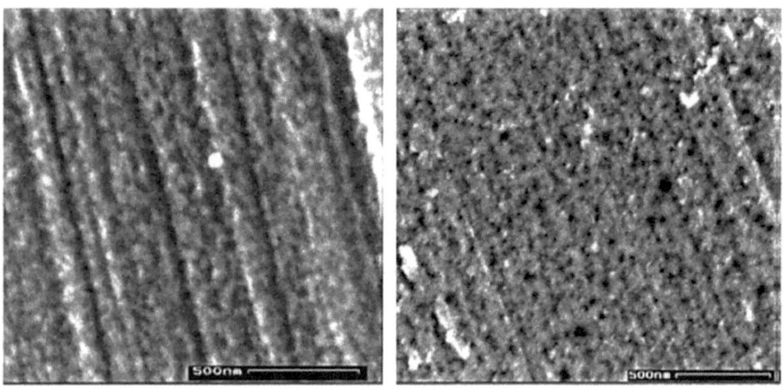

Fig. 6.19 HRSEM of copper exposed to a solution of 3 vol.% H_2O_2 with 10 g/L of Na_2SO_4, buffered at pH 4 for 5 min at a potential of 0.3 V (*left*) and 0.4 V (*right*). The image on the *left* shows a tightly formed oxide layer, while that on the *right* contains numerous pits (visible as *small black dots*). These results are consistent with the chronoamperometric study performed by the same authors and shown in Fig. 6.18 [30]

determining whether a chemical is appropriate for use as a CMP oxidizer or not. In contrast to ammonium hydroxide, chronoamperometric studies of copper in hydrogen peroxide solutions, such as the one shown in Fig. 6.18, demonstrate that this oxidizer can develop an oxide film without severe intergranular corrosion. High-resolution scanning electron micrographs (HRSEM) are used to confirm the quality of the surface film, as shown in Fig. 6.19. In this figure, the passive film is adherent and continuous at an applied potential of 0.3 V, in good agreement with the chronoamperometric profile shown in Fig. 6.18.

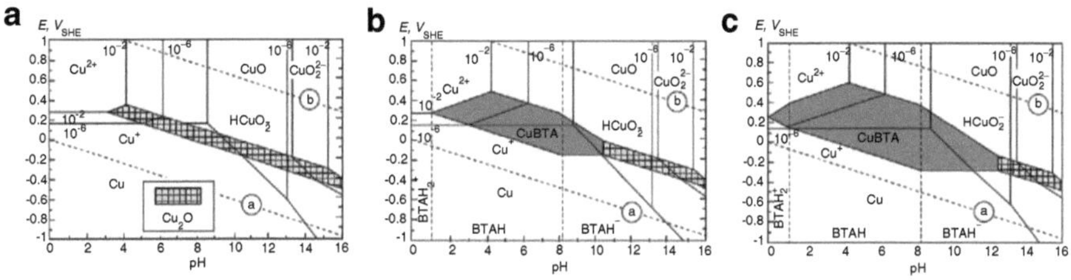

Fig. 6.20 Pourbaix diagrams for (**a**) copper in water, (**b**) copper in water with a total activity of BTAH of 10^{-4}, and (**c**) copper in water with a total activity of BTAH of 10^{-2}. (Originally published in [34], reproduced here from redrawn figure in [35])

Similarly, the breakdown of the film indicated by the chronoamperometric profiles in Fig. 6.18 at an applied voltage of 0.4 V corresponds with the presence of pits at the same voltage in Fig. 6.19. In this example, chronoamperometric characterization can confirm that passivation occurs quickly and that the resulting oxide film is passivating, two essential requirements for use in CMP. The stability of the film formed can be enhanced further when corrosion inhibitors are used in conjunction with oxidizers such as hydrogen peroxide. These are discussed in the next section.

6.3.3 *Corrosion Inhibitors*

Corrosion inhibitors increase the stability of the passive film initiated by oxidation through adsorption on the oxide surface. Benzotriazole (BTAH) is one of the most commonly used corrosion inhibitors for copper and has been used as a corrosion inhibitor in cooling water systems for close to 50 years [33]. Alternative inhibitors based on surfactants, such ammonium dodecyl sulfate (ADS), have also been investigated.

6.3.3.1 Benzotriazole

Thermodynamic analysis indicates that benzotriazole is an effective corrosion inhibitor for copper over a wide pH and potential range. This is shown graphically in the Pourbaix diagrams in Fig. 6.20; in this figure, the addition of BTAH induces a passive CuBTA region in addition to the passivation of copper by its own oxides, shown in Fig. 6.13a for comparison. The range of the passive CuBTA region expands as BTAH concentration is increased, up to a BTAH concentration of around 0.01 M, beyond which no further expansion is observed.

Surface characterization techniques such as surface-enhanced Raman spectroscopy have been used in conjunction with thermodynamic analysis to determine the

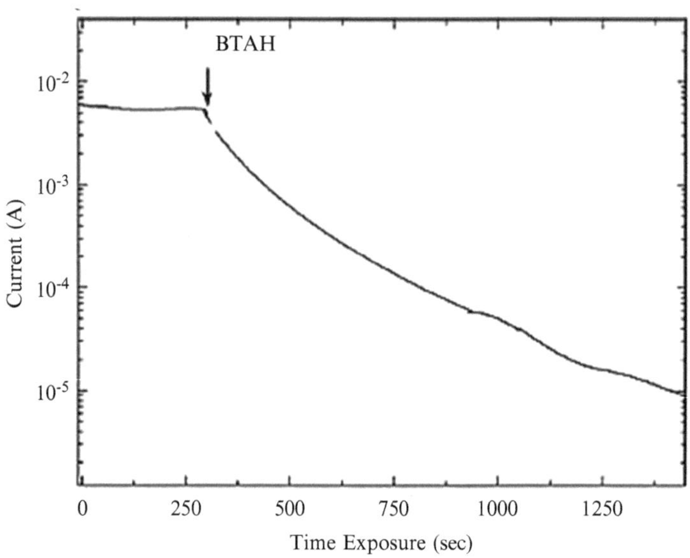

Fig. 6.21 Anodic current of copper over time at an applied potential of 0.1 V in a solution of 3 wt% nitric acid. At the point indicated, 0.02 M BTAH is added. As can be seen from the figure, the BTAH immediately but gradually reduced the transient current, indicating the slow formation of a passive film [30]

morphology of the CuBTA film. The film is thought to form initially by absorption of BTAH molecules on to the copper surface and then by reduction of oxygen:

$$\mathrm{Cu} + \mathrm{BTAH} \rightarrow \mathrm{CuBTA} + \mathrm{H}^+ + \mathrm{e}^- \quad E^\circ{}_\mathrm{a} = -0.0946\ \mathrm{V} \tag{6.7}$$

$$\mathrm{O_2} + 4\mathrm{H}^+ + 4\mathrm{e}^- \rightarrow 2\mathrm{H_2O} \quad E^\circ{}_\mathrm{c} = 1.229\ \mathrm{V} \tag{6.8}$$

This gives the overall reaction:

$$4\mathrm{Cu} + 4\mathrm{BTAH} + \mathrm{O_2} \rightarrow 4\mathrm{CuBTA} + 2\mathrm{H_2O} \quad E^\circ{}_\mathrm{cell} = 1.6074\ \mathrm{V} \tag{6.9}$$

BTAH is the dominant species in solution at pH 2 and applied potential of −0.7 V. Under these conditions, surface-enhanced Raman spectroscopy has been used to show that the CuBTA film forms when the lone pair of the BTAH nitrogen atoms in the triazole ring bond to the copper surface. The adsorbed molecules are oriented perpendicularly to the copper surface and pack closely, forming a continuous adsorbed film.

Although thermodynamically and sterically favorable, BTAH has relatively slow reaction kinetics which may limit its suitability for CMP. In the presence of chloride ions, CuBTA films form rapidly. However, chloride ions can cause pitting of the copper surface, so great pains are taken to ensure that they are eliminated from CMP process streams, slowing the BTAH reaction kinetics. In a study, shown in Fig. 6.21, the anodic current generated by copper in nitric acid free of chloride ions was recorded

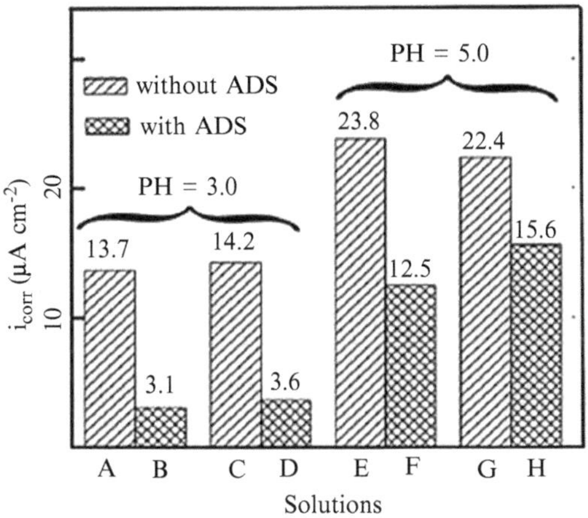

Fig. 6.22 Corrosion currents for copper rotating disc electrodes in various solutions containing 0.13 M of acetic acid. Solutions A, B, C, and D are at pH = 3 while E, F, G, and H are at pH = 5. Additionally, solutions A, B, E, and F contain 1 wt% H_2O_2 while the remaining solutions (C, D, G, and H) contain 5 wt% H_2O_2. ADS is added to one of each type of solution and substantially decreases the corrosion current in all cases [36]

while BTAH was added to the solution. While a passive film did form, as indicated by the reduction in anodic current, the process occurred slowly and several minutes elapsed before the anodic current dropped by an order of magnitude. Despite these kinetic concerns, the ready availability of and body of knowledge surrounding BTAH mean that it is still the most widely used corrosion inhibitor in Cu CMP.

6.3.3.2 Surfactants as Corrosion Inhibitors

Due to these kinetic concerns, and concerns regarding its environmental impacts, a number of surfactants have been investigated as alternatives to BTAH. Surfactants are able to interact with the copper surface, allowing these chemicals to also behave as corrosion inhibitors in CMP. One such substance, ADS, is shown in experiments with rotating disc electrodes over a range of acidic pH values to reduce the corrosion current by a factor of 2 or more. These results, shown in Fig. 6.22, indicate that surface passivation is enhanced when ADS is present in solution.

The rate at which passivation occurs is also important in determining if a surfactant is effective as a CuCMP slurry additive. Two studies with ADS used in conjunction with low concentrations of BTAH suggest that ADS does in fact have faster reaction kinetics than BTAH in isolation. In one study [37], the anodic current of copper in solutions of hydrogen peroxide, 3 mM ADS, and varying concentrations of BTAH (shown in Fig. 6.23) is recorded over a period of 5 min. The results indicate that both BTAH and ADS effectively passivate the copper surface. However, the onset of passivation is slightly more rapid for ADS, with or without BTAH, and

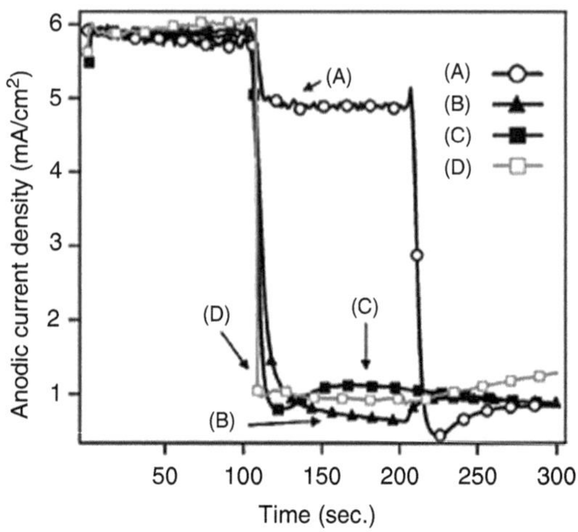

Fig. 6.23 The anodic current density transient for copper in solutions with 1 wt% glycine, 5 wt% H_2O_2, pH = 4, and varying amounts of BTAH and ADS. Solution A contains 0.5 mM of BTAH and 3 mM of ADS, added at 100 and 200 s, respectively. Solution B contains the same concentrations of the inhibitors, added in reverse order. Solution C also contains the same inhibitor concentrations, both added at 100 s. Solution D contains 10 mM of BTAH added at 100 s, followed by 3 mM of ADS added at 200 s [37]

the degree of passivation is greater. The proposed mechanism for these varying passivation rates is that ADS hemi-micelles adhere strongly to the copper surface, preventing further reaction, while BTAH adsorbs between the ADS hemi-micelles. According to this model, ADS and BTAH are in competition for adsorption sites; this is supported by the results shown in Fig. 6.23, where a slight increase in anodic current is observed when either ADS or BTAH is added to a solution already containing the other chemical, regardless of the order in which they are added.

In a second study [38], in situ measurements of the open circuit potential (OCP) were made while a copper electrode was polished, held stationary, and then polished again in a slurry containing ADS and other chemicals. The results, shown in Fig. 6.24, show that the OCP increases by 50 % or more while the electrode is held stationary compared to during polishing. This supports the model of Cu CMP as a process of passivation, film removal, and re-passivation. As in the first study described here, the rate of passivation is greater for solutions containing ADS than for BTAH (Fig. 6.24, curves e and f).

The quality of the copper surface produced after polishing with ADS has been assessed by atomic force microscopy such as those shown in Fig. 6.25. In this example, the copper surface was polished in a slurry containing ADS as well as hydrogen peroxide, glycine, and silica particles. The resulting surface roughness is very low, with a root-mean-square surface roughness (R_a) of 5.1 Å.

A number of other surfactant-type corrosion inhibitors, including Triton X-100 and dodecyl trimethyl ammonium bromide (D-TAB), have been characterized for use in CMP. The outcomes of these and similar studies can be found in the literature.

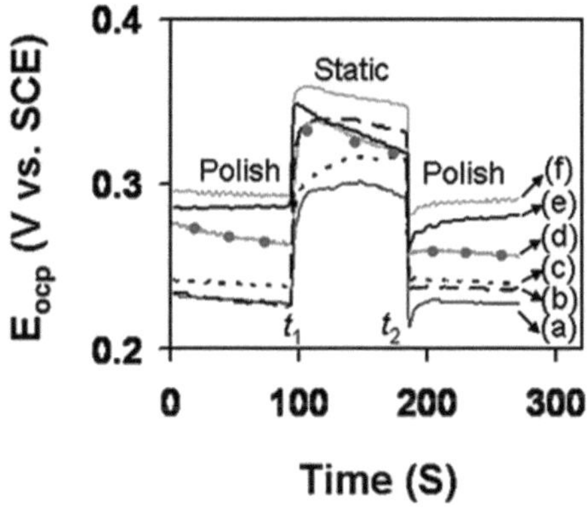

Fig. 6.24 Open circuit potentials for copper discs in polishing and static conditions. Polishing was carried out in a solution of 1 wt% glycine and 5 wt% H_2O_2 at pH = 4 (curve a) with additions of (b) 1 mM BTAH, (c) 1 mM BTAH + 1 mM ADS, (d) 10 mM BTAH, (e) 1 mM ADS, and (f) 10 mM ADS [38]

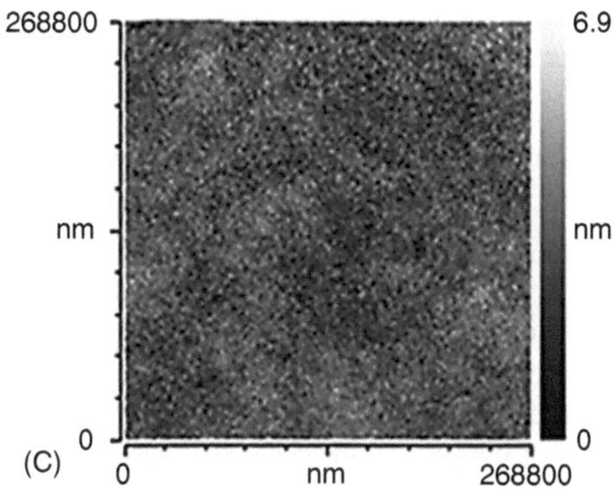

	Image
Mean	-0.00
Sq	0.65
Sa	0.51
Peak/Valley	6.93
Skewness	0.02
Kurtosis	3.05

Fig. 6.25 Copper surface after polishing in a slurry containing 0.4 mM of ADS, 3 wt% fumed silica particles, 1 wt% glycine and 5 wt% H_2O_2 at pH 4 [38]

6.3.4 Efficacy of Evaluating Copper CMP Through Electrochemical Techniques

Significant progress has been made in identifying and testing of chemical solutions suitable for CMP. The progression from nitric acid to ammonium hydroxide to hydrogen peroxide has been shown to have a strong electrochemical justification, both theoretical and experimental, and has resulted in improved polishing outcomes for copper. Polish quality has been further enhanced by the use of the corrosion inhibitors benzotriazole and ADS. The suitability of these additives also has a firm basis in electrochemical theory and has moreover been demonstrated experimentally.

Table 6.1 Data for several buffer systems

Buffering agent	pK_a	Useful pH range
Citric acid	3.06	2.1–7.4
	5.75	3.0–6.2
Acetic acid	4.74	3.8–5.8
Ammonia	9.26	8.2–10.2
Phosphate	2.12	1.1–3.1
	7.21	6.2–8.2
	12.32	11.3–13.3

However, CMP is a complex process. As well as the electrochemical considerations discussed here, the success or otherwise of the technique depends on the chemical interaction with the abrasive particles and the pressure and velocity of polishing. Although some workers have attempted to electrochemically characterize various chemistries in polishing-like conditions, the full effects of slurry transport to and from copper surface has not been examined and remains a key unanswered question in understanding copper CMP.

6.3.5 Buffers in CMP slurries

Control of slurry pH is important for CMP process stability since pH affects removal rate and the formation of defects. The pH of slurries can drift globally or locally due to exposure to air or a redox reaction that occurs during metal CMP, for example. High pH slurries absorb CO_2 from the air forming carbonic acid which lowers the pH over time. Reactions of metals with slurries can locally generate H^+ or OH^- ions that will locally lower or raise the pH, respectively.

The variation of pH can be controlled by adding buffers to the slurry system. For acidic slurries, an acid buffer is required which is a mixture of a weak acid and its salt, which is formed by the reaction of the weak acid with a strong base. A good example of this is the citric acid buffer system, citric acid and potassium citrate. Potassium citrate is formed from the reaction of citric acid and potassium hydroxide. Basic slurries need a basic buffer system, a mixture of a weak base and its salt which is formed by the reaction of a weak base and a strong acid. An example of a basic buffer is ammonium hydroxide and ammonium chloride. Ammonium chloride is formed by the reaction of ammonium hydroxide with hydrochloric acid. Values for pK_a (negative logarithm of the dissociation constant) and pH range for several buffer systems are shown in Table 6.1. Note that the useful pH is typically close to pK_a.

6.3.6 Polishing of Highly Doped Silicon

In this chapter the reaction of metals with solutions has been discussed. Semiconductors such as silicon and germanium are another group of materials that can display

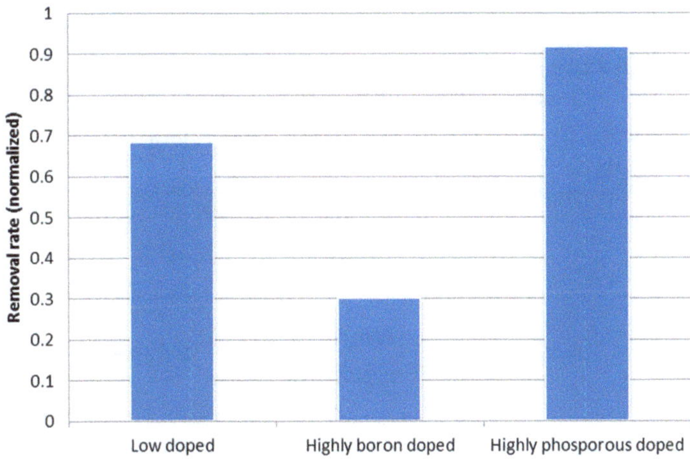

Fig. 6.26 Effect of doping on polish rate of polysilicon, after 20 min of polish (3 psi pressure and 40 rpm in a colloidal slurry at pH = 10.4) [40]

metallic like behavior and which are used in integrated circuit manufacturing. Integration schemes for these materials often require them to be removed by CMP. It has been shown that doping of semiconductors affects the CMP polish rate [39–43].

Polycrystalline silicon (polysilicon) is an important semiconductor which is used for filling vias and making interconnections in integrated circuits. Doping polysilicon with phosphorous increases the CMP removal rate while boron doping decreases the removal rate as shown in Fig. 6.26. This effect is not fully understood. One model postulates that dopant atoms compete with silicon in reacting with the slurry [40, 41]. However, the dopant concentration is significantly less than the silicon concentration ($\ll$1 at.%) and it is unlikely that a competitive reaction could explain the polish rate changes. Another hypothesis is that the difference between the silicon and the dopant atom size generates stress in the material, thereby influencing the polish rate [41, 42]. This is also unlikely since various experiments have shown that the polish rate is more dependent on the charge of dopant than the size. The third model, shown schematically in Fig. 6.27, relates the removal rate change to the influence of the electric field formed by dopants. It has been shown that weakening the Si–Si bonds increases the polysilicon polish rate. Strong polarizers such as hydroxyl anions at high pH weaken Si–Si bonds and can increase the removal rate [41]. Impeding or assisting the transportation of hydroxyl anions on the polysilicon surface directly affects the polish rate. On the other hand, dopants form a depleted layer on the wafer surface, B− ions repelling OH− while the P+ ions attract OH− to the silicon surface and thereby influence the OH transportation at the surface [41, 43].

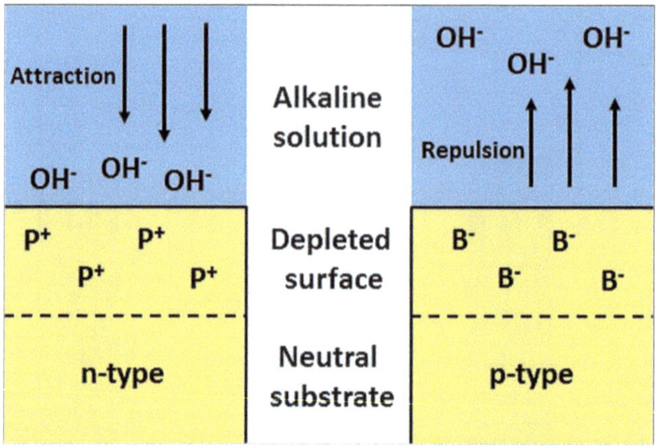

Fig. 6.27 Schematic illustration of the model which describes the electric effect of dopants on polysilicon polish [41]

6.4 Emerging Applications: CMP of Noble Metals

Unique features of CMP make it popular in many areas where smoothness and high surface quality are required. Polishing of noble materials such as platinum, ruthenium, and iridium is another important application of CMP. These materials are widely used in electrical industry because of their high work function, conductivity, and chemical stability. For example, iridium and platinum are known as noticeable candidates for gate electrodes of p-channel devices and as bottom electrodes for capacitors [44–47]. The application and effect of CMP on these noble materials are briefly explained below.

6.4.1 Platinum and Iridium

High work function of materials such as platinum and iridium makes them useful as diffusion barrier capacitors for gate electrodes and molecular device substrates [45, 48]. However, there are many issues in polishing these noble materials. The main problem relates to their inert properties. Platinum is a strong corrosion-resistant material and therefore it shows polish rates below 10 nm/min. Iridium is also difficult to react and is almost inert to all acids and even strong oxidizers at room temperature. There are other problems such as the hardness of iridium which make the abrasive choices limited. The polish rate of iridium is less that 50 nm/min and takes long time to planarize the surface. There have been studies working on increasing the polish rate by using slurries containing abrasives like alumina or iridium oxide and chemicals such as high PH solutions and oxidizer additives [45, 49, 50]. Because of the low polish rate and difficulty in patterning these noble materials, more study is required to commercialize the role of platinum and iridium.

6.4.2 Ruthenium

Ruthenium is a popular noble bottom electrode. This material shows low current leakage and compatibility to high dielectric materials. Like other noble materials, CMP of ruthenium is difficult due to the high hardness and corrosion resistance. Oxide formation can improve the polish rate because it is soluble in aqueous solutions. An example of slurry for ruthenium polishing is highly acidic ceric ammonium nitrate which acts as a strong oxidizer to ruthenium and helps to increase the polish rate. However, the low PH environment is not preferred for polishing due to corrosion and this solution is not popular for ruthenium polishing [47, 48]. It has been shown that alumina or silica abrasive in solutions such as NIO_4 and KIO_4 is another interesting candidate for ruthenium polishing. Maximum polish rate was achieved at PH $\sim$ 6 and was lower than 100 nm/min [46, 48, 51, 52].

6.5 Electrochemically Induced Defects in CMP

The mechanism of metallic contamination can be classified into two types: electrochemical and surface chemical [53]. This is based on electrons being transferred, whether there exist oxidation/reduction reactions and the possibility of valence changes of the elements during the contamination process. In this section we will only concentrate on the electrochemically induced defects.

Cu CMP is a typical example of metal that is susceptible to electrochemical defects. Cu is not chemically stable and it is not self-passivated (that is, it doesn't grow a protective layer) which signifies that Cu will interact with other chemical reagents during the wet CMP process and becomes susceptible to chemically induced defects [54]. The chemically induced defects in Cu CMP are derived from classical corrosion processes. They can be characterized into three types [55]: pitting corrosion, galvanic corrosion, and chemical etching which are discussed in the following sections.

6.5.1 Galvanic Corrosion and Photo-Induced Defects

The Cu dual damascene process, shown in Fig. 6.9, is used in the fabrication of copper interconnect structures, in which chemical mechanical polish, CMP, is used to planarize and remove electroplated copper overburden [56–60]. Prior to the electrodeposition of Cu, a thin diffusion barrier layer material typically Ta and/or TaN and a Cu seed layer are deposited onto the substrate, specifically into the trenches and vias created by reactive ion etching [56, 57]. In the final stage of Cu CMP, both Cu and the barrier layer (Ta and/or TaN) are subjected to CMP and are exposed to polishing slurries [56, 57]. Cu (more noble cathode) and the barrier metal (less noble anode) form a galvanic couple while immersed in the polishing slurry and may induce preferential corrosion of either the barrier material or Cu by galvanic corrosion [56, 57, 59–61]. Shortly afterwards, "fangs" are usually observed at the Cu-barrier interface [59].

CMP slurry provides the conductive pathway between the more noble cathode and the less noble anode [59]. The most common oxidant in slurries used for copper CMP is hydrogen peroxide; however, it has a disadvantage that it decomposes easily decreasing its oxidizing strength [57]. Many studies have been carried out on the influence of slurry composition and pH on the nature and extent of such galvanic corrosion. Tamilmani et al. [57] in their study showed through electrochemical polarization measurements that galvanic corrosion between abraded Cu and Ta is a function of pH of the two slurry systems used (hydroxylamine and peroxide solutions) and was found to be the lowest at pH 4. Brusic et al. [58] in their study also used polarization curves to predict the likelihood of galvanic corrosion between Cu and Ta when immersed in an aqueous electrolyte at different pH values. Assiongbon et al. [59] investigated the extent of galvanic corrosion between Cu and Ta on a partially Cu-covered Ta disk by combined electrochemical impedance spectroscopy with potentiodynamic polarization and galvanic current measurements in two CMP slurry solutions.

During the CMP process, galvanic corrosion of metals is determined commonly by potentiodynamic polarization measurement where the anodic branch of the Ta polarization curve intersects the cathodic branch of Cu forming a galvanic couple [56–59]. In galvanic couples, the less noble metal forms the anode (typically Ta or TaN) and the other acts as the anode (Cu). Lin et al. [56] used potentiodynamic polarization curves of Cu and the barrier material TaN exposed to commercial slurries to show a galvanic couple when the anodic branch of TaN polarization curve intersects the cathodic branch of Cu in the slurries used. Also, it should be noted that the galvanic potential and the galvanic current density can be estimated from the point of intersection of the anodic portion of the TaN curve and the cathodic portion of the Cu portion.

In the estimation of the galvanic current, the anodic portion of the polarization curve represents the metal corrosion reaction while the cathodic portion represents the reduction reaction, which is responsible for metal corrosion due to its chemistry [56, 57]. When abrasion occurs during CMP, galvanic corrosion is increased during which the anodic reaction is the oxidation of the active metal (Ta) and the cathodic reaction is the reduction of the noble metal ions (Cu) in the solution [56, 57]. Tamilmani et al. [57] showed that Ta corrosion is enhanced when Ta and Cu form a couple and are abraded. Furthermore, it was proven that galvanic current density increased when subjected to abrasion in all chemistries. Galvanic current density during abrasion went as high as 500 $\mu A/cm^2$ in hydroxylamine solutions and 120 $\mu A/cm^2$ in hydrogen peroxide at pH 6 and 8, respectively [57].

Faradaic reactions are observed when measuring the corrosion currents [59]. After metallic CMP, the most likely area of the substrate that is exposed is the oxide layer. An oxide film is most likely to be formed on the polished Ta surface when the sample is introduced into an aqueous electrolyte [53]. The anodic oxidation (corrosion) reaction of Ta surface partially covered in oxide occurs in the following three steps [59, 62]:

$$5H_2O = 10H^+ + 5O^{2-} \tag{6.10}$$

$$2Ta = 10e^- + 2Ta^{5+} \tag{6.11}$$

$$2Ta^{5+} + 5O^{2-} = Ta_2O_5 \tag{6.12}$$

with the net reaction as follows [59, 63]:

$$2Ta + 5H_2O = Ta_2O_5 + 10H^+ + 10e^- \tag{6.13}$$

Electro-dissolution of the Ta oxide occurs as follows [59, 62]:

$$3Ta_2O_5 + 8OH^- = Ta_6O_{19}{}^{8-} + 4H_2O \tag{6.14}$$

when peroxide slurries are used, the OH^- ions present in reaction 14 are provided as a result of the electro-reduction (catalytic breakdown) of H_2O_2 [59]. The solution pH determines the intermediate steps of H_2O_2 reduction. In alkaline solutions, H_2O_2 reduces in the following two steps [59–61]:

$$2H_2O = 2H_2O + O_2 \tag{6.15}$$

$$O_2 + 2H_2O + 4e^- = 4OH^- \tag{6.16}$$

with the net reaction written as [61]:

$$2H_2O_2 + 4e^- = 4OH^- \tag{6.17}$$

The intermediate reaction 16 involves the chemisorption of oxygen for the reduction of H_2O_2. This takes us to the corresponding cathodic reaction that originates from hydrogen evolution or O_2 reduction [59]:

$$O_2 + 4H^+ + 4e^- = 2H_2O \tag{6.18}$$

Reaction (6.18) is likely to occur in acidic media while reaction (6.16) is more dominant in a neutral or alkaline environment. In both cases, however, the alkalinity increases at the cathode.

Pertaining to the copper electrode (cathode), Cu dissolution/oxidation occurs in the following steps [59]:

$$Cu = Cu^{2+} + 2e^- \tag{6.19}$$

$$2Cu + 2OH^- = Cu_2O + H_2O + 2e^- \tag{6.20}$$

$$Cu + H_2O_2 + 2H^+ = Cu^{2+} + 2H_2O \tag{6.21}$$

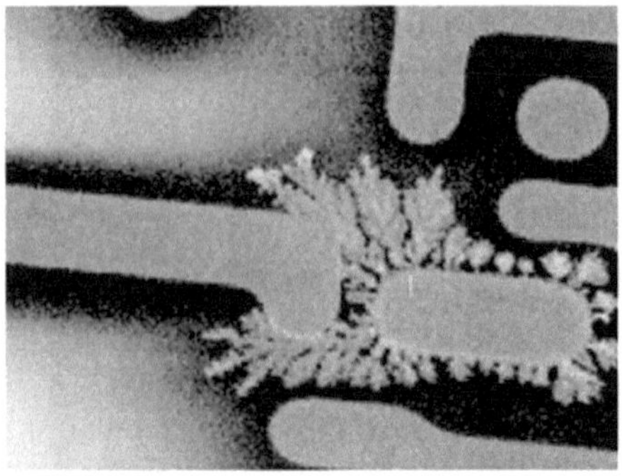

Fig. 6.28 Light-induced copper dendrite growth during post-CMP cleaning [55]

So far, we have concentrated on galvanic corrosion between Cu and the barrier metal Ta and its nitrides; however, the integration of Ru as a barrier material is being considered as an alternative material for Ta/TaN in the fabrication of Cu interconnect structures. The benefits of switching to Ru are its lower resistivity (7 μΩ cm) compared to Ta (14 μΩ cm) and TaN (200 μΩ cm) and Ru is more electrically conductive than Ta and its nitrides [60, 61]. Ru possesses advanced barrier characteristics such that it can support direct electro-deposition of Cu lines without using seed layers [60, 61]. However, due to the noble metal characteristics of Ru, it is difficult to integrate completely Ru barrier films in Cu interconnects due to the fact that Ru induces galvanic corrosion on the adjacent Cu lines during the CMP process of the multilayer components [60, 61]. Due to this, in the wet CMP environment, it is rather difficult to remove material in Ru-CMP without inducing surface defects as Ru forms a strong galvanic couple with Cu, and in this case, it is typically the Cu lines of the Ru–Cu structure that corrodes as a result of the coupling [60, 61]. In this galvanic coupling, Ru acts as the cathode while Cu acts as the anode.

Many studies have been carried out on the control of this galvanic corrosion of Cu during CMP of Ru. It can be concluded that there exists a large difference between the OCPs (E_{oc}) of Ru and Cu exposed to reactive CMP slurry; this large difference of E_{oc} is the cause of the galvanic corrosion of Cu [60, 61]. The predominant oxidation anodic reactions for Ru and Cu are as seen below [60, 61]:

$$Ru + 3OH^- = Ru(OH)_3 + 3e^- \tag{6.22}$$

$$2Cu + 2OH^- = Cu_2O + H_2O + 2e^- \tag{6.23}$$

where both $Ru(OH)_3$ and Cu_2O are passivating species at the anode. However reaction (6.22) is followed by a chemical conversion of $Ru(OH)_3$, where Ru is converted to RuO_2 by H_2O_2 in the solution [58]. The cathodic branch is caused by reaction (6.18).

Photo-induced electron is another cause of galvanic corrosion [55]. Figure 6.28 shows a light-induced copper dendrite growth on the surface of a patterned wafer. This unique corrosion mechanism usually occurs during the post-CMP cleaning steps [55].

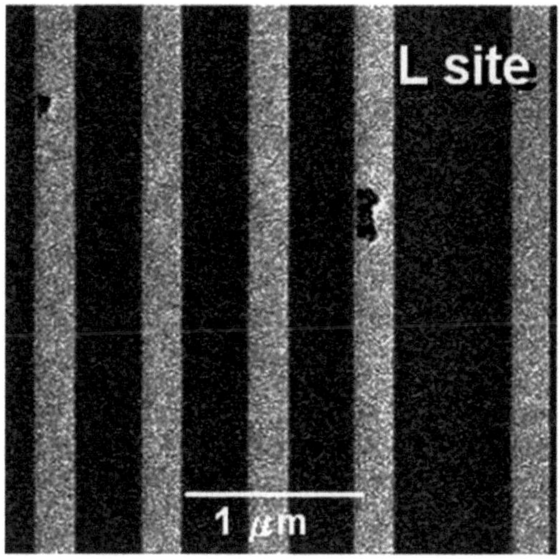

Fig. 6.29 SEM image of the via feature which has pits on the surface [64]

This occurs when the Cu is connected to *p–n* junctions, and is exposed to light. Within this *p–n* junction, there exist excited electron–hole pairs. The electrons are shifted toward the *n*-type material while the holes move toward the *p*-type material by existing electric fields. Copper connected to the *p*-type and *n*-type material forms the anode and cathode, respectively. The result is a depletion of copper from the anode (Cu is oxidized and the electrons are canceled by photo-induced holes) and an accumulation of Cu at the cathode (Cu is reduced by the photo-induced electrons) [55]. The dendrite defect can be controlled by avoiding incident illumination during post-CMP cleaning process.

6.5.2 Pitting

Pitting corrosion is a localized form of corrosion which forms small holes on the surface of metals. The higher corrosion rate increases the pitting on the surface and consequently the surface roughness. Figure 6.29 shows an example of pitting defect on the surface of copper interconnects. To control the pitting behavior the corrosion mechanism should be understood and the polish environment needs to be controlled carefully [65].

In copper the pitting occurs in both acidic and basic solutions with the presence of oxidizers. By using the Pourbaix diagram it can be realized that in acidic solutions the copper corrodes to Cu^{2+} and a cathodic reaction happens before the passive film forms to protect the surface. Defects such as scratches and grain boundaries, and impurities such as halides, $SO^{2-}{}_4$, $PO^{3-}{}_4$, $NO^{-}{}_3$, and $HCO^{-}{}_3$, act as copper pitting agents and help initiating localized galvanic corrosion on the surface. These anions specifically prevent any passive film formation by reacting

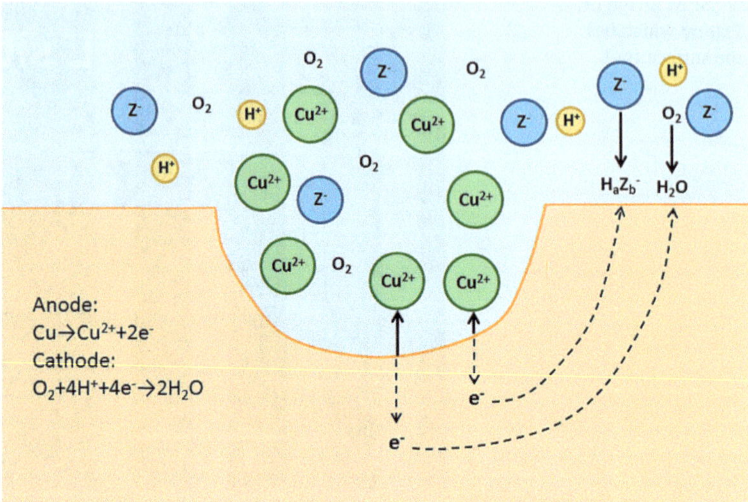

Fig. 6.30 The schematic mechanism of pitting corrosion on copper CMP in acid solution (6.4a). Z^- represents anion ions such as F^- and NO_3^-

with the metastable Cu_2O and releasing Cu^{2+} ions into the solution. Figure 6.30 schematically illustrates the pitting behavior in acidic environment. In this process the potential gradient and the electromigration occur between the localized anodes which form inside and the cathodes which from outside the pit. In basic solutions, Cu_2O passive film formation hinders the corrosion and pitting occurs when the oxide is removed mechanically [65, 66].

Adding corrosion inhibitors and additives such as benzotriazole (BTAH) in the acidic regime can reduce the pitting effect on metals. This happens due to the formation of a nonnative protective film on the surface of metal. Another useful method to decrease the pitting effect is to use a cathodic protection such as tantalum or titanium [64, 65].

6.5.3 Redeposition of Material

As mentioned earlier, chemical mechanical polishing of materials–especially metals–is mainly based on the etching of the surface. Chemical reaction and etching of the surface lead to the release of anions into solution which can be redeposited again on the surface. This undesired phenomenon forms a rough surface with a dendritic texture. Copper polishing is an example of the redeposition as shown in Fig. 6.31. In this figure, the dendrite structure is formed due to the redeposition of $Cu(OH)_2$ precipitation on the polished surfaces [67, 68]. Redeposition of silica abrasives is another unwanted post-CMP issues which can leave a rough surface

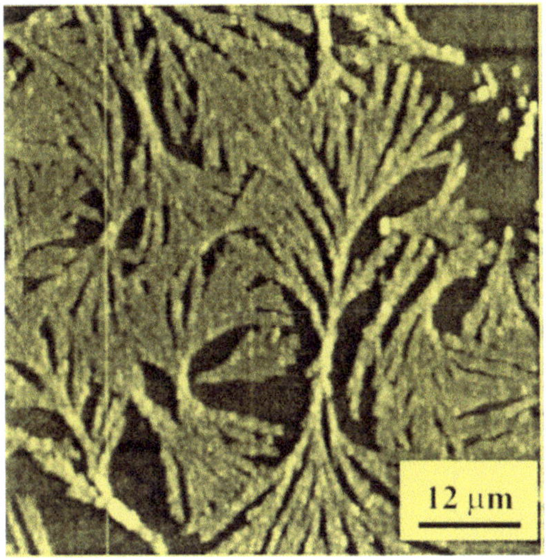

Fig. 6.31 AFM topography of redeposited cupper salt dendrites [65]

after polishing [69]. The redeposition of alumina abrasives after copper polishing is an example of the abrasive redeposition. There have been many cleaning studies aimed at preventing redeposition and it has been shown that the electrostatic repulsion of anions from the wafer surface can be a helpful cleaning method. If the polished surface and the anions carry the same charge repulsion occurs and if they have opposite charge attraction takes place [69, 70]. The repulsion and consequently the redeposition prevention occur by using the suitable chemical conditions such as changing the pH and solution zeta potential.

6.5.4 Etching and the Development of Texture

When two dissimilar metals are immersed in an electrolyte and are placed in contact, one metal will undergo oxidation and the other reduction, leading to etching of the first metal and redeposition of the second metal [55, 64]. Usually, etching solutions are made up of inhibitors and surface active compounds; due to this Cu etching is usually followed by deposition of corrosion products, Cu (I) compounds, on the surface [71].

During Cu CMP, the use of low pH slurries in the absence of corrosion inhibitors results in high static etch rates leading to enhanced surface texturing. This is shown in Fig. 6.32, where etching and surface texturing occurred in patterned wafers exposed to a Cu peroxide-based slurry and a static etch rate of 400 A/min [55, 64]. This type of texturing occurs in wide structures that are coated during copper electroplating in which the Cu at the middle of the structure is thinner than the Cu thickness at the edge. Since the center is thinner it is initially unpolished since the polishing pad is supported by the edges. The center undergoes chemical etching. Excess Cu loss

Fig. 6.32 Texturing defect induced by a high static etch slurry [64]

and surface roughening can occur if the unpolished center of advances below the ILD surface while there is Cu in the field regions [55, 64]. Cu^{2+} ions or a cupric complex is the most common etching reaction product of Cu CMP [64]. Cu^{2+} in the presence of complexing agents such as glycine enhances peroxide activity [72] contributing to increasing localized etching and texturing. However, an increasing peroxide concentration will lead to a decline in both the wet etch rate and polish rate of Cu which can be explained by the stability of the passivating layer [55]. Static etch rates can be controlled by increasing the pH of the slurry and implementing inhibitors like BTAH.

6.6 Conclusion and Future Directions

Chemical mechanical polishing is an essential enabling technology in modern semiconductor manufacturing. Electrochemical techniques have furthered the technology and are capable of characterizing the process with increasing accuracy. Topics of future research in CMP are closely related to the future direction of the semiconductor manufacturing industry as a whole. Some specific areas of progress are outlined below.

As described earlier in this chapter, galvanic interactions can influence the CMP process. As the range of metals involved in devices expands, the range of possible galvanic effects also expands. Using electrochemical research techniques can assist with assessing the effect of these interactions on CMP as part of the materials selection process.

Away from the semiconductor industry, CMP is also used to create atomically smooth surfaces for use (as an example) as substrates for molecular devices. While there is less commercial interest in such applications, CMP can be a significant enabling technology in academic research. Academic users employ a very wide variety of materials, including noble metals such as platinum [73], to produce ultrasmooth substrates for biological devices; hard and brittle ceramics such as transparent neodymium-doped yttrium aluminum garnet (Nd:YAG), used in lasers [74]; and polymeric materials as substrates in MEMS and bio-MEMS devices [75]. CMP of these surfaces is often conducted by the researcher themselves on an ad hoc basis to enable their own research goals, leading to a strongly multidisciplinary body of research. This is a valuable resource for CMP researchers of all persuasions and a source of innovation in the field.

Critical dimensions for chip components are reducing to the extent that they are commensurate with the size of the defects generated by the CMP process. This trend is likely to continue in the future, according to Moore's Law [76]. Hence, the pressure to reduce defects is one of the primary drivers in CMP as it allows critical dimensions to continue to shrink.

Leading-edge silicon processing is carried out on 300 mm wafers. Within the next several years, 450 mm processing is expected to become the standard [77]. This is likely to greatly impact CMP as many polishing metrics, such as removal rate and fluid pressure, are radially dependent. Current research on scaling to 450 mm wafers is focused on slurry usage and dissipation of the frictional heat generated [78].

References

1. Beyer K (1999) IBM MicroNews 5:46
2. ITRS Roadmap: http://www.itrs.net/Links/2012ITRS/Home2012.htm Table FEP14
3. Tripathi A, Suni I I, Li Y, Doniat F, McAndrew J (2009) J Electrochem Soc 156:H555-H560
4. Plummer J D, Deal M D, Griffin P B (2000) Silicon VLSI Technology, Prentice Hall, Upper Saddle River, New Jersey
5. Lin F, Nolan L, Xu Z, Cadien K (2012) J Electrochem Soc 159:H482-H489
6. Lu Z, Lee S-H, Gorantla V R K, Babu S V, Matijevic E (2003) Mater Res 18:2323-2330
7. Bielmann M, Mahajan U, Singh R K (1999) Electrochem Sol-Stat Lett 2:401-403
8. Luo J, Dornfeld D A (2003) IEEE Trans Semi Manuf 16:469
9. Nolan L, Cadien K (2103) Wear 307:155
10. Steigerwald J M, Murarka S P, Gutmann R J (2004) Chemical Mechanical Planarization of Microelectronic Materials. Wiley-VCH, Weinheim, Germany
11. Bastaninejad M, Ahmadi G (2005) J Electrochem Soc 152:G720
12. Greenwood J (1984) In: Proceedings of the Royal Society of London. Series A, Mathematical and Physical Sciences 393:133-157
13. Philipossian A, Mitchell E (2003) J App Phys 42:7259-7264
14. Coppeta J, Rogers C, Racz L, Philipossian A, Kaufman F (2000) J Electrochem Soc 147 (5):1903-1909
15. Preston F. W (1927) J Society of Glass Technology XI :214-256
16. Luo Q, Ramarajan S,.Babu S. V (1998) Thin Solid Films 335:160-167
17. DeNardis D, Rosales-Yeomans D, Borucki L, Philipossian A (2010) Thin Solid Films 518:3910–3916
18. Lee S-Y, Lee S-H, Park J-G (2003) J Electrochem Soc 150:G327-G332
19. Mazaheri A, Ahmadi G (2003) J Electrochem Soc 150: G233-G239

20. Li J, Lu X, He Y, Luo J (2011) J Electrochem Soc 158: H197-H202
21. Rogers C, Coppeta J, Racz L, Philipossian A, Kaufman, F.B, Bramono D (1998) J Electronic Materials 27:1082-1087
22. Seok J, Sukam C, Kim A, Tichy J, Cale T (2003) Wear 254 :307–320
23. Lu Y. S, Li N, Wang J, Zhang T, Duan M, Xing X.L (2011) Adv Mat Res 215:217-222
24. Wijekoon K, Lin R, Fishkin B, Yang S, Redeker F, Amico G, Nanjangud S (1998) Solid State Technology 41(4)
25. Cook LM (1990) J. Non-Crystalline Solids 120:152-171
26. Ziomek-Moroza M, Miller A, Hawk J, Cadien K, Li D.Y (2003) Wear 255:869-874
27. Lim G, Lee J-H, Son J-W, Lee H-W, Kim J (2006) J Electrochem Soc 153:B169-B172
28. Hernandez J, Wrschka P, Oehrlein GS (2001) J Electrochem Soc 148:G389-G397
29. Aksu, S (2005) Mater Res Soc Symp Proc 867:W1.6.1 - W1.6.6
30. Carpio R, Farkas J, Jairath R (1995) Thin Solid Films 266:238-44
31. Ein-Eli Y, Abelev E, Rabkin E, Starosvetsky D (2003) J Electrochem Soc 150:C646-C652
32. Ein-Eli Y, Abelev E, Starosvetsky D (2004) J Electrochem Soc 151:G236-G240
33. Aiken JK, Howard DK, Popplewell AF (1965) Inhibiting Copper Corrosion. British Patent 994,409
34. Tromans D (1998), J Electrochem Soc 145: L42-L45
35. Cadien KC, Nolan L (2012) Handbook of Thin Film Deposition. Seshan, K (Ed.), Elsevier, Amsterdam
36. Zheng JP, Roy D (2009) Thin Solid Films 517:4587-4592
37. Hong Y, Devarapalli VK, Roy D, Babu SV (2007) J Electrochem Soc 154:H444-H453
38. Hong Y, Patri UB, Ramakrishnan S, Roy D, Babu SV (2005) J Mat Res 20:3413-3424
39. Yang W L, Cheng C, Tsai M, Liu D, Shieh M (2000) IEEE Elect Dev Lett 21: 218–220 doi:10.1109/55.841301
40. Forsberg M, Keskitalo N, Olsson J (2002) Microelectronic Engineering 60:149–155. doi:10.1016/S0167-9317(01)00591-3
41. Liu D-G, Tsai M S, Yang WL, Cheng C-Y (2001) J Electronic Materials 30:53–58. doi:10.1007/s11664-001-0214-9
42. Senna JR, Smith RL (1995). Paper presented at the 8th Int Conf on Solid State Sensors and Actuators, and Eurosensors IX, Stockholm, Sweden, 25-29 June 1995.
43. Watanabe J, Yu G, Eryu O, Koshiyama I, Izumi K, Nakashima K, Kodama K (2005) Precision Engineering 29(2):151–156. doi:10.1016/j.precisioneng.2004.06.006
44. Evans D (2002). The Future of CMP. In: MRS Bulletin. http://www.mrs.org/publications/bulletin
45. Kim I-K, Cho B-G, Park J-G, Park J-Y, Park H-S (2009) J Electrochem Soc 156:H188 doi:10.1149/1.3058594
46. Lee W, Park H, Lee S, Sohn H (2004) Applied Electrochem 34:119–125
47. Peethala BC, Babu SV (2011) J Electrochem Soc 158:H271. doi:10.1149/1.3528942
48. Lee S-H, Kang Y-J, Park J-G, Lee S-I, Lee W-J Chemical Mechanical Planarization of Ruthenium for Capacitor Bottom Electrode in DRAM Technology. http://www.electrochem.org/dl/ma/203/pdfs/0446.pdf
49. Mainka G, Beitel G, Schnabel RF, Saenger A, Dehm C (2001) J Electrochem Soc 148:G552. doi:10.1149/1.1396339
50. Kim N-H, Ko P-J, Kang S K, Lee W-S (2007) Microelectronic Engineering 84:2702–2706. doi:10.1016/j.mee.2007.05.027
51. Kim I-K, Kang Y-J, Kwon T-Y, Cho B-G, Park J-G, Park J-Y, Park H-S (2008) Electrochem and Solid-State Letters 11: H150. doi:10.1149/1.2901544
52. Cui H, Park J-H, Park J-G (2012) J Electrochem Soc 159:H335–H341 doi:10.1149/2.103203jes
53. Zhang LM, Raghavan S, Weling M (1999) J Vac Sci Technol B 17:2248-55
54. Choiu WC, Chen YC, Lee SN, Jeng SM, Jang SM, Liang MS (2004) IEEE 8:127–129

55. Miller AE, Feller AD, Andryushchenko TN, Cadien KC (2003) ASM Handbook. vol 13A ASM International, p 164–9
56. Lin JY, Wang YY, Wan CC, Feng HP, Cheng MY (2007) Electrochem Solid-State Lett 10: H23-H6
57. Tamilmani S, Huang W, Raghavan S (2006) J Electrochem Soc 153:53-9
58. Brusic V, Kistler R, Wang SM, Hawkins J, Schmidt C (1998) Elec Soc S 98:119-25
59. Assiongbon KA, Emery SB, Gorantla VRK, Babu SV, Roy D (2006) Corrosion Science 48:372-88
60. Peethala BC, Roy D, Babu SV (2011) Electrochem and Solid-State Lett 14:306-10
61. Turk MC, Rock SE, Amanapu HP, Teugels LG, Roy D (2013) J Sol Stat Sci and Technol 2: P205-P13
62. Assiongbon KA, Emery SB, Pettit CM, Babu SV, Roy D (2004) Mater. Chem. Phys. 86:347
63. Kerrec O, Devilliers D, Groult H, Chemla M (1995) Electrochim. Acta 40:719
64. Miller AE, Fischer PB, Feller AD, Cadien KC (2001) Chemically induced defects during copper polish. In: Proceedings of the IEEE 2001 International Interconnect Technology Conference, Piscataway, NJ, USA, 4-6 June 2001
65. Tsai T-H, Yen S-C (2003) Appl Surf Sci 210:190–205. doi:10.1016/S0169-4332(02)01224-2
66. Bentz DN, Jackson KA (2002) Mat Res Soc Symp Proc 697:1–6
67. Xu G, Liang H, Zhao J, Li Y (2004) J Electrochem Soc 151:G688. doi:10.1149/1.1787497
68. Vogt MR, Polewska W, Magnussen OM, Behm RJ (1997) J Electrochem Soc 144:L113-L6
69. Fyen W, Vos R, Teerlinck I, Lagrange S, Lauerhaas J, Meuris M, Mertens P, Heyns M (2000) The Ninth International Symposium on Semiconductor Manufacturing (P-55):415–418
70. Xu K, Vos R, Vereecke G, Doumen G, Fyen W, Mertens P W, Kovacs F (2005) J Vac Sci & Technol B: 23:2160. doi:10.1116/1.2052713
71. Ein-Eli Y, Starosvetsky D (2007) Electrochimica Acta 52:1825–1838
72. Hariharaputhiran M, Zhang J, Ramarajan S, Keleher JJ, Li Y, Babu SV (2000) J Electrochem. Soc 147:3820–3826
73. Islam MS, Jung GY, Ha T, Stewart DR, Chen Y, Wang SY, Williams RS (2005) Appl Phys 80:1385-1389
74. Li J, Zhu Y, Chen TC (2008) Key Engineering Materials 278:375-376
75. Zhong ZW, Wang ZF, Zirajutheen BMP, Tan YS, Tan YH (2005) CMP of PC, PMMA and SU-8 Polymers. Paper presented at Polytronic 2005. 5th International Conference on Polymers and Adhesives in Microelectronics and Photonics, pp.58-62, 23-26 Oct 2005
76. Nigam T, Yiang K-Y, Marathe A (2013) Microelectronics to Nanoelectronics; Devices & Manufacturability. In: Kaul, Anupama (Ed.), CRC Press, Boca Raton, USA
77. Davis J (2013) 450 mm – It's bigger than you think. In: Solid State Technology. Available via DIALOG. http://www.electroiq.com/articles/sst/2013/06/450mm-_-it_s-bigger-than-you-think.html
78. Borucki L, Philipossian A, Goldstein M (2009) Solid State Technology, pp 10-13

Chapter 7
Metallization of Semiconductors and Nonconductive Surfaces from Aqueous Solutions

Stojan S. Djokić and Luca Magagnin

7.1 Introduction

Deposition of metals and alloys from aqueous solutions onto semiconductors and nonconductive surfaces, e.g., polymers, ceramics, and textiles, can be challenging. Direct electrodeposition of metals from aqueous solutions onto nonconductive surfaces is not practically successful. Using the physical vapor deposition (PVD) or chemical vapor deposition (CVD) in order to deposit various metals on semiconductors or nonconductors is usually quite successful. However, PVD and CVD processes are very expensive and require complicated equipment.

Specific pretreatment of the nonconductive surface is required in order to deposit metals from aqueous solutions. Once when the semiconductor, polymer, or ceramic is properly activated, it may act as a catalytic surface for the autocatalytic electroless deposition of a desired metal. To achieve a specific thickness or surface morphology of the deposited film, electrodeposition may be applied as well. Activation of the nonmetallic surfaces is practically achieved by using the $SnCl_2$/$PdCl_2$ processes [1–3]. A further autocatalytic deposition of metals such as Ni, Cu, Ag, Au, etc., on the activated nonconductive surfaces from appropriate aqueous solutions containing mentioned metallic ions and specific reducing agents can be very successful. These processes are used on industrial scale.

This chapter describes the achievements related to the metallization of semiconductors, polymers, and ceramics from aqueous solutions.

S.S. Djokić (✉)
Elchem Consulting Ltd, 15511-103 Street NW, Edmonton, AB, Canada T5X 6B3
e-mail: sdjokic@telus.net

L. Magagnin
Dipartimento di Chimica, Materiali e Ingegneria Chimica 'Giulio Natta',
Politecnico di Milano, Via Mancinelli, 7, 20131 Milano, Italy

S.S. Djokić (ed.), *Electrodeposition and Surface Finishing: Fundamentals and Applications*, Modern Aspects of Electrochemistry 57,
DOI 10.1007/978-1-4939-0289-7_7,

7.2 General Concepts of Metallization of Semi-Conductive and Nonconductive Surfaces

Semi-conductive and nonconductive surfaces must be appropriately activated prior to deposition of metals and/or alloys from aqueous solutions. Due to differences among the semiconductors and nonconductive polymers and ceramics, different approaches are applied in order to deposit specific metal onto their surfaces. Consequently they are analyzed separately in the present chapter.

Practically, any nonconductive surface can be activated for the deposition of metals from aqueous solutions using the $SnCl_2/PdCl_2$ procedure [1, 2]. A general outline of the specific steps involved with the metallization of the nonconductive surfaces from aqueous solutions is schematically presented in Fig. 7.1.

Of course, it is not necessary to apply every step outlined in Fig. 7.1 in order to achieve a successful metallization of the nonconductive surface. An omission of some of these steps could still lead to a successful deposition, which depends on the material. In some cases an introduction of additional steps may be required in order to achieve the successful results as will be demonstrated in the following sections. Similar steps as those outlined in Fig. 7.1 can be used for the electroless deposition of some metals onto other metallic surfaces which are not susceptible to the deposition of the desired metal.

Degreasing and cleaning are aimed to remove the light soils such as fingerprints, dirt, and other debris from the surface. The cleaning and degreasing are usually achieved by using the organic solvents, e.g., acetone, ethanol, and similar. Rinsing

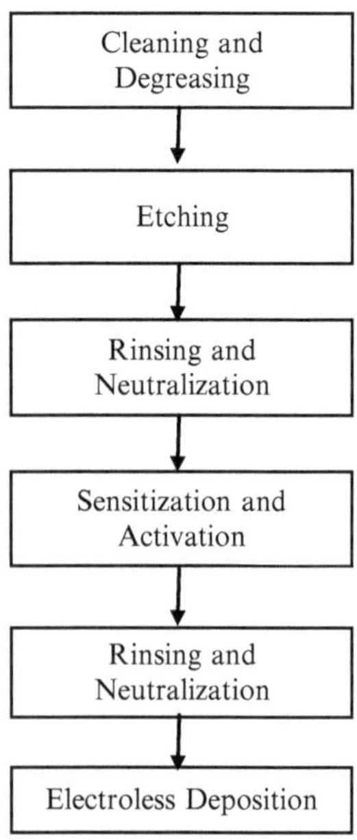

Fig. 7.1 An outline of the surface preparation prior to electroless deposition

and neutralization are carried out using the deionized water. A choice of the etching solution depends on the nature of the surface. It is obvious that various polymeric, ceramic, or semiconductor surfaces would require different etching agents. After all required steps, e.g., cleaning, degreasing, etching, etc., are performed, the surface is further activated using appropriate solutions for the autocatalytic electroless deposition [1].

7.2.1 Surface Sensitization and Activation

Prior to metallization, clean nonmetallic surfaces must be activated for the electroless autocatalytic deposition. The surface activation for the electroless deposition depends on the material and a specific metal aimed to be deposited. For example, after degreasing and cleaning, glass surface is usually treated with the $SnCl_2$ solution [1]. It is believed that upon immersion of glass into $SnCl_2$ solution, the Sn^{2+} ions are adsorbed onto the surface. After rinsing, when immersed into an Ag (I) containing solution in the initial stages the adsorbed Sn^{2+} ions at the glass surface act as a reducing agent of Ag^+ ions according to the following reaction:

$$Sn^{2+} + 2Ag^+ \rightarrow 2Ag + Sn^{4+} \tag{7.1}$$

If an appropriate reducing agent, e.g., formaldehyde, hydrazine, or similar, is present in the solution, then Ag produced according to reaction (7.1) provides active sites for the further growth of silver film onto a glass surface via the autocatalytic electroless deposition.

For other materials, the most widely used method for the surface sensitization/activation includes the classical approach based on the $SnCl_2$ and $PdCl_2$ [1]. This is an early version of the "two-step" procedure. A treatment of the nonmetallic surfaces with $SnCl_2$ is known as a sensitization step [1]. Using the acidic $SnCl_2$ solution during the sensitization step the Sn^{2+} ions are adsorbed at the treated surface, which is described in the existing literature as a "first step" [1]. In the following step the substrates are immersed into the $PdCl_2$ solution, which is known as a "second step." During the "second step" the Pd^{2+} ions are reduced at the spots of the adsorbed Sn^{2+} ions according to the following reaction:

$$Sn^{2+} + Pd^{2+} \rightarrow Sn^{4+} + Pd^{o} \tag{7.2}$$

The reduced Pd atoms, formed during the activation step, act as catalytic sites for the autocatalytic deposition of the desired metal, e.g., Ni, Cu, Ag, Au, or similar.

In some cases the sensitization and the activation steps are combined as one step. In other words, $SnCl_2$, $PdCl_2$, and HCl are mixed together representing one solution which is further used for the treatment of nonconductive surfaces prior to electroless deposition.

Other methods for the surface activation for the electroless deposition of metals onto the nonconductive surfaces were also described in the literature. They may include PVD or CVD of specific metals, e.g., Pd, Ag, Au, or similar, which may act as the catalytic surfaces for the autocatalytic electroless deposition or as a conductive surface for the electrodeposition. One published communication reports on PVD of thin Pd film (approximately 20 nm thick) onto alumina surface, on which Ti film of approximately similar thickness was deposited using the PVD [4]. A plasma treatment process for palladium chemisorption onto polycarbonate substrates before the autocatalytic electroless deposition was also recommended [5]. An immersion of polymeric or ceramic surfaces into organosilane solutions, followed by an exposure to the palladium catalyst solution, was also disclosed as an activation step for the further metallization [6, 7]. Other processes recommended for the activation of nonconductive surfaces for the metallization from aqueous solutions are carbon/graphite systems, conductive polymers, and non-formaldehyde electroless processes [8].

7.2.2 Galvanic Displacement and Semiconductors

The semiconductors such as silicon or germanium can be successfully metallized in a way as described. However, it is believed that an easier approach may be applied for these materials taking the advantage of the galvanic displacement reaction. Standard electrode potentials for Si^{4+}/Si and Ge^{4+}/Ge couples are -1.2 and -0.124 V, respectively [9]. The values of these standard electrode potentials suggest that many noble metals, e.g., Cu, Ag, Au, Pd, Pt, etc., can successfully be deposited via the galvanic displacement reaction onto silicon and germanium substrates. Galvanic displacement reaction is generally based on the oxidation of the substrate and a consequent reduction of the ions of a more positive element. For semiconductors, if a proper solution containing the ions of a more positive element is chosen then due to oxidation of Si or Ge reduction of ions such as Cu(II), Ag(I), Au(I), or similar may occur at the surface.

The metallization of silicon and germanium via the galvanic displacement reaction can be quite challenging due to the existence of the naturally formed oxides such as SiO_2 and GeO_2 at the surface of the substrate. The naturally formed thin SiO_2 film prevents the oxidation of the silicon substrate and the reduction of more positive ions. Consequently, the deposition of the desired metal does not occur. This further suggests that a proper composition of an aqueous solution must be selected in order to deposit a desired metal onto the surface of a semiconductor (Si or Ge). For the case of silicon it is successfully realized using the fluoride-based solutions [10]. It is very well known that fluoride-based solutions or hydrofluoric acid can dissolve the SiO_2 film according to the following reaction:

$$SiO_2 + 4HF \rightarrow SiF_4 \uparrow + 2H_2O \tag{7.3}$$

As this reaction describes, the SiO_2 is etched away, producing the SiF_4 gas. Upon the dissolution of the SiO_2 film, a further oxidation of the silicon substrate can be described with the following half reaction:

$$Si + 6F^- \rightarrow SiF_6{}^{2-} + 4e \quad (7.4)$$

with $E^o(Si^{4+}/Si) = -1.20$ V.

As reaction (7.4) clearly shows, a reduction of any more positive element should be possible if the fluoride-based solutions contain these ions, e.g., Cu(II), Ag(I), Pd(II), Au(I), Pt(IV), or similar. This type of deposition has been experimentally confirmed by many researchers [9–13].

For example, the deposition of Pt onto silicon semiconductor from $PtCl_6{}^{2-}$ containing solutions can be explained by the following reaction [10]:

$$PtCl_6{}^{2-} + Si + 6F^- \rightarrow Pt + SiF_6{}^{2-} + 6Cl^- \quad (7.5)$$

Reaction (7.5) is obtained as a combination of the cathodic reaction,

$$PtCl_6{}^{2-} + 4e \rightarrow Pt + 6Cl^- \quad (7.6)$$

with $E^o(Pt^{4+}/Pt) = 0.74$ V,

and the anodic reaction (7.4), described above.

A similar process for the deposition of Pd onto Si substrates based on the $PdCl_2$–HF–NH_4F solution [14] can be described by the reaction:

$$Si + 6F^- + 2Pd^{2+} \rightarrow SiF_6{}^{2-} + 2Pd^o \quad (7.7)$$

It seems, based on the experimental observations, that the metallization of silicon requires the fluoride-based solutions.

Similar is applicable to the germanium semiconductor substrates. However, experimental observations [15] have shown that in the case of germanium fluoride-based solutions are not necessarily required to deposit the noble metals on these surfaces via the galvanic displacement reaction. This further suggests that if a proper fluoride-free solutions are selected, GeO_2 formed at the surface of the germanium substrate can be dissolved and a reduction of desired metallic ions, e.g., Cu(II), Ag(I), Au(I), or similar, may occur [17]. The details of the metallization of semiconductor substrates will be discussed in the later sections of this chapter.

7.2.3 Autocatalytic Electroless Deposition

When the nonconductive or semiconductor surfaces are properly activated by the methods described in Sect. 7.2.1 or 7.2.2, for the further deposition and an increase in the thickness of deposited metal, autocatalytic electroless deposition or electrodeposition methods are recommended.

Description of the autocatalytic electroless processes, which use specific reducing agents, depending on the metal being deposited is thoroughly described in the existing literature. For comprehensive reviews, the reader is referred to the publications [1–3, 16, 18] and references therein.

7.2.4 Electrodeposition

The electrodeposition methods can be very useful if thicker deposits are required for a particular application. In this case, after the surface activation (Sects. 7.2.1 and 7.2.2) in many cases a direct electrodeposition may be applied. If this approach may not work, then the methods described in Sect. 7.2.3 should be used. Methods of the electrodeposition of many different metals and/or alloys are readily available in the published literature.

7.3 Metallization of Semiconductors

Integration of metals with semiconductors has been a crucial milestone for many technologies, from integrated circuit and microsystem technologies to nowadays alternative energy systems, e.g., batteries and fuel cells. The need for all wet metallization processes of semiconductor is gaining great interest, proving the feasibility of these approaches in many fields. A lot of literature has been written of this topic and fundamental description of the wet-based processes has been discussed in many dedicated chapters or books; see [1, 2, 11, 19]. In this section, a review of the more recent literature is provided, showing new applications and routes for wet-based metallization processes of semiconductors.

Galvanic displacement can be considered a type of electroless deposition: the main difference between autocatalytic plating and galvanic displacement is that the former uses reducing agents in the solution to deposit metal ions on the substrate, whereas displacement corrodes the underlying substrate to reduce the metal ions onto the substrate. Galvanic displacement of noble metals and platinum group elements is a well-known process, already proven for many technological applications [11]. Relatively new is the use of galvanic displacement for the deposition of two different metals. The selective and conformal nature of the electroless deposition was used to fabricate very long (up to 1 mm in length), slender (5 μm × 5 μm) metalized micro-cantilever beams with a very simple, one-mask fabrication process, combining the galvanic displacement of copper on silicon with the galvanic displacement/autocatalytic deposition of gold in a cyanide electrolyte [20]. Copper was deposited in a plating solution based on a promoting adhesion formulation by Magagnin et al. [21] and is comprised mainly of copper sulfate, ammonium fluoride, ascorbic acid, polyethylene glycol, and methanol to improve the film quality and adhesion [22]. Cu adhesion to Si with a simple Cu-silicide (Cu_3Si) formation step was improved. Since active Cu surface oxidizes rapidly, it needs to be capped immediately with a noble metal, such as Au. Direct immersion gold is a relatively new approach that has been formulated to meet the industry's need for defect-free surface finishes for second-level interconnect technology [23]. The method uses a combination of galvanic displacement and autocatalytic deposition to form relatively pore-free selective Au thin films on Cu. This method was used to fabricate metalized micro-cantilever beams with selective metallization [20].

Although Au–Si systems prepared by galvanic displacement have been extensively studied, according to our knowledge, the preferential deposition on certain crystallographic facets has not been reported. Sayed et al. reported the preferential deposition of gold nanoparticles on the Si(110) rather than the Si(111) facets of silicon nanowires (NWs) grown along the <112> growth direction [24]. A solution of $KAuCl_4$ and HF was used to grow gold nanoparticles on Si NWs. It was demonstrated that the galvanic displacement of gold salts on silicon NWs results in gold nanoparticles interfaced directly to the NWs. In Si NWs with growth directions in the <112> direction, Si(110) planes were found to be the preferred surfaces for gold deposition by galvanic displacement. The (110) surfaces are more reactive than the (111), which agrees with the surface properties and energies, and lattice structures of the (111) and (110) facets.

The electroless deposition of metallic nanoparticles (NPs) onto semiconducting surfaces by galvanic displacement has been demonstrated to be a powerful, simple, and fast method to produce metallized surfaces with a high density of NPs [25]. It has been demonstrated that fluoride-free electroless deposition by galvanic displacement combined with a UV photolithography method has been used for micrometer scale patterning of metallic NPs on Si(100)–H surfaces. Using this simple and versatile process, Fabre et al. have demonstrated the ability to fabricate single Au and bimetallic Au and Pt NP micropatterns [25]. Such metallic NP micropatterned surfaces were used as photocathodes for H_2 evolution from water, exhibiting much higher electrocatalytic performance than that of non-metallized Si (100)–H, both in the absence of light and under illumination. Cerruti et al. reported on the galvanic deposition of Pt on Si from solutions containing $PtCl_2$ and different concentrations of HF, proving that the Pt and PtSi films formed are able to catalyze the formation of Si nanowires arrays formed via vapor–liquid–solid process [26]. For low [HF]/[Pt] ratios, no Pt deposition was observed, and only a thin layer of PtSi was formed. For intermediate values of [HF]/[Pt], an increase in the rate of Pt deposition was observed; large clusters of Pt were observed only for high [HF]/[Pt], which did not cover the Si substrate as densely as the smaller ones observed at lower [HF]/[Pt] ratios. Ohmic contacts with a palladium (Pd) diffusion barrier were formed on GaAs substrates [27]. The Pd films were deposited on GaAs substrates by the displacement method: the chemical bath was a mixture of deionized water, HF, and $PdCl_2$. Pd films were deposited by immersing the substrates for a period of 10–120 s. The authors suggest the following mechanism:

$$GaAs + 3Pd^{2+} \rightarrow 3Pd + Ga^{3+} + As^{3+} \quad (7.8)$$

with formation of $GaCl_3$ and AsH_3. The inclusion of the Pd film in the semiconductor–metallic–alloy interface was proved to act as diffusion barrier. A remarkable result of this study was that the Pd film inhibited the formation of metal oxide films at the interface structure.

It has been shown by Romann et al. that bismuth films with up to few hundred nm thickness can be deposited onto silicon wafer using bismuth(III)oxide and hydrofluoric acid as reagents [28]. XPS data showed that pure bismuth on silicon was deposited by galvanic displacement, but no silicon(IV)oxide or bismuth silicide was detected.

It deserves to be mentioned the introduction of the classical zincate process by galvanic displacement on aluminum into the solar cell technology. Sufficient adhesion on PVD aluminum layers on silicon was demonstrated by zincate processes with subsequent plating of nickel, copper, and silver on both random pyramid and damage etched surfaces. Small-size back-contact back-junction solar cells with an aluminum or aluminum–silicon seed layer were successfully processed in this way [29].

Cutting-edge interconnection processes now have electrochemical deposited copper lines of critical dimensions below 60 nm set within inter-metal dielectric layers. The linewidth is expected to shrink further as technology nodes advance, thus raising several issues. Chen et al. developed an alloying and direct patterning process for fabricating Cu–Mn nanostructured thin films using electroless deposition in conjunction with a new site-selective seeding process. The manganese alloyed in Cu has concentrations of only 0.4 at.%, but can be segregated to form a Mn-bearing oxide barrier at the SiO_2/Cu interface [30]. The use of Ni/Cu double layers as front contact on commercial c-Si solar cells for low concentration application is proposed by Chaudhari and Solanki [31]. The front contact consists of Ni and Cu layers, which are deposited using electroless deposition and electroplating, respectively. These double layers of metals help in reducing the series resistance of solar cells. Ni is deposited on Si wafers, using Ni salt and sodium hypophosphite reducing agent in an electroless bath. Further the Ni-deposited wafers are annealed between temperatures 400 and 430 °C, to obtain NiSi (nickelsilicide), which lowers the contact resistivity between Ni and Si. Electroless deposition of CoNi alloy thin films on a Si substrate was investigated by Kim et al. in order to transfer electrons from metal interconnections to the Si substrate by an ohmic contact [32]. The quality of the film was strongly influenced by the concentration of the electrolyte and the presence of NH_4F. A quality adhesive CoNi thin film was formed when the concentration of $NiSO_4{\cdot}6H_2O$ exceeded 0.03 M, and this is attributed to self-activation at higher Ni ion concentrations. The presence of NH_4F in the electrolyte also influenced the crystallinity of the film and the deposition rate. Electroless deposition of barrier layers is gaining more and more interest in the last few years for integrated circuits and particularly for TSV application. Self-aligned CoWP metal cap has been reported to be helpful in improving degraded electromigration (EM) reliability.

It was found by Chen et al. that CoWP could induce a severe degradation of low-k dielectric if its process is not optimized, and its impacts on dense low-k and porous ultra low-k dielectrics could be different [33]. Electroless technology offers a low-temperature and low-cost diffusion barrier process for fine diameter and high aspect ratio TSVs. A thin continuous diffusion barrier was formed using electroless Co–W–B plating with a Pd–NP catalyst in fine TSV by Inoue et al. [34]. The high-density adsorption of Pd–NPs assisted the efficient formation of a thin and continuous electroless Co–W–B layer. It was shown that the adhesion strength of 40-nm-thick electroless Co–W–B films exceeded 80 MPa. The adhesion strength of the electroless Co–W–B films exceeded 70 MPa after annealing at 200 °C, even for a film thickness of 80 nm. In addition, the Co–W–B film has good barrier properties against Cu diffusion, even after annealing at 300 °C. Tao et al. consistently proved that, according to the experimental results, with W in the layer the barrier performance is improved. Although the resistance of NiWP is higher than that of NiP, the resistance of a stacked film of Si/NiWP/Cu is close to that of SiNiP/Cu [35].

Recently, electroless deposition has been receiving interest as a Cu-filling method because this technique does not require a Cu seed layer. Kawano et al. developed an electroless Ni plating process for TSVs and applied it to 3D stacked DRAM. Electroless Ni was plated onto W pads from the back of the Si, and void-free bottom-up filling was achieved. The adhesion strength between the Ni and W increased during thermal annealing of over 100 °C. The plating rate uniformity was significantly improved by light shielding during plating, because this prevented the occurrence of photoelectric currents at the p–n junctions [36]. Han et al. successfully filled a 41-nm trench pattern using SAM-modified electroless copper deposition [37]. Santagata et al. demonstrated the possibility to use Cu electroless plating on the chemically activated TiN surfaces to uniformly fill the TSVs in a very short time. TSVs, 300 μm deep and 20 μm wide, were filled in less than 2 h [38].

The interface between Si and Au has been studied for many decades primarily due to interest in its electronic properties. Akhtari-Zavareh et al. investigate the electrical properties of Au/Si diodes prepared using aqueous electrodeposition under specific conditions of polarization to obtain epitaxial Au (111) layers that are free of visible grains and smooth on the atomic scale. The plating solution was a $HAuCl_4$ solution (0.1 mM) in a supporting acidic electrolyte [pH 4, K_2SO_4 (0.1 M) KCl (1 mM) H_2SO_4 (1 mM)] [39]. The superior stability in air of epitaxial electrodeposited Au/Si interfaces compared to their vacuum-evaporated Au indicated the presence of a uniform, submonolayer thick, interdiffusion barrier (such as residual impurities) at the Si interface. The nanocrystalline magnetic Ni films were grown on n-Si(111) substrate by pulsed electrodeposition in nonaqueous $NiCl_2$ + methanol solution by Lee et al. [40]. Ni nanocrystallites at initial deposition stage nucleated through 3D instantaneous nucleation followed by diffusion-limited growth. The maximum values of Hc perpendicular and Hc parallel were 200 Oe and 203.5 Oe, respectively.

A three-dimensional selenium solar cell with the structure of Au/Se/porous TiO_2/compact TiO_2/fluorine-doped tin oxide-coated glass plates was fabricated by an electrochemical deposition method of selenium, which can work for the extremely thin light absorber and the hole-conducting layer, by Nguyen et al. [41]. The crystallinity of the selenium layer after annealing at 200 °C for 3 min in the air was significantly improved, and the band gap became narrower in comparison to the sample both with and without annealing at 100 °C. The obtained results and PV photovoltaic efficiencies were interesting for PV researchers because the fabrication method for this kind of solar cells is quite simple.

7.4 Metallization of Polymers

The fabrication of metallic patterns on functional polymeric materials is becoming extremely important for manufacturing processes of microelectronic devices and flexible electronics. The conventional metallization is typically achieved by plasma activation, laser-induced deposition, or vapor phase deposition.

Unfortunately, the poor adhesion strength of the metal directly deposited onto polymers, which is a consequence of their low free surface energy, has to be overcome. Plasma technology, corona discharge, and bombardment with charged particles have been used in order to improve the wettability, printability, biocompatibility, and other related surface properties of the polymeric materials [42]. Low-pressure plasma treatment is a well-known method to modify the surface of polymers, presenting advantages in contrast to chemical treatments. It is a fast treatment, it is suitable for the treatment of complex geometric samples, and it modifies the surface chemistry producing an increase in the adhesion properties [42]. Pretreatment of polymers with the purpose to change adhesion properties can be classified in two main methods: chemical surface modifications and physical surface modifications. The effects caused on the polymer surface by these techniques are the incorporation of functional groups, changes in the surface morphology, and alteration in the chain structure. Cho et al. applied microwave MW plasma to the surface of polyimide (PI) films as a treatment to enhance the adhesion of electroless copper deposition layers [43]. The results showed that N_2 MW plasma was successfully used for achievement of chemical modification on the PI surface. The N_2 MW plasma produces chemical change to the PI surface through nitrogen insertion on the PI surface.

Following electroless copper by free formaldehyde method using glyoxylic acid as a reducing agent for 40 min at 40 °C was characterized by high adhesion to the PI substrate. Borris et al. developed an environmentally friendly plasma amination process for the activation of polymers prior to electroless metallization using dielectric barrier discharges (DBD) at atmospheric pressure [44]. Using XPS spectroscopy it was found that a DBD treatment of polyimide (PI) films with mixtures of N_2 and H_2 leads to considerably higher Pd surface concentrations than on untreated reference samples or foils treated in air-DBD. Typical thicknesses of the electroless metal deposits were estimated to be in the range of 50–200 nm. Galvanic reinforcement of electrolessly metallized samples to coating thicknesses of up to approximately 35 mm was carried out using conventional acid copper baths. Results obtained so far show that also adherent nickel layers could be deposited on PI using electroless nickel baths [44].

Metallization on fluorinated polymers can be achieved by electroless metallization after plasma activation in order to increase adhesion. Pascu and coworkers have modified in a controlled manner the surface properties of polyvinylidene fluoride (PVDF) films, by treatment in a radiofrequency nitrogen plasma, and a microwave one, at different treatment powers [42]. It was proved that both plasma treatments significantly improved the adherence and the quality of the deposited metal layer. The RF plasma treatment seems to be less efficient in respect to the microwave plasma one, for small powers, but the obtained results are similar at discharge powers higher than 50 W. The amount of the deposited copper and the thickness of the metal layer were significantly higher for the treated PVDF samples in respect to the untreated one.

Innovative fabrication methods to tune the polymer surface characteristics and enhance the adhesion between metal and polymeric substrate are recently studied and developed based on the grafting of the polymer with monolayers. Recently, Akamatsu et al. [45, 46] reported site-selective surface modification for the generation of copper micropatterns on a polyimide surface. Masuda and coworkers [47] fabricated a micropattern of copper thin film on a poly(ethylene terephthalate) substrate by electroless deposition using a self-assembled monolayer patterned with different functional groups (SH and OH terminal groups) as a template. Carmichael and coworkers [48] oxidized the various polymeric substrates to form surface-bound carboxylic acid groups, patterning of an aluminum porphyrin monolayers to bind a Pd/Sn colloidal catalyst that subsequently initiated the selective deposition of copper in an electroless plating solution. Yang et al. demonstrated a novel, simple, and universal method for the site-selective deposition of copper granular layer patterns on a variety of flexible polymeric substrates such as poly (imide) (PI), poly(ethylene naphthalate), and poly(ethylene terephthalate), as in Fig. 7.2. The method relies on the use of grafted polymer template incorporating palladium catalysts, which can be acted as adhesive interlayers for fabricating precise copper patterns on flexible polymers [46].

A supercritical carbon dioxide ($scCO_2$)-assisted electroless Ni–P plating technique on a polymer substrate was modified and improved for polyolefin polymers by Tengsuwan and Ohshima [49]. The technique consists of two steps: the first step is the $scCO_2$-assisted impregnation of the polymer substrate with $Pd(hfa)_2$ and the second step is the electroless plating reaction. A the Ni–P metal layer was successfully formed by electroless deposition onto polypropylene PP with an average adhesive strength of 7.9 ± 0.5 N/cm to the polymeric substrate, as shown in Fig. 7.3.

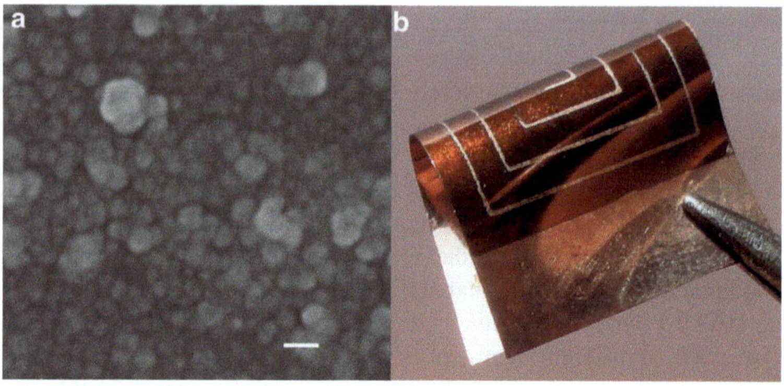

Fig. 7.2 (**a**) SEM image of copper patterns on PI substrate. The scale bar = 200 nm. (**b**) Optical photograph of copper circuit patterns with the lines of 200 μm width. Reprinted with permissions from Elsevier (Yang et al. [46])

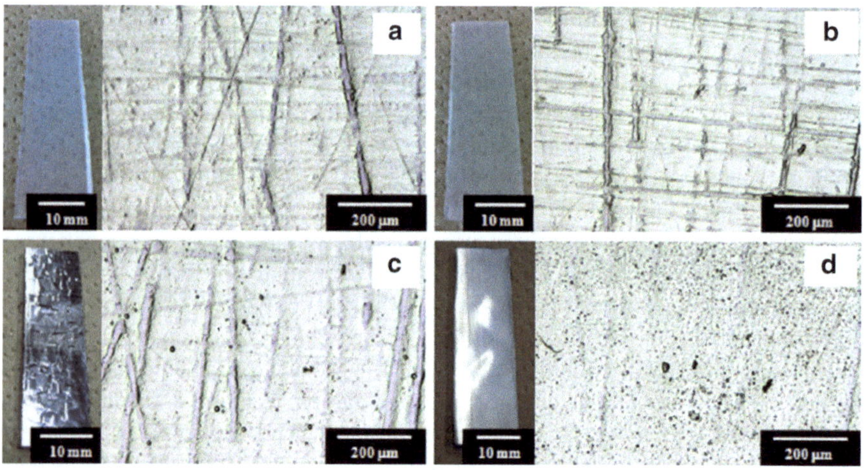

Fig. 7.3 Images of a digital camera (*left*) and an optical microscope image (*right*) of the surface of (**a**) a non-plated neat PP substrate, (**b**) a non-plated PP-modified substrate, (**c**) a Ni–P plated PP substrate, and (**d**) a Ni–P plated PP-modified substrate. Reprinted with permissions from Elsevier (Tengsuwan and Ohshima [49])

7.5 Metallization of Ceramics

The ceramic substrates can successfully be metallized via electroless deposition from aqueous solutions. However, the ceramic surfaces do not have catalytic activity for the electroless deposition and, as such, require sensitization and activation. In addition, a significant attention must be paid to achieve an acceptable adhesion of the metal coating to the ceramic substrate. This is usually realized

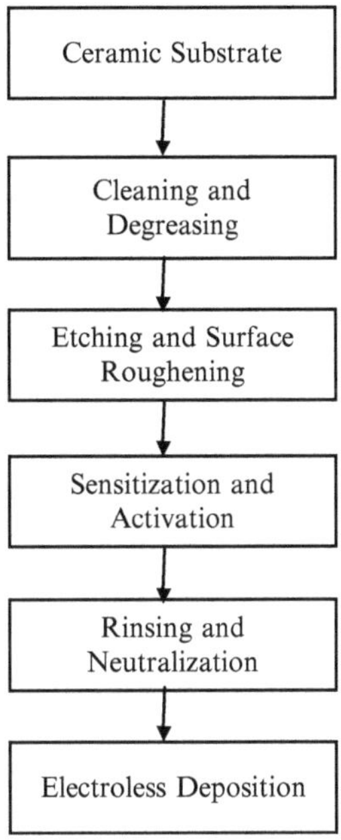

Fig. 7.4 An outline of steps required for the metallization of the most ceramic substrates via electroless deposition

by the micro- or nano-roughening of the substrate using the mechanical or chemical methods.

Metallization of ceramics via electroless deposition in general terms can be carried out as illustrated in Fig. 7.4. Of course, one should keep it in mind that the scheme given in Fig. 7.4 is too general. Many concepts of the ceramic substrate such as origin of the sample, composition, surface roughness, etc., must be known in order to successfully metallize its surface.

Procedures for cleaning, degreasing, etching, and activation can be found in the literature [1, 2]. Usually, the ceramic substrates, prior to etching, are cleaned and degreased using the detergents, alkaline solutions (e.g., NaOH or KOH), and various organic solvents (e.g., acetone, ethanol, ethyl acetate, or similar). Rinsing is carried out using the distilled water.

The successful etching and roughening of the surface depend on the composition of a specific ceramic material. Substrates, such as AlN and Al_2O_3, can be etched using alkaline solutions [50]. Glass, silicon-based, and alumina-based ceramics can be efficiently etched using the hydrofluoric acid due to affinity of fluoride ions for silicon or aluminum. These materials are etched with solutions containing HF, NH_4F, and similar, usually at room temperature in the plastic containers. These operations must be very carefully carried out with all safety precautions and in fumehood. Other ceramics, e.g., barium and zirconium titanates, and similar are usually etched with fluoroboric acid (48 %, for 10 min at room temperature) [1]. As mentioned earlier, the composition of a particular ceramic is very important and this will determine the proper solution and the conditions for the etching of a specific surface.

The etched ceramics are further treated with sensitizing and activating agents to provide the active sites or, as commonly known in the practice, a *seed layer* for the future electroless deposition of a desired metal, e.g., Ni, Cu, Ag, Au, Co, etc., or their alloys. The most frequent method used for the activation of ceramics for the electroless deposition is based on the $SnCl_2$ and $PdCl_2$ pretreatment steps as described earlier. Other methods such as pretreatment with $AgNO_3$ or $CuSO_4$ and similar solutions, further photosensitization, or treatments with corresponding reducing agent solutions can lead to quite successful results.

When the ceramic surface is activated, further electroless deposition of nickel, copper, silver, gold, or other desired metal or alloy from aqueous solutions proceeds successfully if a proper solution, reducing agent, and temperature are correctly chosen.

7.5.1 Developments in the Metallization of Ceramics

The metallization of ceramics was investigated in the past predominantly due to their applications in the microelectronics packaging, sensors, and aerospace applications. Electroless deposition of copper, silver, or nickel was used to achieve a good electrical contact in the tin oxide film resistors and integrated circuits [51].

De Luca and McCormack described a process for the metallization of ceramics [52]. In this process, a flat black ceramic containing 90 % Al_2O_3 and 10 % other oxides were used as substrates for a further metallization for the electronics applications. After degreasing and cleaning procedures in alkaline solutions the black ceramic substrates were immersed into 760 g/L NaOH solution and then removed without washing or rinsing with water. These samples were then dried at 170 °C for 10 min in order to remove water in the sodium hydroxide film on the substrate. After drying, ceramic samples were heat treated in an air atmosphere for 15 min at 450 °C. The heat treatment at 450 °C was carried out in order to fuse the sodium in the ceramic substrate, to increase the surface roughness and to consequently promote a good adhesion of the deposited metal. After cooling, the ceramic substrate is rinsed with water and 20 % H_2SO_4 to neutralize the residual NaOH at room temperature. These authors claimed that similar results can be obtained using any type of alkali metal hydroxides, nitrates, or carbonates. Upon the treatments with molten alkali metal compounds and rinsing with water and H_2SO_4 solution, ceramics were further immersed into chloride (>4 M HCl) or bromide (>5 M NaBr, pH $<$ 1) solutions in order to achieve a complete coverage of the surface with the metallic film.

In the next step, the ceramic substrates were sensitized and activated using the usual $SnCl_2/PdCl_2$ procedures, and further electroless deposition of copper from commercial solutions is carried out. The adhesion of the deposited copper was excellent and, consequently, De Luca et al. recommended it for use in the microelectronics packaging.

Aluminum nitride ceramics attracted the electronic industry as packaging substrates due to AlN higher thermal conductivity increase in the circuit operating temperatures. Prior to electroless deposition of NiP, it was suggested that AlN ceramics should be etched with NaOH [50, 53].

Oxide ceramics, such as $SrTiO_3$, $PbTi_{1-x}Zr_xO_3$ (PZT), and Al_2O_3, were activated by UV excimer laser irradiation in the presence of air prior to electroless deposition of copper or nickel on these substrates [54, 55]. The deposition of metal is restricted to the ablated areas, which results in a selective metallization of the ceramic surfaces. Even the decrease in the surface roughness was observed [56]; the deposited metal showed good adherence to the surface. It appeared that the laser ablation creates catalytic sites onto the ceramic substrates which are responsible for the electroless deposition. The authors attributed this catalytic activity to the formation of oxygen vacancies on the laser irradiated samples, which changes the reactivity of the surface. They also stated that after the deposition of a few monolayers of metal, further deposition proceeds autocatalytically.

An interesting method for the activation of AlN ceramics prior to electroless deposition was described by Yanagimoto et al. [57]. This method intentionally avoids the pretreatments with $PdCl_2$ and/or $SnCl_2$ solutions. The active catalytic sites for the electroless deposition on the AlN surface are produced as follows. The AlN substrates were first coated with a composite film consisting of Cu_2O nanoparticles (about 3.5 nm in size) and NH_2-terminated poly(ethylene oxide) matrix. The coated substrates were then heat-treated at 1,000 °C for 120 min in air. During the heat treatment at 1,000 °C the authors observed the transformation of Cu_2O into CuO. After cooling, samples were additionally heat treated in a separate furnace with a flowing reducing gas composed of H_2 and N_2. During this process, the CuO is reduced to Cu, as evidenced by their XRD data. Finally, on the obtained Cu/AlN samples, Cu film was electroless deposited using commercial solutions. Yannagamoto et al. reported an excellent adhesion of the electroless deposited Cu onto AlN substrates, which was attributed to the chemical interactions between Cu particles and the interfacial layer consisting of $CuAl_2O_4$ and α-Al_2O_3.

The activation of low-temperature co-fired ceramic substrates prior to electroless deposition was achieved through silver nanoparticle inject printing [58]. This solvent-based silver ink was printed onto ceramic substrate with an aim to produce a seed layer of silver which can act as a catalyst for the electroless deposition. Silver-printed ceramic substrates were then cured at 200 °C for about 1 h. Prior to electroless deposition of nickel, silver-coated ceramic substrates were first cleaned with sulfuric acid solution and then activated with palladium containing solution.

Although other approaches for the activation of nonconducting substrates for the electroless deposition of various metals are described in the literature, it seems that on the industrial scale the most commonly used method is based on the $SnCl_2/PdCl_2$ treatments. Other methods of the surface treatments on nonmetallic surfaces showed good results, but significant future work is required, prior to a real application on the industrial scale.

7.6 Conclusions

The results thus far have shown that the metallization of nonconductive surfaces, e.g., ceramics, polymers, and glasses, or semiconductors, e.g., Si or Ge, from aqueous solutions can be quite successful. These surfaces must be properly activated in order to initiate the electroless deposition. Further growth of metallic films is possible through autocatalytic deposition using appropriate reducing agents or electrodeposition.

References

1. Mallory O, Hajdu JB (1990) Electroless Plating: Fundamentals and Applications, AESFS, Orlando, Florida
2. Djokić SS (2002) Electroless Deposition of Metals and Alloys In: Conway BE, White RE and Bockris JO'M (eds), Modern Aspects of Electrochemistry, chap.2, no 35, Kluwer Academic/ Plenum Publishers, New York p.51
3. Djokić SS, Cavallotti PL (2010) Electroless Deposition: Theory and Applications In: Djokić SS (ed) Electrodeposition: Theory and Practice, Modern Aspects of Electrochemistry, chap. 6, no 48, Springer, New York p.251
4. Sverin JW, Hokke R, van der Wel H, de With G (1993) J Electrochem Soc 140: 611
5. Charbonnier M, Alami M, Romand M (1996) J Electrochem Soc 143: 472
6. Dulcey CS, Georger Jr. JH, Krauthamer V, Stenger DA, Fare TL and Calvert JM (1991) Science 252: 551
7. Brandow SL, Dressick WJ, Marian CRK, Chow G-M and Calvert JM, J. Electrohem. Soc., 142 (1995) 2233
8. Deitz KH (1995) Plat and Surf Finish 82 (7): 60
9. Scudiero L, Fasasi A, Griffiths PR (2011) Appl Surf Sci 257:4422
10. Gorostiza P, Diaz R, Servat J. Sanz F, Morante JR (1997) J Electrochem Soc 144:909
11. Corraro C, Maboudian R, Magagnin L (2007) Surf Sci Rep 62(12):499.
12. Magagnin L, Bertani V, Cavallotti PL, Maboudian R, Carraro C (2002) Microelectr Eng 64:479
13. da Rosa CP, Maboudian R, Iglesia E (2008) J Electrochem Soc (155) (6):E70
14. Karmalker S, Sridhar D, Banerjee (1997) J Electrochem Soc 144:1696
15. Aizawa A, Cooper AM, Malac M, Buriak JM (2005) Nanolett 5:815
16. Djokić SS, Djokić NS (2011) J Electrochem Soc 158: D204
17. Djokić SS, Djokić NS, Guthy C, Thundat T (2013) Electrochim Acta 109: 475
18. Djokić SS (2012 Production of metallic Powders from Aqueous Solution Without an External Current Source In: Djokić SS (ed) Electrodeposition of Metal powders from Aqueous Solution, chap. 6, no 54, Springer, New York p.369
19. Magagnin L, Cojocaru P, Secundo F (2012) Electroless Synthesis of Metallic Nanostructures for Biomedical Technologies In: Djokić SS (ed) Biomedical Applications, chap. 2, no 55, Springer, New York p.73.
20. Uncuer M, Koser H (2012) J. Micromech. Microeng. 22:015003
21. Magagnin L, Maboudian R, Carraro C (2001) Electrochemical And Solid-State Letters 4:C5
22. Carraro C, Maboudian R, Magagnin L (2003) Thin Solid Films 434:100

23. Dongwook K, Bhimaraj P, Watts N, Isao Y, Kumar C, Youren X (2006) Evaluation of DIG (direct immersion gold) as a new surface finish for mobile applications In: 56th Proc. Electronic Components and Technology Conf., (Stoughton, WI: The Printing House, Inc.) p. 5
24. Youssef Sayed S, Wang F, Malac M, Li P, Wang D, Buriak J (2012) Cryst Eng Comm 14:5230
25. Fabre B, Hennous L, Ababou-Girard S, Meriadec C (2012) ACS Appl. Mater. Interfaces 5:338
26. Cerruti M, Doerk G, Hernandez G, Carraro C, Maboudian R (2009) Langmuir 26:432
27. Galvan-Arellano M, Diaz-Reyes J, Pena-Sierra R (2010) Vacuum 84:1195
28. Romann T, Anderson E, Kallip S, Mändar H, Matisen L, Lust E (2010) Thin Solid Films 518:3690
29. Kamp M, Bartsch J, Cimiotti G, Keding R, Zogaj A, Reichel C, Kalio A, Glatthaar M, Glunz S (2013) Solar Energy Materials and Solar Cells, in press. http://dx.doi.org/10.1016/j.solmat.2013.05.035
30. Chen GS, Chen ST, Lu YL (2010) Electrochemistry Communications 12:1483
31. Chaudhari VA, Solanki CS (2010) Solar Energy Materials & Solar Cells 94:2094
32. Kim D, Shanmugam R, Choi M, Yoo B (2012) Electrochimica Acta 75:42
33. Chen F, Shinosky M, Li B, Christiansen C, Lee T, Aitken J, Badami D, Huang E, Bonilla G, Ko T, Kane T, Wang Y, Zaitz M, Nicholson L, Angyal M, Truong C, Chen X, Yang G, Law SB, Tang TJ, Petitdidier S, Ribes G, Oh M, Child C, Sawada H, Kolics A, Rigoutat O, Gilbert N (2010) Comprehensive investigations of CoWP metal-cap impacts on low-k TDDB for 32 nm technology application In: Reliability Physics Symposium (IRPS), IEEE International (Anaheim, California) p.566
34. Inoue F, Shimizu T, Miyake H, Arima R, Ito T, Seki H, Shinozaki Y, Yamamoto T, Shingubara S (2013) Microelectronic Engineering 106:164
35. Tao Y, Hu A, Hang T, Peng L, Li M (2013) Applied Surface Science 282:632
36. Kawano M, Takahashi N, Komuro M, Matsui S (2010) Low-cost TSV process using electroless Ni plating for 3D stacked DRAM In: Electronic Components and Technology Conference (ECTC), Proceedings 60th (Las Vegas, USA) p. 1094
37. Han WK, Hwang GH, Hong SJ, An HH, Yoon CS, Kim JH, Lee MJ, Hong G, Park KS, Kang SG (2010) Materials Chemistry and Physics 123:401
38. Santagata F, Farriciello C, Fiorentino G, Wvan Zeijl H, Silvestri C, Zhang GQ, Sarro PM (2013) J. Micromech. Microeng. 23:055014
39. Akhtari-Zavareh A, Li W, Maroun F, Allongue P, Kavanagh KL (2013) Journal Of Applied Physics 113:063708
40. Lee JD, Kim HS, Jeong SY, Kim KH, Lee JJ, Kim JE (2010) Current Applied Physics 10:249
41. Nguyen D, Tanaka S, Nishino H, Manabe K, Ito S (2013) Nanoscale Research Letter 8:8
42. Pascu M, Nicolas D, Poncin-Epaillard F, Vasile C (2006) Journal Of Optoelectronics And Advanced Materials 8:1062
43. Cho SJ, Nguyen T, Boo J-H (2011) J. Nanosci. Nanotechnol. 11:5328
44. Borris J, Dohse A, Hinze A, Thomas M, Klages C-P, Mobius A, Elbick D, Weidlich E-R (2009) Plasma Process. Polym. 6-S258
45. Akamatsu K, Ikeda S, Nawafune H, Yanagimoto H, (2004) Journal of the American Chemical Society 126:10822
46. Yang F, Li P, Li X, Huo L, Chen J, Chen R, Na W, Tang W, Liang L, Su W (2013) Applied Surface Science 274:241
47. Sawada S, Masuda Y, Zhu P, Koumoto K, Langmuir (2006) 22-332
48. Miller MS, Filiatrault HL, Davidson GJE, Luo M, Carmichael TB (2010) Journal of the American Chemical Society 132:765
49. Tengsuwan S, Ohshima M (2012) J. of Supercritical Fluids 69-117
50. Djokić S et al. (1998) Electroless/electrolytic methods for the preparation of metallized ceramic substrate, US Patent 5,849,170A
51. Dearden J (1976) Electrocomp Sci Technol 3: 103
52. De Luca MA, McCormack JF (1986) Metallization of ceramics, US Patent 4,604,249

53. Osaka T, Asada T, Nakajima E, Koiwa I (1998), J Electrochem Soc 135: 2578
54. Shafeev GA (1993) Adv Mater for Opt and Electr 2: 183
55. Shafeev GA, Autric M, Marine V (1997) Adv Mater for Opt and Electr 7: 249
56. Nicholas G, Autric M, Marine, V Shafeev GA (1997) Appl Surf Sci 109-110: 128
57. Yanagimoto H, Akamatsu K, Deki S, Gotoh H (2003) J Electrochem Soc 150: C260
58. Gian PW, Shan X, Liang YN, Lok BK, Lu CW, Ooi BL (2008) EDA Publishing ISBN:978-2-35500-006-5

Index

S.S. Djokić (ed.), *Electrodeposition and Surface Finishing: Fundamentals and Applications*, Modern Aspects of Electrochemistry 57,
DOI 10.1007/978-1-4939-0289-7,

MIX
Papier aus verantwortungsvollen Quellen
Paper from responsible sources
FSC® C105338

If you have any concerns about our products,
you can contact us on
ProductSafety@springernature.com

In case Publisher is established outside the EU,
the EU authorized representative is:
Springer Nature Customer Service Center GmbH
Europaplatz 3, 69115 Heidelberg, Germany

Printed by Libri Plureos GmbH
in Hamburg, Germany